Acta Physica Austriaca

Supplementum XI

Proceedings of the
XII. Internationale Universitätswochen für Kernphysik 1973
der Karl-Franzens-Universität Graz
at Schladming (Steiermark, Austria)
5th February—17th February 1973

Sponsored by
Bundesministerium für Wissenschaft und Forschung
International Centre for Theoretical Physics, Triest
Steiermärkische Landesregierung
Sektion Industrie der
Kammer der gewerblichen Wirtschaft für Steiermark

1973

Springer-Verlag
Wien New York

Recent Developments in Mathematical Physics

Edited by Paul Urban, Graz

With 39 Figures

1973

Springer-Verlag
Wien New York

Organizing Commitee

Chairman

Prof. Dr. Paul Urban
Vorstand des Institutes für Theoretische Physik
der Universität Graz

Committee Members

Prof. Dr. Ludwig Streit
Fakultät für Physik
der Universität Bielefeld

Dr. H.-J. Faustmann

Dr. H. Latal

Dr. L. Pittner

Secretaries

M. Krautilik

I. Primschitz

Library of Congress Catalog Card Number 73-13322
Printed in Austria by Bors & Müller, A-1010 Wien

ISBN 3-211-81190-7 Springer-Verlag Wien-New York
ISBN 0-387-81190-7 Springer-Verlag New York-Wien

CONTENTS

Acta Physica Austriaca, Suppl.XI, 1–2 (1973)

OPENING ADDRESS

BY

P. URBAN
Institut für Theoretische Physik
Universität Graz

In the past years the Schladming conferences were devoted to various aspects of elementary particle physics, ranging over the wide spectrum from weak over electromagnetic to strong interaction processes. Most of the topics were prompted by experimental progress - as e.g. the topic "Vectormesons" of our first meeting in 1962, or by the now historical neutrino experiments at CERN in 1963/64 -, or the advent of new phenomenological machinery - as examples I quote the Regge poles of 1963 and again of 1968, the concept of higher symmetries of 1964 and of the following years.

During the past decade or so and until rather recently, however, the development of mathematical physics and the attempts to explain things starting from a dynamical ansatz have been less impressive. One feature that made the field so uninviting was the fact that the landscape of mathematical physics was studded with a lot of little unconnected ivory towers. These offered various confusing views according to which the world appeared to be one-dimensional, or a quark oscillator, or an infraparticle, or travelling on the light cone, or even off into imaginary time - to name only a few of the dizzying sights.

In the recent past, however, things have changed dramatically. All of a sudden interrelations between various approaches have become visible - so e.g. between the previously rather disjoint pursuits of, among others, statistical mechanics, constructive quantum field theory, the infrared problem, conformal quantum field theory, light cone expansions, and the singularities of phase transition points. In this way, through a new and vigorous dialogue the previously separate lines of thought have become intertwined, they all have experienced accelerated progress and have, in stepping out of their one-time isolation, become more relevant to our understanding of the material world.

Now at last theoreticians of various specialities talk to each other, and with the mathematicians. There have been some, however, who kept insisting that mathematical physics was both worthwhile and in the long run relevant. In this context I think it is most appropriate to thank Res Jost for the encouragement and advice which he had for the scientific organizers of this meeting.

Exchange of ideas, concepts, techniques between the various subfields of mathematical physics is the hallmark of present progress. To this we hope to contribute through this year's lectures and discussions and I wish the school to be a profitable and pleasant experience for all of us.

Acta Physica Austriaca, Suppl.XI, 3–116 (1973)

DYNAMICAL PROBLEMS OF THE RELATIVISTIC QUARK MODEL*

BY

M. BÖHM
Physikalisches Institut der
Universität Würzburg

H. JOOS and M. KRAMMER
Deutsches Elektronen-Synchrotron
DESY, Hamburg

A. BASIC QUESTIONS

1. Introduction.
   - 1.1. Why does field theory seem to be a good choice for a frame work of hadron dynamics?
   - 1.2. How does field theory look like? (A glossary of general field theory.)
   - 1.3. How many fundamental fields? Why quark fields?
   - 1.4. What are the dynamical problems in quark field theory posed by phenomenology?
   - 1.5. Is our scheme too restrictive?

---

*Lectures given at XII. Internationale Universitätswochen für Kernphysik, Schladming, February 5 - 17, 1973.

A. BASIC QUESTIONS

1. Introduction

In our lectures we deal with the specific dynamical problems arising in the formulation of the relativistic quark model in the framework of field theory[1]. After the discussion of some basic questions occurring in this particular attempt towards a foundation of hadron dynamics, we develop extensively the methods for the description of mesons as quark-antiquark bound-states and give some illuminating phenomenological applications. In order to outline the spirit of our approach, let us first answer some general questions.

1.1. Why does field theory seem to be a good choice for a framework of hadron dynamics?

Certainly the great success of QED, i.e. the quantum

theory of the coupled Dirac-Maxwell equations of electrons and photons, seduces many physicists to believe also in the success of field theory of hadron dynamics (HD). But in HD, the correct field equations are not known and phenomenological success is not evident. The strongest argument for hadron field theory is, that the combination of general principles formulated in the "Axioms of General Quantum Field Theory" represents still the only known closed scheme for the successful relativistic reaction theory ("relativistic quantum mechanics"):

Related to a field $\psi(x)$ and the particles described by it, there is an S-matrix showing the following features:

Relativistic kinematics for the particles, including the relation of spin and statistics;
The validity of the conservation laws for the reactions;
Cluster property and macro-causal one-particle structure, guaranteeing independence of reactions in distant space time regions;
Dispersion relations and crossing symmetry;
Threshold behaviour and asymptotic limits of scattering amplitudes, etc..

Besides the S-matrix aspect of quantum field theory, we have a consistent description of the physical phenomena by local observables, like electromagnetic currents.

For further discussion we shall give a short glossary of these well-known facts[2].

1.2. How does field theory look like? (A Glossary of General Field Theory).

1.2.1. "Axioms"

(a) Quantum mechanical description. The field $\psi(x)$ is a distribution valued operator in a Hilbert space H. $\psi(x)$ generates a complete operator system in H.

For fixed x, $\psi(x)$ has a finite number of components $\psi_A(x)$ including the hermite conjugates. In case of the quark field theory (QUARFT)*:
$\psi(x) = \{\psi_{a,\alpha}(x), \bar{\psi}_{b,\beta}(x)\}$, a,b = 1,...,4 Dirac indices,
$\alpha = p,n,\lambda$, SU(3) indices in quark notation.

(b) Relativistic covariance. There is a unitary representation of the Poincaré group $P^{\uparrow}_{+}$, (resp. $SL(2,C) \circledS t_4$) defined in H: $(\Lambda,a) \to U(\Lambda,a)$ under which the field transforms covariantly:

$$U(\Lambda,a)\ \psi(x)\ U^{-1}(\Lambda,a) = S^{-1}(\Lambda)\ \psi(\Lambda x+a),$$

which is extended by charge conjugation and antiunitary time reversal:

$$U(C)\ \psi(x)\ U^{-1}(C) = C\ \psi(x);\ \bar{U}(\tau)\ \psi\ (x)\ \bar{U}^{-1}(\tau) = T\psi(\tau x)$$

[In QUARFT: $S(\Lambda)$ spinor representation of $L^{\uparrow}$: $(\frac{1}{2},0) \oplus (0,\frac{1}{2})$ for ψ and $\bar{\psi}$. $U(C)\psi(x)U^{-1}(C) = -(\bar{\psi}(x)C)^T$, $C = i\gamma_o\gamma_2$; $\bar{U}(\tau)\psi(x)\ \bar{U}^{-1}(\tau) = C^{-1}\gamma_5\psi(\tau x)$ etc.]

---

*The specifications given for QUARFT define our notations.

(c) Spectrum condition.

(i) There is a unique vacuum state $|0\rangle \varepsilon H$: $U(\Lambda,a)|0\rangle = |0\rangle$.

(ii) The spectrum of the energy momentum operator P_μ of $U(\Lambda,a)$ lies in the forward light cone. ($P_\mu P^\mu > m_0^2 > 0$ except $P_\mu|0\rangle = 0$).

(d) Locality.

$$[\psi(x), \psi(y)]_\mp = 0 \quad \text{for} \quad (x-y)^2 < 0.$$

Commutators or anti-commutators for Bosons or Fermions respectively.

(e) Internal symmetry. ("Conservation of isospin I, baryon number B and hypercharge Y")

There is a unitary representation of the group

$$SU_I(2) x U_B(1) x U_Y(1) \quad \text{IBY in H: } \exp[\vec{\beta}\,\vec{I} + \beta_0 B + \beta_8 Y] \rightarrow U(\beta)$$

$$U(\beta)\ \psi(x)\ U^{-1}(\beta) = D^{-1}(\beta)\ \psi(x);\ U(\beta)|0\rangle = |0\rangle\ .$$

[in QUARFT: $D(\beta) = \exp -i(\frac{1}{2}\vec{\beta}\vec{\lambda} + \frac{\beta_8}{\sqrt{3}}\lambda_8 + \frac{1}{\sqrt{6}}\beta_0\lambda_0)$ for $\psi(x)$, $D^*(\beta)$ for $\bar{\psi}(x)$, - λ_i the Gell-Mann matrices, - .]

1.2.2. Description of particles

(a) Single particle states.

States $|\emptyset\rangle \varepsilon H$ of single particles are normalizable eigenstates of the mass operator $P_\mu P^\mu|\emptyset\rangle = M_i^2|\emptyset\rangle$. It follows from "relativistic covariance" and "internal symmetry" that $|\emptyset\rangle$ is an element of an irreducible sub-space of H invariant under the transformations of the total symmetry group: $P^C \otimes$ IBY. An (improper) base of these states is denoted by

$$\left|\begin{matrix} M & S^{\Pi C}; & I, & Y, & B \\ P & S_3 & ; & I_3 & \end{matrix}\right\rangle$$

normalized according to $\langle P|P'\rangle = 2P_o\ \delta(\vec{P}-\vec{P}')$.

It is assumed that there is no further accidental degeneracy of the mass eigenvalues M^2. (Models with higher symmetry yielding more degeneracy might be considered).

(b) Single particle amplitudes.

A condition for a particle with a given set of quantum numbers to be described by the field $\psi(x)$ is that[3]

$$\langle O|\psi(x_1)\ldots\psi(x_n)\left|\begin{matrix} M & S^{\Pi C}; & I, & Y, & B \\ P & S_3 & ; & I_3 & \end{matrix}\right\rangle \neq O \qquad (1.1)$$

for some n. A necessary condition for this is that the product $\psi(x_1)\ldots\psi(x_n)$ "carries" the quantum numbers of the particle: the product representation of $P^C \otimes$ IBY defined by the covariance axioms 1.2.1. (b,c) in the space spanned by the states $\psi^+(x_1)\ldots\psi^+(x_n)|O\rangle$ must contain the irreducible representation of the symmetry group characterized by (M^2, $S^{\Pi C}$, I, Y, B). For practical discussions one assumes that this condition is also sufficient; the vanishing of $\langle O|\psi(x_1)\ldots\psi(x_n)|{}^{M}_{P}\ldots\rangle$ would be accidental otherwise. The internal structure of the particles is described by amplitudes like those of Eq.(1). A particularly important role plays in this respect the Bethe-Salpeter amplitude and its Fourier transform:

$$(2\pi)^{-3/2}\chi_{[{M \atop P}]}(p_1\ldots P_n)\ \delta(p_1+\ldots+p_n-P) = \qquad (1.2)$$

$$= (2\pi)^{-4}\iint dx_1\ldots dx_n\ e^{i(p_1x_1+\ldots+p_nx_n)} \langle 0|T\psi(x_1)\ldots\psi(x_n) \left| \begin{matrix} M, & S^{\Pi C}; & IYB \\ P, & S_3 & ; I_3 \end{matrix} \right\rangle$$

$$(2\pi)^{-3/2}\bar{\chi}_{[{M \atop P}]}(p_1\ldots p_n)\ \delta(p_1+\ldots+p_n+P) =$$

$$= (2\pi)^{-4}\iint dx_1\ldots dx_n\ e^{i(p_1x_1+\ldots+p_nx_n)} \left\langle \begin{matrix} M, & S^{\Pi C}; & IYB \\ P, & S_3 & ; I_3 \end{matrix} \right| T\psi(x_1)\ldots\psi(x_n)|0\rangle .$$

T denotes time ordering, including the conventions on signs for Fermion fields etc. We shall omit often in our notation variables which are not important for the special point under consideration; or we introduce some appropriate abbreviation. In Eq. (1.2.) $[{M \atop P}]$ denotes all particle variables and $P_\nu = (p_\nu, A_\nu)$ combines the field momenta and internal variables.

1.2.3. Green's functions and S-matrix elements

(a) Green's functions.

The relation between the fields $\psi(x)$ and the S-matrix may be described with help of the Green's functions, i.e. the vacuum expectation values of time ordered products of fields and their Fourier transforms:

$$\tilde{\tau}(x_1 \ldots x_n) = \langle 0 | T\psi(x_1) \ldots \psi(x_n) | 0 \rangle$$

$$\tau(p_1 \ldots p_n) = (2\pi)^{-4} \int \ldots \int dx_1 \ldots dx_n \, e^{i(p_1 x_1 + \ldots p_n x_n)} \tilde{\tau}(x_1 \ldots x_n) . \tag{1.3}$$

(b) Cluster decomposition.

The vaccuum structure[4] of the Green's functions follows from the covariance of the fields with respect to space time translations, and from the spectrum condition, particularly from the existence of $|0\rangle$

$$\tau(p_1 \ldots p_n) = \delta(p_1 + \ldots + p_n) \eta(p_1, \ldots p_n)$$

$$+ \sum_Z \delta(p_{i_1} + p_{i_2} + \ldots) \eta(p_{i_1}, p_{i_2}, \ldots) \ldots \delta(p_{j_1} + p_{j_2} + \ldots) \eta(p_{j_1}, p_{j_2}, \ldots) . \tag{1.4}$$

(The sum is to be taken over all possible partitions Z of the indices 1,...n in distinct classes $i_1, i_2, \ldots ; j_1, j_2 \ldots ;$ ...., where $\eta(p_1 \ldots p_k)$ does not contain any singularities of the type $\delta(\Sigma p_i)$).

(c) Single particle singularities.

Single particle intermediate states lead to poles of the

τ-functions[4] whenever partial sums of the momenta lie on the mass shell:

$$\tau(p_1,..p_n) = \lim_{\varepsilon\to 0} \frac{\{(k^2-M^2)\tau(p_1...p_n)\}}{(k^2-M^2+i\varepsilon)} \quad , \quad k=p_{i_1}+...+p_{i_\nu} \tag{1.5}$$

where the residue $\{(k^2-M^2)\tau(p_1,...p_n)\}$ is finite and "factorizes" after separation of the δ-functions in BS-amplitudes:

$$\{(k^2-M^2)\tau[p_1...p_n]\}_{\substack{k^2=M^2\\ k_o>0}} = i\,\delta(p_1+...+p_n)$$

$$\sum_{S_3 I_o} \chi_{\left[{M \atop k}\right]}(p_{i_1}...p_{i_\nu})\,\bar{\chi}_{\left[{M \atop k}\right]}(p_{i_{\nu+1}},...p_{i_n})$$

or (1.6)

$$\{(k^2-M^2)\tau[p_1...p_n]\}_{\substack{k^2=M^2\\ k_o<0}} = i\,\delta(p_1+...+p_n)$$

$$\sum_{S_3 I_o} \chi_{\left[{M \atop k}\right]}(p_{i_{\nu+1}}..p_{i_n})\,\bar{\chi}_{\left[{M \atop k}\right]}(p_{i_1}...p_{i_\nu})$$

Similar expressions hold for the residues of the poles of the η functions. In that case the δ-functions of energy-momentum conservation are already separated and the connected parts of the BS amplitudes of the one-particle states $|\emptyset>$ have to be inserted instead of χ.

Apart from their slightly more complicated definition the expressions with complete removal of the vacuum singularities are theoretically and practically more appropriate for most of the discussions.

(d) S-matrix elements.

The S-matrix elements of the reactions $\left(\begin{matrix} M_1 \ldots M_r \\ K_1 \ldots K_r \end{matrix} \middle| \; S \; \middle| \begin{matrix} M'_1 \ldots M'_{r'} \\ K'_1 \ldots K'_{r'} \end{matrix}\right)$

$$[M'_1] + [M'_2] + \ldots [M'_{r'}] \rightarrow [M_1] + \ldots [M_r]$$

are related to the one particle poles. Let $k'_1 \ldots k'_{r'}$ and $k_1 \ldots k_r$ be the 4-momenta of the incoming and outgoing particles, then we have

$$(2\pi)^4 (-i)^{r+r'} \{ (k_1^2 - M_1^2) \ldots (k_r^2 - M_r^2)(k_1'^2 - M_1'^2) \ldots (k_{r'}'^2 - M_{r'}'^2) \tau(p_1 \ldots p_n) \}_{\substack{k_i^2 = M_i^2 \\ k_j'^2 = M_j'^2}}$$

$$= (2\pi)^{\frac{3}{2}(r+r')} \sum_{S_{3i} I_{oi}} \chi_{\left[\begin{smallmatrix} M_1 \\ K_1 \end{smallmatrix}\right]}(p_{i_1}) \ldots \chi_{\left[\begin{smallmatrix} M_r \\ K_r \end{smallmatrix}\right]}(p_{i_r}) \left(\begin{matrix} M_1, \ldots M_r \\ K_1, \ldots K_r \end{matrix} \middle| \; S \; \middle| \begin{matrix} M'_1, \ldots M'_{r'} \\ K'_1, \ldots K'_{r'} \end{matrix}\right) \cdot$$

$$\cdot \bar{\chi}_{\left[\begin{smallmatrix} M'_1 \\ K'_1 \end{smallmatrix}\right]}(p'_{j_1}) \ldots \bar{\chi}_{\left[\begin{smallmatrix} M'_{r'} \\ K'_{r'} \end{smallmatrix}\right]}(p'_{j_{r'}}) \qquad (1.7)$$

The arguments P_j of the normalized BS amplitudes (3.(5c)) denote the momenta contained in the partial sums $k_1 = p_{i_1} + \ldots p_{i_\ell}$, $k_j' = -(p_{j_1} + \ldots + p_{j_{\ell'}})$; $p_1 \ldots p_n$ appears in precisely one partial sum k_i, k_j'.

The vacuum structure and the single particle poles of the τ-functions lead to the cluster property of the S-matrix and its one particle structure. These imply the correct description of spatially separated independent processes and the macro-causal sequence of consecutive reactions.

1.2.4. Currents and local observables.

From a practical point of view, the S-matrix contains the most important information on measurable quantities, like cross sections, life times etc. Fields are, in general, not directly measurable quantities (f.i. Fermi fields). A thorough analysis of the concepts of field theory has to discuss which quantities can really be observed. - Since the fact, that quark particles have not been found, puts some serious problems on QUARFT, we would like to mention this problem. - The elements of the algebra of observables may be considered as local functions of the fields. For a more precise definition we refer to the work of Haag, Araki and Kastler.[5] Important observable fields are the sources of the e.m. and weak interactions, i.e. the e.m. and weak currents. The expression of these currents in terms of field operators is therefore an important part of the interpretation of a particular field theory following the line of QUARFT. (We discuss this in Section 3.3).

1.3. How many fundamental fields? Why quark fields?

A theorem by Thirring[6] states, that hadron dynamics (HD) must be based on a minimum of six independent Majorana fields. We repeat Thirring's arguments in the language of general field theory. For this we assume that in HD (isolated from e.m. and weak interactions) the pseudo-scalar meson octet (π, K, η) and the baryons (N, Σ, Λ, Ξ) are described as single particle states (Sect.1.2.2.(a)) with the well known quantum numbers. In order that these particles are contained in HD, there must be single particle amplitudes of the fundamental field $\psi_A(x)$ which are not identical zero (Sect.1.2.2.(b)). For this it is necessary that a certain product of the $\psi_A(x)$ carries the quantum numbers of the stable particles.

The internal quantum numbers of $\psi_A(x)$ are determined by the representation of IBY in the space spanned by the field components (Sect.1.2.1.(e)). In particular, if the field components are diagonal with respect to the additive quantum numbers $Q = I_3 + \frac{1}{2} Y$, Y, B:

$$[Q,\psi_A(x)] = Q'_A\psi_A(x),\ [Y,\psi_A(x)] = Y'_A\psi_A(x), [B,\psi_A(x)] = B'_A\psi_A(x) \tag{1.8}$$

the state $\psi_{A_1}(x) \ldots \psi_{A_n}(x)|0\rangle$ is an eigenstate of Q,Y,B with the eigenvalues

$$Q' = \sum_{i=1}^{n} Q'_{A_i} \; ; \; Y' = \sum_{i=1}^{n} Y'_{A_i} \; ; \; B' = \sum_{i=1}^{n} B'_{A_i} .$$

Since the three quantum numbers (Q', Y', B') of the

stable particles $\pi^+: (1,0,0)$, $K^0: (0,1,0)$, $\Lambda: (0,0,1)$ are linear independent, we must have at least 3 independent fields with respect to the internal quantum numbers. Charge conjugation C anticommutes with Q,Y,B, QC = -CQ etc. therefore:

$$\psi_A{}^C = C\psi_A C^{-1}; \; [Q,\psi_A{}^C(x)] = - Q_A'\psi_A{}^C(x) \text{ etc.}$$

Because of C-invariance of HD, we have to introduce these fields as independent field components. The fundamental fields of HD must therefore contain 3 pairs of charge conjugate fields, which corresponds to 6 independent self conjugate fields ("Majorana fields").

There are further conditions on the quantum numbers of the fields, which we shall discuss for 3 fields with their charge conjugates ("fundamental triplet fields"):

(α) The existence of half integer isospin multiplets implies that the fundamental triplet field consists of an $I = 1/2 \oplus 0$ field.

(β) The existence of the baryons with spin 1/2 implies, that one of the fundamental fields must have spin 1/2.

(γ) The signature of spin and isospin $\varepsilon_S = (-)^{2S}$; $\varepsilon_I = (-)^{2I}$ of the fields are multiplicative quantum numbers. Since half integer spins are independent ($\pi \rightarrow (\varepsilon_S, \varepsilon_I) = (1,1)$, $K \rightarrow (1,-1)$, $\Xi \rightarrow (-1,1)$, $= (-1,-1)$) the assignment of spin-isospin to the fundamental triplets must be:
(A) $(\varepsilon_S, \varepsilon_I)$: $(-1,-1)$, $(-1,1)$;
(B) $(\varepsilon_S, \varepsilon_I)$: $(1,-1)$, $(-1,-1)$
the assignments $(1,-1)$, $(1,1)$ or $(-1,-1)$, $(1,1)$ being excluded

The assumption of integer charges and baryon numbers, and the restriction to spin zero and 1/2 fields lead to two possible models:
(Using the notation of existing particles to describe the quantum numbers of the fields (Q', Y', B'; ε_S, ε_I) - omitting the charge conjugate fields)

(A)	P	N	Λ
	(+1,+1,+1; -1,-1)	(0,+1,+1;-1,-1)	(0,0,+1;-1,1)
			(Sakata[7], Wess[8])
(B)	K^+	K^0	Λ
	(+1,+1,0;1,-1)	(0,1,0;1,-1)	(0,0,+1;-1,1)
			(Thirring[6])

It was the phenomenological success of (approximate) SU (3) symmetry with the eightfold way assignment of the ground state baryons which suggests to give up the restriction to integer Q', B' of the fields, in favour of considering the fundamental fields as an SU(3) triplet field, the quark fields

(A')	p	n	λ
	(2/3,1/3,1/3;-1,-1)	(-1/3,1/3,1/3;-1,-1)	(-1/3,-2/3,1/3;-1,1)
			(Gell-Mann, Zweig[9])

This model allows a SU(3)-symmetric description of the mesons by amplitudes $\langle \bar{q}(x)\, q(x') | \text{Mes} \rangle$ and of the baryons by $\langle q(x) q(x') q(x'') | \text{Baryon} \rangle$. The experimental absence of particles and resonances, which cannot be described by such amplitudes ("exotic particles"[10]) may be considered as an additional, phenomenological hint in favour of fundamental quark fields.

Therefore our answer to the question of this section is:

"The quark fields form a set of fundamental fields, with a minimal number of components, which is favoured over other triplet fields by the phenomenological experience of approximate SU(3) symmetry and the absence of exotic states."

We are aware of the fact that, by the assumption of the quark fields as the fundamental fields of hadron dynamics, we are confronted with the question of the existence of free particles with the quantum number of the quark fields (quarks). We consider this as a first dynamical problem in field theory created by the relativistic quark model and we shall discuss it in Section 2.3.

1.4. What are the dynamical problems in quark field theory posed by phenomenology?

The application of quark field theory to phenomenological questions becomes more complicated with increasing complexity ("number of arguments") of the Green's functions describing the reactions. We shall give a short sketch of the content of our lecture along this line.

1.4.1. Two-point problems

The spectrum of the states: $\int f(x)\bar{\psi}(x)|0\rangle\, dx \equiv \bar{\psi}(f)|0\rangle$, which carry the quantum numbers of the quarks, may be studied by the spectral representation of the two-point function:

$$\tau(p,\bar{p}) = \frac{1}{(2\pi)^4} \iint dx\, d\bar{x}\, \langle O|T^{*}\psi(x)\bar{\psi}(\bar{x})|O\rangle\, e^{i(px-\bar{p}\bar{x})}$$

$$= \eta(p)\ \delta(p-\bar{p})$$

(Against our definition of the Fourier transform in Eq.(1.3), we have changed the sign of $\bar{p}$, in order to be consistent with the usual description of Dirac fields). Therefore we consider the question of the existence of quarks as a two-point problem; we shall discuss it in the next section.

1.4.2. Four-point problems.

The four-point function

$$\tau(p_1p_2;\bar{p}_1\bar{p}_2) = (2\pi)^{-4} \iint dx_1dx_2d\bar{x}_1d\bar{x}_2\langle O|T^{*}\psi(x_1)\psi(x_2)\bar{\psi}(\bar{x}_1)\bar{\psi}(\bar{x}_2)|O\rangle\ e^{i(p_1x_1+p_2x_2-\bar{p}_1\bar{x}_1-\bar{p}_2\bar{x}_2)}$$

$$= \eta(p_1p_2;\bar{p}_1\bar{p}_2)\cdot\delta(p_1+p_2-\bar{p}_1-\bar{p}_2)+\eta(p_1\bar{p}_1)\eta(p_2\bar{p}_2)\delta(p_1-\bar{p}_1)\delta(p_2-\bar{p}_2)$$
$$+\eta(p_1\bar{p}_2)\eta(p_2\bar{p}_1)\delta(p_1-\bar{p}_2)\delta(p_2-\bar{p}_1).$$

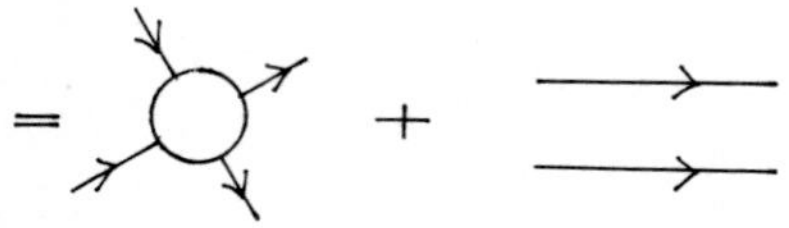

(compare Eq.(1.4))

would describe the scattering of quarks if there would be one.

Our main interest is, however, the single particle

structure in the $q\bar{q}$-sector (compare Sect.1.2.3.c).

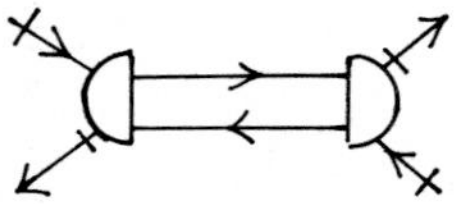

$$= \quad i\,\frac{\sum_r \chi_P^r(p_1,\bar{p}_1)\,\bar{\chi}_P^r(p_2,\bar{p}_2)}{P^2-M^2+i\varepsilon}$$

which contains the meson spectrum and the meson B S-amplitudes. This is the main topic of our lecture and it is treated in Chapter B.

The e.m. and weak currents shall have the form $j_\mu(x) = \bar{\psi}(x)\,\gamma_\mu\,Q^i\psi(x)$. Therefore:

(i) The leptonic decays of the mesons

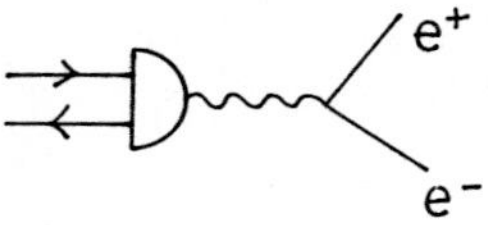

: $\rho^o,\omega,\phi \rightarrow e^+e^-$; and $\pi^+ \rightarrow \mu^+\nu$ etc.

(ii) The total cross section for $e^+e^- \longrightarrow$ hadrons

$\sigma_{tot} \approx \mathrm{Im}$

(iii) The e.m. quark formfactor

are also four-point problems. We shall discuss these as simple applications of our B S model in Sect.7.

1.4.3. Six-point problems.

In the same way as the meson spectrum is related to the four-point function, the baryon spectrum is a six-point problem:

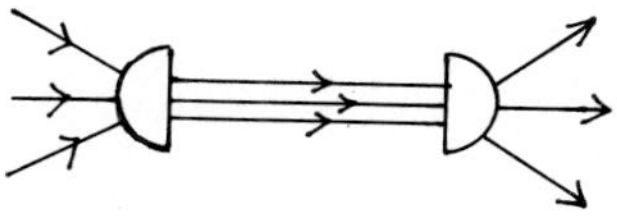

$$= \quad i \frac{\sum_r \chi(p_1p_2p_3)\bar{\chi}(p_1p_2p_3)}{p^2-M^2+i\varepsilon} \quad .$$

In Section 9.1. we report on some first attempts to discuss the baryons as a relativistic three-body problem.

The mesonic coupling constants described by the graph are calculated with our model in Sect.8.

Other six-point problems are:
The e.m. formfactors of mesons:

Quark production by meson-meson interaction:

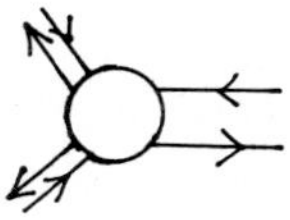

Mesons and baryons are described in QUARFT by the same field. In the non-relativistic quark model, the additivity assumption allows to relate the properties of mesons and baryons. In relativistic QUARFT meson properties and baryons should match by some analytic continuation of the six-point function. (Take this as a speculative remark!)

We could mention "more-points" like: meson-meson scattering, deep inelastic electron meson scattering, meson baryon coupling constants (eight-point problems), pion-nucleon scattering, deep inelastic electron-proton scattering,[11]... (10-point problems), nucleon-nucleon scattering,... (12-point problem). But besides a short remark on a resonance excitation model for meson meson scattering (Sect.9.2.), we restrict ourselves to six-points, because beyond that we cannot offer, at the moment, more than pure speculations.

1.5. Is our scheme too restrictive?

We are aware of many serious objections against our approach:
The spin statistics problem raised by the baryon spectrum in the non-relativistic quark model may suggest para-statistics[12] instead of Fermi statistics or some of the generalizations of the quark model suggested by Han and Nambu,[13] Gell-Mann[14] and others.

In view of the unsolved problem of the existence of quarks, and the possibility of gauge fields, the assumption of local fields in a positive metric Hilbert space may be too restrictive.[15]

The ansatz of a finite component field theory may be opposed to attempts with infinite component fields suggested by the operator formalism of duality[16] and so on. We shall comment on some of these problems in due course.

It is our aim to incorporate phenomenological experience in the straight forward field theory scheme[17], in order to see if such objections are serious arguments against this scheme. Lagrangean field theory models,[18] like gluon models are outside of the scope of this lecture.

2. The existence of quarks.

The experimental discovery of quarks would be one of the greatest triumphs of theoretical physics! It would change the credibility of the quark field approach to hadron dynamics by orders of magnitude. However, up to now the quarks did escape the observation of the experimentalists. Therefore we have to ask what this means (A) from the phenomenological, (B) from the theoretical point of view.

2.1. The phenomenological view on the negative result of quark hunting.

For quarks was searched in accelerator experiments up to cms-energies of 50 GeV's, in cosmic rays, and in solid material. No fractionally charged particles were found. The accelerator experiments[19] give upper bounds for the cross-section of quark production in elementary particle reactions (Table I). In order to judge the significance of these results, one has to compare the upper bound of the cross-sections with some model calculations for these quantities. For this end, one uses nowadays the statistical theory of Hagedorn and Maksimenko[20], which gives for the cross section of the reaction $p + p \to p + p + q + \bar{q} + \ldots$:

$$\sigma(q,\bar{q}) = C \left[\frac{m_q}{T_o}\right]^3 e^{-2m_q/T_o} \qquad (2.1)$$

with $C = 250$ mb; $T_o = 160$ MeV.

This formula reproduces the cross-sections of $p\bar{p}$, $\Sigma\bar{\Sigma}$, $\Sigma^*\bar{\Sigma}^*$, $d\bar{d}$, $He_3\bar{He}_3$ production reasonably well.

	CERN-PS[19a]		SERPUKHOV[19b]	ISR[19c]	
m_q	≤ 2 GeV		4.5 - 5 GeV	≤ 13GeV	≤ 22GeV
Q	$-\frac{2}{3}$	$-\frac{1}{3}$	$-\frac{2}{3}$	$(\pm)\frac{2}{3}$	$(\pm)\frac{1}{3}$
$\sigma\,(cm^2)$	$\leq 1.7\cdot 10^{-37}$	$\leq 2\cdot 10^{-38}$	$\leq 4.10^{-37}$	$\leq 6\cdot 10^{-34}$	$\leq 3\cdot 10^{-34}$

Table I: Examples of upper limits of quark production cross-sections.

From Table I we see that quarks with a mass of $m_q \geq 3$ - 4GeV could not have been observed. Cosmic ray and chemical experiments do not add to this information.[21] According to Eq.(2.1.), high quark masses offer always an excuse for not finding quarks. But, of course, a more detailed dynamical quark theory based on heavy quarks must reproduce the estimates relying on this formula.

All quark search relies on the assumption that, among all states with fractional charge, there should be one with lowest mass and this one should be a state of a stable particle. It was pointed out by de Swart[22] that this stable particle might be a quark compound with a high fractional charge, like 4/3 or so. It would be difficult to find such a particle in accelerator experiments. No such particles have been found.

It is a puzzle that a theory is rather successful on an elementary level, but its basic ingredient is missing. In view of this "Quark Puzzle", the following possibilities may be considered from the field theoretical point of view:

(a) There exist no free quarks.[23]
(b) Free quarks cannot be produced by interactions of mesons and baryons.[24,25]
(c) Quark field dynamics is based on the existence of heavy quarks.[26]

We have to discuss the implications of these possibilities on QUARFT.

2.2. Analysis of the two-point function.

The states $\int dx\, f(x)\psi(x)|0\rangle = \psi(f)|0\rangle$, $\bar{\psi}(g)|0\rangle$ are the simplest states, which carry the quantum numbers of the quarks. They determine the spectral decomposition of the two-point function[27]:

$$\langle 0|\psi_{a\alpha}(x)\ \bar{\psi}_{b\bar{\beta}}(x')|0\rangle = (2\pi)^{-3}\int dp\ e^{-ip(x-x')}\ \theta(p_o)$$
$$(\omega_1^{(\alpha)}(p^2)+\gamma^{\mu}_{ab}\ p_{\mu}\ \omega_2^{(\alpha)}(p^2))\,\delta_{\alpha\bar{\beta}} \qquad (2.2)$$

where the spectral functions ω_1, ω_2 are defined by sums over intermediate states with fixed four-momentum p:

$$\sum_{\eta}\langle 0|\ \psi_{a\alpha}(0)|p\eta\rangle\langle p\eta|\ \bar{\psi}_{b\bar{\alpha}}(0)|0\rangle =$$
$$(2\pi)^{-3}(\omega_1^{(\alpha)}(p^2)\ \delta_{ab} + \gamma^{\mu}_{ab}\ p_{\mu}\ \omega_2^{(\alpha)}(p^2)) \ . \qquad (2.3)$$

From isospin invariance it follows $\omega_i^{(p)} = \omega_i^{(n)} \equiv \omega_i^{(I)}$;

but we may have $\omega_i^{(I)} \neq \omega_i^{(\lambda)}$ by SU(3)-breaking.

Disregarding subtractions, we get for the Green's function of the time ordered product

$$<0|T\psi(x)\ \bar{\psi}(x')|0> = i(2\pi)^{-4} \int dp\ e^{-ip(x-x')} \int_0^\infty ds\ \frac{\omega_1(s) + \gamma p\ \omega_2(s)}{p^2 - s + i\varepsilon} \tag{2.4}$$

$$= (2\pi)^{-4} \int dp\ e^{-ip(x-x')}\ S_F'(p).$$

A free quark state is described in QUARFT by a single quark particle state

$$\left| \begin{matrix} m_\alpha & 1/2 & \alpha \\ P & S_3 & \end{matrix} \right\rangle$$

which is an eigenstate of the discrete spectrum of the mass-operator (Sect. 1.2.2.). Its amplitudes

$$<0|\psi(0)\left| \begin{matrix} m_\alpha & 1/2 & \alpha \\ p & S_3 & \end{matrix} \right\rangle = \frac{\sqrt{2\ m_\alpha}}{(2\pi)^{3/2}}\ U^{S_3}(p);$$

$$\left\langle \begin{matrix} m_\alpha & 1/2 \\ p & S_3 \end{matrix} \right|\ \bar{\psi}(0)|0> = \frac{\sqrt{2\ m_\alpha}}{(2\pi)^{3/2}}\ \bar{U}^{S_3}(p) \tag{2.5}$$

($\sum_{S_3} U^{S_3}(p)\ \bar{U}^{S_3}(p) = (2\ m_\alpha)^{-1}\ (m_\alpha + \gamma p)$ contribute a δ-function to the l.h.s. of Eq.(2.3.):

$$\omega_1^{(\alpha)}(p^2) = m_\alpha\, \delta(p^2 - m_\alpha^2) + \bar{\omega}_1^{(\alpha)}(p^2)$$

$$\omega_2^{(\alpha)}(p^2) = \delta(p^2 - m_\alpha^2) + \bar{\omega}_2^{(\alpha)}(p^2) \ . \tag{2.6}$$

The states of the continuous mass spectrum contribute to $\bar{\omega}_i(p^2)$. In an asymptotic complete field theory these states would decay in outgoing quark plus conventional particle states. Hence the support of these functions would be:

$$\text{Support} \quad \omega_i^{(\alpha)}(p^2): \quad p^2 \geq m_{oi}^2 > m_\alpha^2 \tag{2.7}$$

On the other hand, an isolated δ-function in the spectral decomposition Eq.(2.2) would imply that a state $\bar{\psi}(f)|0>$, where the support of the Fourier transform of f is concentrated around the positive quark mass shell, is a normalizable state, and hence a single particle state with the internal quantum numbers of the quarks.

Therefore our three possible solutions of the quark puzzle show up in the two-point function in the following form:

(a) No free quarks: The $\omega_i(p^2)$ have no δ-singularity.

(b) No quark production by conventional particles: There may be a δ-function singularity in $\omega_i(p^2)$, but the higher Green's functions do not have all the quark poles required by the single particle structure of Green's functions. (We shall discuss a model of this type proposed by K. Johnson[24] in Sect.2.3).

(c) Heavy quarks: There is a single particle contribution to the two-point function, but the support of the spectral function is well above the conventional single particle hadron masses.

Eq.(2.5) together with our normalization of single particle states (Sec.1.2.2.) defines our quark fields as <u>renormalized</u> fields. The vacuum expectation value of the equal time anticommutator determines the renormalization constant Z_2:

$$\langle 0|[\psi(x),\bar{\psi}(x')]_+|0\rangle_{x_0=x_0'} = \gamma_0\delta(\vec{x}-\vec{x}')\cdot Z_2^{-1} \tag{2.8}$$

$$(Z_2^{\alpha})^{-1} = \int_0^{\infty}\omega_2^{\alpha}(s)\,ds = 1+\int_{m_0^2}^{\infty}\bar{\omega}^{(\alpha)}(s)\,ds \quad . \tag{2.9}$$

If QUARFT could be based on canonical field theory, Z_2 would be the proportionality constant between unrenormalized fields ${}^u\psi$, satisfying canonical commutation relations and the renormalized fields: $\psi(x) = \sqrt{Z_2}\,{}^u\psi(x)$. Positive metric in Hilbert space implies a positivity condition for the spectral functions:

$$3\sqrt{s}\,\omega_2(s) \geq \omega_1(s) > \sqrt{s}\,\omega_2(s) \geq 0 \quad . \tag{2.10}$$

Therefore $0 \leq Z_2 \leq 1$. These considerations are completely formal and depend on the high energy behaviour of $\bar{\omega}^{(\alpha)}(s)$ i.e. the convergence of the integral (2.9) and necessary subtractions in Eq.(2.4). In the case of heavy quarks we are interested in an expansion of the two-point Green's function for small p^2, because this describes the behaviour of strongly bound quarks with small "kinetic energy p^2":

$$S_F'(p) = i \int_{m_q^2}^{\infty} ds \, \frac{\omega_1(s) + \gamma p \, \omega_2(s)}{p^2 - s + i\varepsilon}$$

$$\approx - i \int \frac{ds}{s} \, \omega_1(s) - i \, \gamma p \int \frac{ds}{s} \, \omega_2(s) - i \, p^2 \int \frac{ds}{s^2} \omega_1(s)$$

$$= - i \, (\omega_{11} + \gamma p \, \omega_{21} + p^2 \, \omega_{12}) + \ldots \qquad (2.11)$$

The inverse propagator has the expansion

$$(S_F')^{-1} = - i \, Z \, (\not{p} - \hat{m} - \frac{1}{\kappa} \, p^2) \qquad (2.12)$$

with

$$Z = \frac{1 + m_q^2 \, \bar{\omega}_{21}}{(1 + m_q \, \bar{\omega}_{11})^2} \, ,$$

$$\hat{m} = m_q \, \frac{1 + m_q \, \bar{\omega}_{11}}{1 + m_q^2 \, \bar{\omega}_{21}} \, , \qquad (2.13)$$

$$\frac{1}{\kappa} = \frac{1}{m_q} \, \frac{2 \, \bar{\omega}_{21} + (m_q \, \bar{\omega}_{11})^2 - m_q \, \bar{\omega}_{12} - m_q^2 \, \bar{\omega}_{11} \, \bar{\omega}_{12} - m_q^{-1} \, \bar{\omega}_{11}}{1 + m_q^2 \, \bar{\omega}_{21} + m_q \, \bar{\omega}_{11} + m_q^3 \, \bar{\omega}_{11} \, \bar{\omega}_{21}}$$

Using the positive norm inequalities (2.10) we get the result

$$\bar{\omega}_{21}=\int_{m_{o2}^2}^{\infty} \frac{\bar{\omega}_2(s)}{s}ds<\int_{m_{o2}^2}^{\infty} \frac{\bar{\omega}_1(s)}{s^{3/2}}ds\leq \frac{1}{m_q}\int_{m_{o2}^2}^{\infty} \frac{\bar{\omega}_1(s)}{s}ds = \frac{1}{m_q}\bar{\omega}_{11} \qquad (2.14)$$

and therefore

$$\hat{m} > m_q \quad . \qquad (2.15)$$

In a similar way one can show that $\frac{1}{\kappa}$ is of the order $\frac{1}{m_q}$. Hence in models working with "free" inverse propagators, these may be interpreted as effective Dirac operators with an effective quark mass, related approximately to the true one by the renormalization constant Z. We want to emphasize that these estimates of $\hat{m}$ and κ depend on the assumption of unsubtracted dispersion relations for the two-point function.[28]

2.3. Theories with no free quarks?

Is it possible to have a solution of the "Quark Puzzle" within QUARFT by assuming

(a) There no single particle quark states?

or

(b) Free quarks can not be produced in collisions of mesons and baryons.

We cannot answer these questions. There are many conceptional and formal problems involved. The following remarks, which we owe mainly to conversations with R.Haag and J.E. Roberts, should shed some light on these questions.

(a) A main conceptional problem in this approach is the quantum mechanical interpretation (Sect. 1.2.1. and 1.2.4.) of states $\bar{\psi}(f)|0\rangle$, or more general, of states with fractional baryon quantum number. We call the subspace of the states with fractional B' the quark sector H_q, the subspace with integer B' the hadronic sector H_h; $H = H_q \oplus H_h$. Of course if the quark sector does not contain single particle states, asymptotic incoming and outgoing states do not form a base of H, such a QUARFT is therefore not asymptotic complete. We do not have the simple particle interpretation of field theory.

Is there a possibility to avoid the prediction of the observation of fractional charges in finite space-time regions, by restricting the set of states and observables to those which describe conventional hadron physics?

In order to fix ideas, let us assume that the algebra A generated by e.m. and weak currents[29]:

$$A = \{J = \int \ldots \int j(x_1)\ldots j(x_n) f(x_1\ldots x_n) dx_1 \ldots dx_n\}$$

would be sufficient to describe conventional <u>hadron states</u> by linear expectation functionals $\omega_h(J)$, $J \in A$. There is a theorem[30] which says that these $\omega_h(J)$ might be approximated by the states $\omega_o(J)$ of the vacuum sector; i.e. those states described by density matrices in the Hilbert space $H_o = A|0\rangle \subset H_h \subset H$. But the same theorem assures also that the expectation functionals of A with respect to states of the total Hilbert space H can be weakly approximated by those of the vacuum sector. As an illustration, let us consider the state $\phi(\lambda) = \bar{\psi}_\lambda(f)|0\rangle \in H_q$ with a test function f being approximately a (normalized) characteristic function of a space time region θ; the "expectation function" $<\phi(\lambda)\ j_\mu^{e.m.}(x)\ \phi(\lambda)>$, $<\phi(\lambda)\ j_\mu^{e.m.}(x)\ j_\nu^{e.m.}(x')\ \phi(\lambda)>$ etc., shows that in this state it is almost certain that there is a charge $-\frac{1}{3}$ in θ. The theorem assures that there is a (hadronic) state in the vacuum sector, for which $\omega_o(j_\mu(x))$ etc. also contains the information that with almost certainty there is a charge $-\frac{1}{3}$ in θ. Therefore, in the framework of QUARFT, it is not possible to circumvent the possibility of the observation of fractional charges on a general level. Of course the theorem says nothing about the difficulty to prepare such a state in the hadron sector, and therefore one may question the practical relevance of this theorem.

(b) One may adopt a pragmatic point of view and consider QUARFT only as a formal frame in which S-matrix elements of hadronic processes are described according to Eq.(1.7). The main problem in such an approach is the unitarity of the S-matrix. Unitarity would be violated, if in hadron—induced collisions any kind of states gets produced, which does not correspond to a set of outgoing hadrons. Recently K.Johnson[24] proposed a model with a mechanism which tries to avoid this difficulty. Because it gives some insight into our problem, we shall have a look at this model. Johnson assumes field equations for scalar quark fields q(x):

$$(\Box + m^2)\ q(x) = -\ I(x)\ q(x). \qquad (2.16)$$

m denotes the mass of free physical quarks, I(x) is a local field operator, possibly a function of q(x), which describes the interaction of the quarks. In the notation of Johnson, states describing incoming our outgoing hadrons are called good states $|G\rangle \varepsilon H_h$, those which contain free outgoing "quarks" are bad states $|B\rangle$. It is consistent with this interpretation to assume:

$$\langle B|q(x)|G\rangle \not\equiv 0; \qquad \langle G|q(x)|G\rangle \equiv 0, \qquad \langle B|q|B'\rangle \equiv 0 \quad (2.17)$$

In order that free quarks are produced in hadron collisons:

$$G \to B + q$$

the corresponding Green's functions (Sect.1.2.3.d), and hence the matrix element $\langle B|q(x)|G\rangle$ must have a pole at the quark mass m. Usually this pole is a consequence of the field equation:

$$(p^2 - m^2)\langle B|q|G\rangle = \sum_{B'} \langle B|I|B'\rangle\langle B'|q|G\rangle \qquad (2.18)$$

where $p^2 = (p_B - p_G)^2$ is the virtual mass of the quark. Johnson argues that like in the Schrödinger equation of the harmonic oscillator:

$$(\vec{p}^{\,2} - 2\, m\, E)\ \psi(\vec{p}) = \omega^2\ \Delta_p\ \psi(\vec{p}) \qquad (2.19)$$

sufficient singular contributions to the r.h.s. of Eq.(18) might compensate the quark pole. Therefore he assumes that $\langle B|I|B'\rangle$ is so singular when $m_B^2 = m_{B'}^2$ and $t = (p_B - p_{B'})^2 \to 0$ that the effect of the integration over $p_{B'}$ in (18) produces a differential operator in p^2. Johnson discusses a special model with

$$\langle B|I|B'\rangle \approx 4\ \pi^2 \gamma\, m_B^5 \frac{\partial}{\partial t}\, \delta''(-\sqrt{t})\,. \tag{2.20}$$

In the case of the harmonic oscillator, the infinitely rising potential prevents dissociation of the bound state at any energy, and therefore the harmonic oscillator wave function has no single particle pole. Similarly the interaction density $I(x)$ in Johnson's model is not a localized field quantity, but has an infinite, "strong" range. For illustration, we may consider the matrix element of $I(x)$ between a localized quark state: $\langle q(f)\ I(0)\ \bar{q}(f)\rangle$. We separate the quark from the interaction center (possibly the position of the other quark in the bound state) by a space-like translation a: $f(x) \rightarrow f(x+a) \equiv f_a(x)$.

It follows from (2.20) that

$$\langle q(f_a)\ I(0)\ \bar{q}(f_a)\rangle \rightarrow \text{const.}|a^2| \quad \text{for} \quad a^2 \rightarrow -\infty,$$

whereas for local interaction densities the cluster decomposition theorem (Sect.1.2.3.b) says:

$$\langle q(f_a)\ I(0)\ \bar{q}(f_a)\rangle \rightarrow \langle 0|I(0)|0\rangle\langle q(f)\bar{q}(f)\rangle = \text{const.for } a^2 \rightarrow -\infty.$$

Therefore we hesitate to accept Johnson's model as a solution of the quark puzzle in the framework of conventional field theory.

In the following we make the assumption of the existence of real heavy quarks. But we are open to the possibility that there may be some limiting theory with infinitely heavy quarks as a QUARFT with no free quarks.

3. A dynamical pattern of the relativistic quark model.

Since we do not have field equations or their equivalents, we have to rely on

(a) phenomenological patterns (models),
(b) the general structure of field theory

for the discussion of hadronic processes. Summing up the results of phenomenology, one may tentatively adopt the standpoint that the properties of (unstable) one particle states determine the dynamical structure of hadronic processes, at least at low and intermediate energies. In the following we try to discuss, how such a dynamical pattern fits into the frame work of general quark field theory.

3.1. Bound states and Bethe-Salpeter equations.

The description of bound states - and resonances - in field theory plays a central role in such a program. We already mentioned in Sect. 1.2.2.b that single particle states are described by B S amplitudes. In the quark model, the simplest non-vanishing B S amplitudes of mesons and baryons are:

$$(2\pi)^{-3/2}\chi_P(q) = \int dx\ e^{iqx} \langle 0 | T\psi(\tfrac{x}{2})\bar{\psi}(-\tfrac{x}{2}) \begin{vmatrix} \text{Mes} \\ P \end{vmatrix}, r\rangle; \quad \begin{matrix} p_1 = \frac{P}{2} + q \\ p_2 = \frac{P}{2} - q \end{matrix}$$

$$(2\pi)^{-3/2}\chi_P(p_1p_2p_3)\ \delta(p_1+p_2+p_3-P) = (2\pi)^{-4} \int dx_1 \dots dx_3 \cdot \\ \cdot\ e^{i(p_1x_1+p_2x_2+p_3x_3)} .$$

$$\cdot \langle 0 | T\psi(x_1)\psi(x_2)\psi(x_3) \Big| \begin{matrix} \text{Baryon} \\ P \end{matrix} \rangle$$

It follows from the many-particle structure analysis (MPSA) of Green's functions[31] that these amplitudes satisfy B S equations, to be described now. Although the setting of MPSA is general quantum field theory, the many-particle structure of Green's functions corresponds to that of perturbation theory and may be described with help of Feynman graphs. For this purpose we introduce a graphical notation:

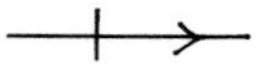

denotes the full two-point Green's function S'

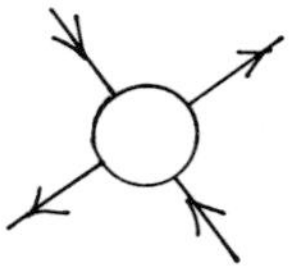

the connected, amputated four-point function $\underline{\eta}$

$\eta(p_1p_2;\bar{p}_1\bar{p}_2) = S'(p_1)\ S'(p_2)\cdot$

$\underline{\eta}(p_1p_2;\bar{p}_1\bar{p}_2)\ \bar{S}'(\bar{p}_1)\ \bar{S}'(\bar{p}_2)$, etc.

Connecting argument lines of higher Green's functions with two-point lines, means multiplication (in matrix sense with respect to spinor and quark indices); one has to integrate over inner lines. In this spirit, the B S kernel

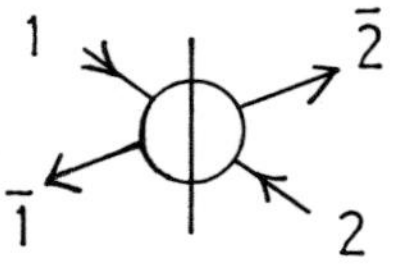

$:K(p_1p_2;\bar{p}_1\bar{p}_2)$ is defined by the graphical equation

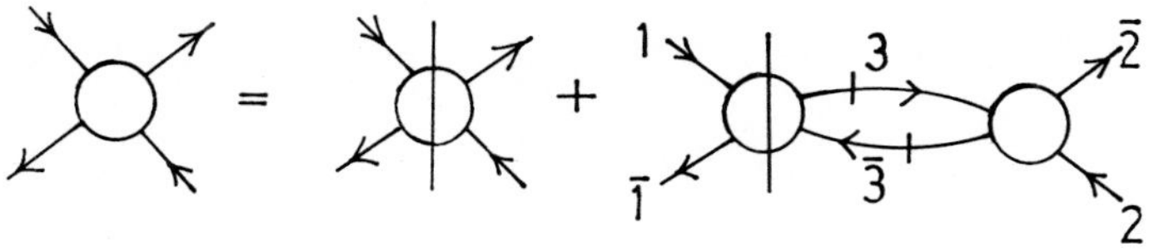

(3.1)

$$\underline{n}(p_1p_2;\bar{p}_1\bar{p}_2) = K(p_1p_2;\bar{p}_1\bar{p}_2) + \iint dp_3d\bar{p}_3K(p_1p_3;\bar{p}_1\bar{p}_3)\cdot$$

$$\cdot\ S'(p_3)\ \underline{n}(p_3p_2;\bar{p}_3\bar{p}_2)\ \bar{S}'(\bar{p}_3)$$

The kernel K is "two-particle irreducible" in the channel $1,\bar{1} \rightarrow (2,\bar{2})$ and is therefore - at least in the studied perturbation theoretic treatment - simpler as the full Green's function. Similarly the "three-particle irreducible" kernel

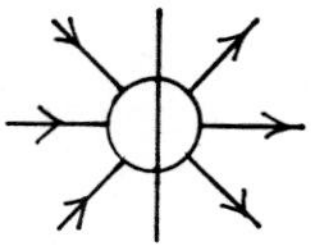

is defined by the connected, amputated six-point function

and the corresponding four- and two-point functions with help of the graphical equation[32]:

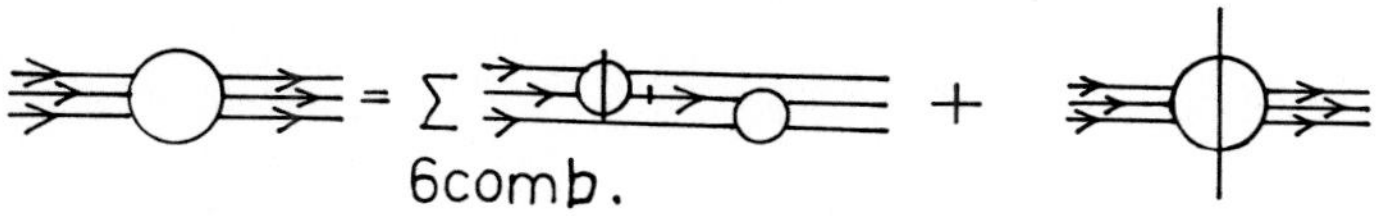

(3.2)

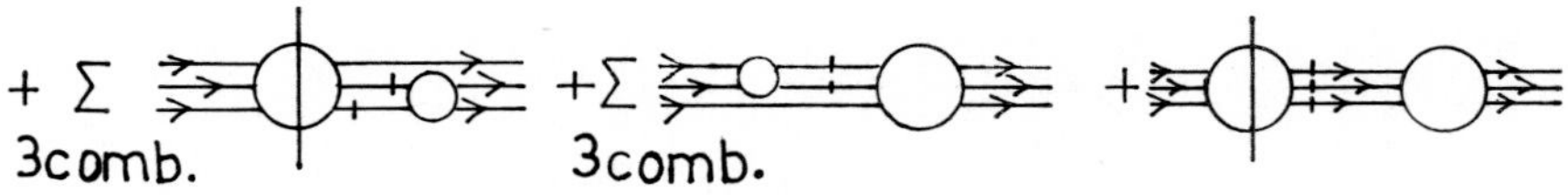

There is a simpler form of the graphical equations (3.1) and (3.2) which indicates how the equations for higher Green's functions look like[33]:

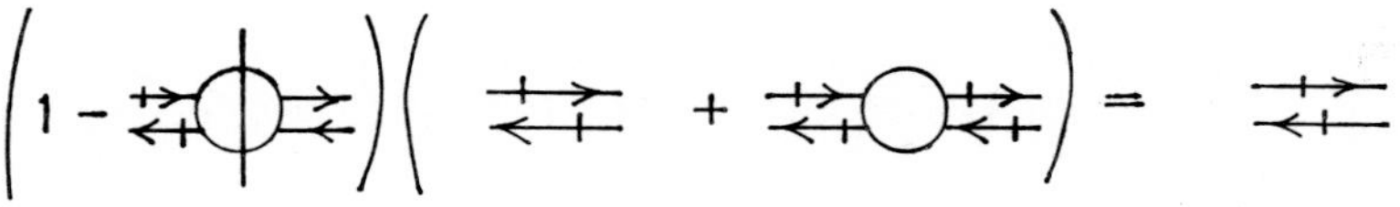

(3.3)

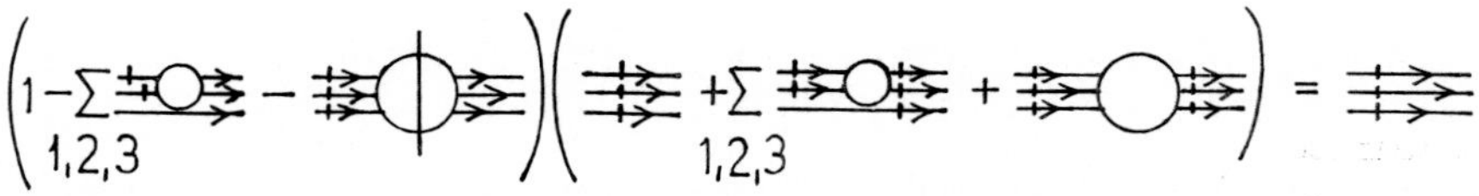

Now we regard the B S equations (3.1), (3.2) in the neighbourhood of the bound state mass shell, where according to Eq.(1.5):

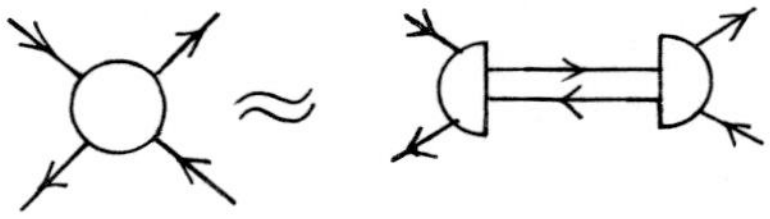

$$= i \frac{\sum_r \underline{\chi}^r(q,P)\,\bar{\underline{\chi}}^r(\bar{q},P)}{P^2 - M^2 + i\varepsilon}$$

$$\text{with } \underline{\chi}^r = S^{-1}(p_1)\,\chi^r\,\bar{S}^{-1}(p_2) \tag{3.4a}$$

and

$$= i \frac{\sum_r \underline{\chi}_P^r(p_1p_2p_3)\,\bar{\underline{\chi}}_P^r(p_1p_2p_3)}{P^2 - M^2 + i\varepsilon}$$

$$\text{with } \underline{\chi}^r(p_1p_2p_3) = S^{-1}(p_1)\; S^{-1}(p_2)\; S^{-1}(p_3)\;\; \chi(p_1p_2p_3)$$

(3.4b)

We insert these expressions into the inhomogeneous B S Eqs.(3.1), (3.2) and expand around $P^2 = M^2$. Comparison of the pole terms on both sides of these equations leads to the homogeneous B S equations of the bound state amplitudes[34]

(3.5a)

and

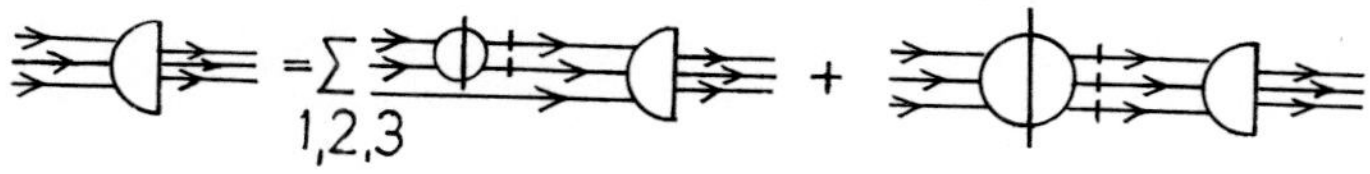

(3.5b)

The zero order term in $(P^2 - M^2)$ gives the field theoretical normalization condition for the B S amplitudes[35].

The discussion of the B S equation of the mesons in the quark model (3.5a)

$$\chi(q,P) = i \int K(q,q',P,\lambda) S^{-1}(p_1') \chi(q',P) S^{-1}(p_2') dq' \tag{3.6}$$

and its solutions in Chapt. B will be one of the main points of these lectures. In the spirit of our approach, we shall regard the form of $K(q,q',P,\lambda)$ and $S(p_i)$ as an input, to be chosen in such a way that the experimental facts should be reproduced under the assumption of heavy quarks (Sect.2.2). As trivially suggested by practical reasons, we will begin with simple forms of Eq.(3.6).

In view of our discussion of the two-point function of heavy quarks, we shall start with propagators of free quarks

$$S^{-1}(p) = (\gamma p - m), \; \bar{S}^{-1}(p) = (\gamma p + m) \tag{3.7}$$

considering modifications according to Eq.(2.12) later. The kernel is assumed to be of convolution type

$$K(q,q',P,\lambda) = \lambda \, K(q - q') \; . \tag{3.8}$$

Under these assumptions Eq.(3.6) may be written as

$$(\gamma p_1 - m)\chi(q,p)(\gamma p_2 + m) = i \; \lambda \int K(q-q')\chi(q',P)dq' \tag{3.9}$$

and the normalization condition gets the simple form

$$(2\pi)^{-4} \; i \; \mathrm{Tr}\int \bar{\chi}^{r}(q,P)(p_1\gamma - m)\chi^{r'}(q,P)(p_2\gamma+m)dq = \lambda\frac{dP^2}{d\lambda}\delta_{rr'} . \tag{3.10}$$

In Sect.9.1. we shall report on some attempts to tackle the relativistic baryon equations.[36]

Of course in discussing more complex phenomena including more refined features of the whole theory (f.i. crossing symmetry) we have to be prepared to consider more complex models. Only when this theoretical development converges after some time, we shall consider our approach to quark field theory as successful.

3.2. Tree graph approximation.

Our approach to the dynamics of more complex physical problems (Sect.1.4.) is motivated mostly by phenomenological experience and relies only vaguely on the general structure of field theory. In the present understanding, hadron dynamics is dominated by resonances as intermediate states. This dynamical pattern, called "duality", was first proposed and analyzed for meson-nucleon scattering and later applied to meson-meson scattering[37], deep inelastic electron scattering[38] and e^+e^- annihilation into hadrons.[39]

The reason for quark concepts being important in these dual dynamical schemes is the fact that intermediate mesons and baryons with exotic quantum numbers seem to play a minor role than those with non-exotic ones. An approximation which uses these ideas most consequently starts with tree graphs, having all the mesons and baryons of the quark model spectrum as internal lines[40].

Of course, such a procedure can only be considered as a very crude approximation. It must be modified, in order to account for the instability of most of the intermediate states, it has to take into account all the general requirements of crossing symmetry, analyticity and unitarity following from field theory. But this approach introduces an hierarchy in the interactions, where the forces responsible for the particle structure ("superstrong forces") are more important than those responsible for their hadronic decays ("mesonic forces"). And as already mentioned, there is some indication for such a hierarchy in nature.

For the tree graph approximation one has to know the hadronic vertices. In the relativistic quark model we shall try to calculate them from the following graph[41]:

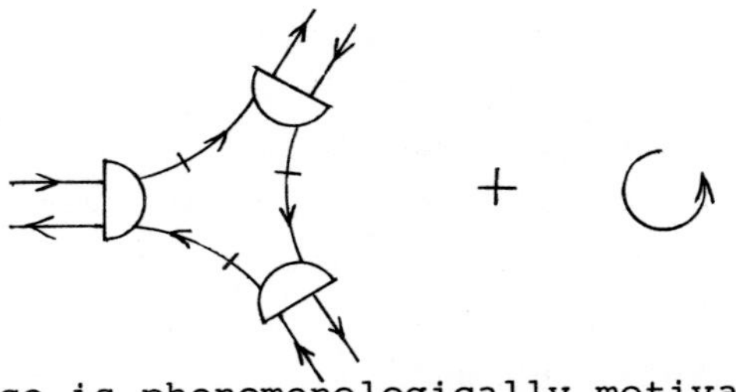

(3.11)

This choice is phenomenologically motivated. We are guided by the successful selection rules derivable from the nonrelativistic quark model[42] or the quark duality diagrams.[43] The justification of this whole scheme with help of field theoretical arguments from MPSA under the assumption of heavy quarks with strong coupling is still to be found. The study of this limit might also contribute to the solution of the quark puzzle (Sect.2.2. - 2.3.). In Sect. 8 we shall discuss the specific problems appearing in the calculation of the hadronic vertices. We shall study some general features of the tree graph approximation with such vertices for the reactions $e^+e^- \to$ hadrons (Sect.7.2.).

3.3. Currents and fields.

An important part of the interpretation of QUARFT relies on the relation of weak, electromagnetic currents to the fundamental quark fields. We assume as a local form of the commutation relations (1.8) the equal time commutators

$$[j_o^Q(x),\ \psi_\alpha(y)]_{x_o=y_o} = Q'_\alpha\ \psi_\alpha(x)\ \delta(x-y)$$

$$[j_o^Y(x),\ \psi_\alpha(y)]_{x_o=y_o} = Y'_\alpha\ \psi_\alpha(x)\ \delta(x-y) \qquad (3.12)$$

The consequences of this for Green's functions are the Ward-Takahashi[44] identities. As a special example we mention for the amputated $j_\mu(x)$ $\bar{\psi}(x')$ $\psi(x'')$ Green's function:

$$\Gamma_\mu(q,q') = S'^{-1}(q)\ \eta(\psi(q)\ \bar{\psi}(q')\ j_\mu(k))\ \bar{S}'^{-1}(q') \tag{3.13}$$

the relation

$$(q - q')^\mu \Gamma_\mu(q,q') = Q'_\alpha\ (S'^{(\alpha)}_F(q) - S'^{(\alpha)}_F(q')), \tag{3.14}$$

where $S'^{(\alpha)}_F(q)$ is the two-point Green's function, discussed in Sect.2.2.[45]

The connection between the currents and quark fields becomes more definite by making the ansatz:

$$j_\mu^{(i)}(x) = (Z_2^i)^{-1}\ [\bar{\psi}(x),\ \gamma_\mu \lambda^i \psi(x)]\ . \tag{3.15}$$

Such an ansatz is a quite general feature of Lagrangean field theory models. It relates n-point functions of quark fields directly to weak and electromagnetic properties. Such applications shall be discussed in Sect.7.2.

It is generally assumed that the currents of quantum field theory should satisfy the relations of "Current Algebra". In our partly phenomenological approach, which relies strongly on the dynamics of bound states, it is not yet clear to what extent we can satisfy this postulate.[46]

B. THE FERMION-ANTIFERMION-BETHE-SALPETER EQUATION

4. Heavy quarks and strong binding.

The nonrelativistic quark model has been very successful in providing a scheme for the classification of the hadrons.[47] Therefore, a relativistic description of the resonances, as boundstates of quarks, should be orientated at the ideas of the phenomenological model.[48] In a field theoretical description, the Bethe-Salpeter equation[49] is the appropriate framework (see Sect.3.1.). Since, however, we do not know the forces between the quarks, we have to use a phenomenological ansatz for the interaction kernel K. If we accept as solution to the "quark puzzle" the existence of heavy quarks (Sect.2.1.), then we are led to look for solutions of the B S equation under the condition of heavy constituents and strong binding. In order to get a feeling for this dynamical situation we shall discuss first a model with scalar quarks.[50,51]

4.1. The scalar B S equation for strong binding of heavy quarks.

The scalar amplitude

$$\tilde{\Phi}(x,P) = (2\pi)^{3/2} \langle O|T(\phi[\tfrac{x}{2}]\ \bar{\phi}\ [\tfrac{-x}{2}])|P,M\rangle \qquad (4.1)$$

M = boundstate mass

satisfies the momentum space B S equation

$$\Phi(q,P) = i\,\Delta_F'\,(q-\tfrac{P}{2})\ \Delta_F'(q+\tfrac{P}{2})\int dq'K(q,q',P)\Phi(q',P). \qquad (4.2)$$

We shall make the assumption that the kernel is of convolution type

$$K(q,q',P) = K(q - q') \equiv V(q - q') \ . \qquad (4.3)$$

We represent these kernels by a superposition of Yukawa-type potentials (generalized ladder type)

$$V(q - q') = \sum_i \frac{\alpha_i}{(q-q')^2 - \mu_i^2 + i\varepsilon} \qquad (4.4)$$

thereby preserving the field theoretically required analyticity structure. The propagators Δ_F' will be assumed to be the free ones.

One advantage of the analyticity structure of the kernel is that we can perform, in the restframe $(P = (M,\vec{0}))$, the Wick rotation[52] in Eq.(4.2), i.e. we analytically continue in the relative energy q_o

$$q_o \to i\, q_4, \qquad q^2 \to - q^2_{\text{Euclid}} \ . \qquad (4.5)$$

We obtain the Wick rotated B S equation

$$[(q^2 + m^2 - \frac{M^2}{4})^2 + q_4^2\, M^2]\,\Phi_M(q_4,\vec{q}) = -\int d^4q' V(q - q')\,\Phi_M(q_4',\vec{q}\,') \ . \qquad (4.6)$$

m = quark mass.

In Ref: 50 a simple choice was made for the superposition of Yukawa's

$$V_k(q-q') = \frac{\lambda}{[(q-q')^2+\mu^2]^k} = \frac{(-1)^{k-1}}{(k-1)!}[\frac{\partial}{\partial\mu^2}]^{k-1}[\frac{\lambda}{(q-q')^2+\mu^2}] \ . \qquad (4.7)$$

Figure 1 shows these "potentials" in Euclidean configuration space

$$(R = \sqrt{x_4^2 + \vec{x}^2}) \quad .$$

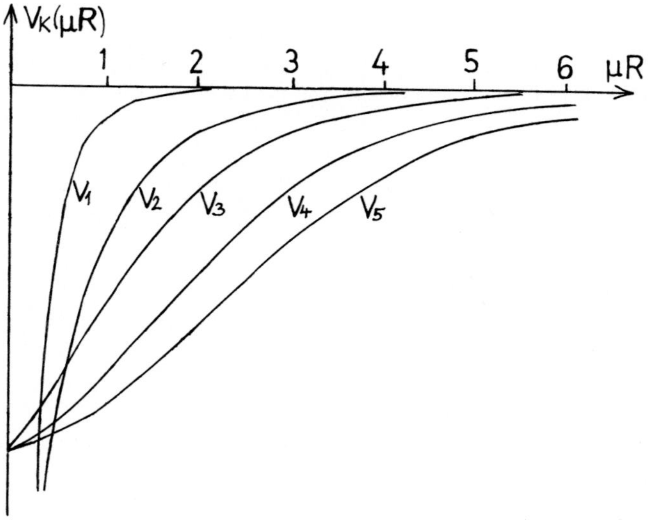

Fig. 1

The assumption of heavy quarks suggests that we use as starting point the solutions of the boundstate mass $M = 0$ equation and then perform a perturbation calculation in the boundstate mass M.

The Wick rotated boundstate mass zero equation is invariant under fourdimensional rotations, i.e. the group O(4). It is therefore natural to expand the amplitude into O(4)-spherical harmonics:[53,54]

$$\Phi(q,\Omega^4) = \sum_{n,\ell,m} \phi^n_{\ell,m}(q^2) Y^n_{\ell,m}(\Omega^4) \tag{4.8}$$

Insertion of this ansatz into Eq. (4.6) leads to a system of coupled equations for the radial functions $\phi^n_{\ell,m}(q^2)$ with the coupling proportional to

$$\varepsilon^{\left|\frac{n'-n}{2}\right|} \quad , \quad (n'-n = \text{even}) ,$$

where

$$\varepsilon = \frac{q^2 + m^2 - \frac{M^2}{4} - \sqrt{(q^2 + m^2 - \frac{M^2}{4})^2 + q^2 M^2}}{q^2 + m^2 - \frac{M^2}{4} + \sqrt{(q^2 + m^2 - \frac{M^2}{4})^2 + q^2 M^2}} \quad . \tag{4.9}$$

For strong binding ($M^2 << 4 m^2$) one obtains

$$|\varepsilon| \approx \frac{q^2 M^2}{4(q^2 + m^2)^2} \leq \frac{M^2}{16 m^2} << 1 \text{ for all } q^2 \quad . \tag{4.10}$$

Up to order ε^2 the coupled system is equivalent to the set of uncoupled integral equations

$$\phi^n_\ell(q^2) = \frac{2(1 + \frac{n(n+2) - 2\ell(\ell+1)}{n(n+2)} \cdot \varepsilon)}{(q^2+m^2 - \frac{M^2}{4})(q^2+m^2 - \frac{M^2}{4} + \sqrt{(q^2+m^2-\frac{M^2}{4})^2 + q^2M^2})} \quad . \tag{4.11}$$

$$\cdot \left[\frac{-2\pi^2 \lambda}{(n+1)}\right] \cdot \sum_i \alpha_i \int_0^\infty dq' \, q'^3 \, \frac{h_i^{n+1}(q,q')}{q\, q'} \phi^n_\ell(q'^2) + O(\varepsilon^2) ,$$

where the h_i result from the expansion of the kernel:

$$\frac{1}{q^2+q'^2+\mu_i^2-2(qq')} = 2\pi^2 \sum_{n''\ell''m''} \frac{h_i^{n''+1}(q,q')}{qq'} \frac{Y^{*n''}_{\ell''m''}(\Omega'^4)\, Y^{n''}_{\ell''m''}(\Omega^4)}{n''+1} \tag{4.12}$$

$$h_i(q,q') = \frac{q^2+q'^2+\mu_i^2-\sqrt{(q^2+q'^2+\mu_i^2)^2-4\,q^2\,q'^2}}{2\,qq'}$$

From the mathematical point of view, Eq.(4.11) is an eigenvalue equation for λ or M^2 and one may apply standard techniques, like variational principles to solve this equation.[55]

For single particle exchange (V_1) one cannot obtain a series of low mass bound states,[56] because of the singular behaviour at the origin. In order to get many levels with a spacing small compared to the quark mass, one needs a smooth potential ($k \geq 3$).

The mass spectrum, which one obtains for $\ell = n$ is shown in Fig. 2 for several values of the range parameter μ and for $m = 4$ GeV, $k = 5$. The q-dependence of the amplitude ϕ_o^o is shown in Fig. 3.

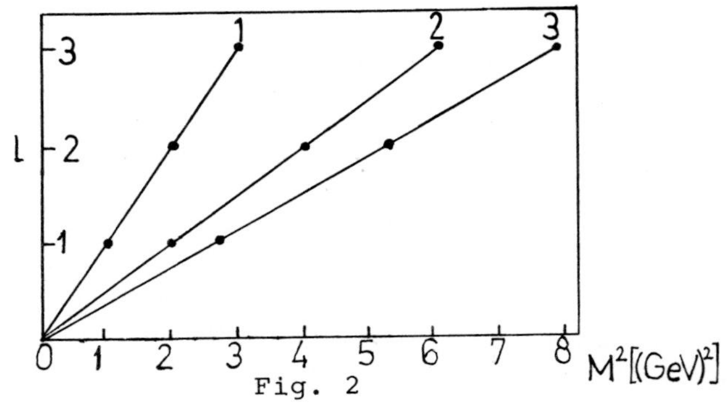

Fig. 2

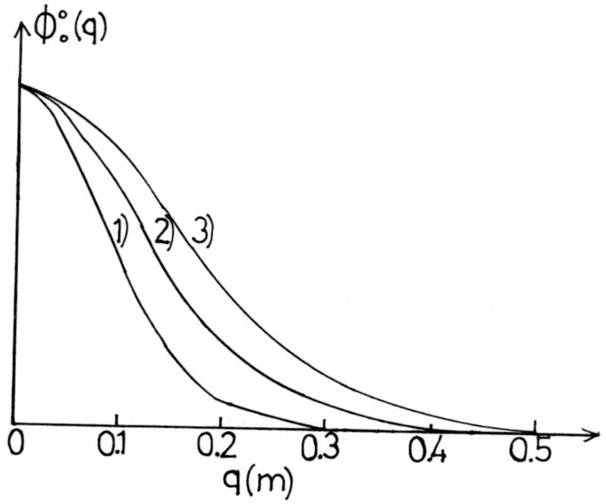

Fig. 3

1) μ = 130 MeV; 2) μ = 250 MeV; 3) μ = 330 MeV;
m = 4 GeV, M_o = 150 MeV;

We obtained with our numerical procedure linear Regge trajectories, and amplitudes decreasing very fast with q_E^2. This means that the quarks are moving in the neighbourhood of the bottom of the potential. We may look for a simpler model, which reproduces these features and can be solved analytically.

4.2. The oscillator approximation.

Consider the Wick rotated B S equation (4.6) for the amplitude (in the c.m.s.):

$$[(q^2+m^2-\frac{M^2}{4})^2 + q_4^2M^2]\Phi(q_4,\vec{q}) = -\int d^4q'\ V(q-q')\Phi(q_4',\vec{q}') \tag{4.13}$$

Since the potential for our meson model has to be smooth around the origin, it can be expanded

$$\begin{aligned} V(R) &= \alpha + \beta R^2 + \ldots \\ V(q) &= \delta^4(q)\ (\alpha - \beta \Box) + \ldots \end{aligned} \qquad (4.14)$$

For the low lying states it is sufficient to keep only terms up to order R^2. Since $M^2 << 4\, m^2$ and $<q^2> << m^2$, we may also neglect the nonleading terms in the inverse propagators on the left hand side. The result is the equation of a fourdimensional oscillator:[57]

$$(2\, q^2 + m^2 - \frac{M^2}{2})\, \Phi(q) = -\left(\frac{\alpha}{m^2} - \frac{\beta}{m^2} \Box\right) \Phi(q)\,. \qquad (4.15)$$

The solutions of this equation are:

$$\Phi(q) = (2\pi)^2 \sqrt{\frac{r!}{(n+r+1)!}}\ \sqrt{\frac{8}{2m^2\beta'}}\left[\frac{q}{\sqrt[4]{\beta'}}\right]^n L_r^{n+1}\left[\frac{q^2}{\sqrt{\beta'}}\right] e^{-\frac{q^2}{2\sqrt{\beta'}}} \cdot Y^n_{\ell,m}(\hat{q}_E)$$

$$\beta' = \frac{\beta}{2m^2}$$

(L_r^{n+1} denotes the standardized Laguerre polynomials[54]) with the eigenvalue

$$\frac{\alpha}{m^2} = -\ (m^2 - \frac{M^2}{2}) - 4\,(n + 2\, r + 2)\ \sqrt{\beta'} \qquad (4.16)$$

$n = 0, 1, \ldots;\quad r = 0, 1, \ldots$

The solutions are characterized by the O(4) quantum

number n and a hyperradial excitation r. In the approximation we made, there is degeneracy with respect to ℓ in an O(4)-multiplet.

The mass spectrum is obtained by inverting the eigenvalue equation:

$$M^2 = 4 \left[\frac{\alpha + m^4}{2 m^2} + 2(n + 2 r + 2) \sqrt{\beta'}\right]. \qquad (4.17)$$

It is seen explicitly that for small bound state masses M the quark mass term has to be compensated by the depth α of the potential.

The solutions (4.16) are normalized according to the field theoretical condition (3.10) which reads:

$$\frac{-1}{(2\pi)^4}\int d^4q\ \bar{\Phi}_{r'}(q,P)\left[\left(\frac{P}{2}+q\right)^2+m^2\right]\Phi_r(q,P)\left[\left(\frac{P}{2}-q\right)^2+m^2\right] =$$

$$=\left(\lambda\frac{\partial P^2}{\partial\lambda}\right)_{P^2=M^2}\delta_{rr'}\ ,\quad \bar{\Phi}(q_4,\vec{q},P) = \Phi^*(-q_4,\vec{q},P) \qquad (4.18)$$

in a Wick rotated scalar model with convolution type kernels and free propagators. In our case

$$\lambda\frac{\partial P^2}{\partial\lambda} = (-m^4)\frac{2}{m^2} < 0$$

and (4.19)

$$\Phi^*(-q_4;\vec{q};P) = (-1)^{n-\ell}\ \Phi^*(+q_4;\vec{q};P)\ .$$

We see that the solutions with negative "relative-time parity", $n - \ell$ = odd, belong to negative norm[58] and have to be omitted on field theoretical grounds.

The characteristics of the spectrum of this relativistic oscillator model are:

(a) Linear trajectories, $M^2 = (a + b\,N)$; $N = n + 2\,r$.

(b) There are oscillations in relative energy, a consequence of the additional relativistic degree of freedom. This leads to a degeneracy of the levels which increases quadratically with N, whereas in the nonrelativistic oscillator the degeneracy increases linearly.

(c) "Odd daughters" are eliminated by our field theoretical framework.

The spectrum is shown in Fig. 4.

The oscillator approximation for the Wick rotated equation simplifies mathematics considerably. However, let us remember that this approximation is good only for $q^2 \ll m^2$. Especially the analyticity structure has changed completely, the B S amplitude has not the one-particle poles at $q^2 = m^2$ and has an essential singularity at infinity. In contrast to this standpoint, Johnson (see the discussion in Sect.2.3.) uses the oscillator also for $q^2 > m^2$, <u>just</u> in order to exclude the appearance of real quarks in hadron-hadron collisions.

A relativistic oscillator model has also been discussed by Feynman, Kislinger and Ravndal.[59] They use it purely phenomenologically in the Lorentz case, encountering the difficulty how to eliminate in a covariant way the exponential increasing factor

$$e^{q_o^2/2\sqrt{\beta'}} .$$

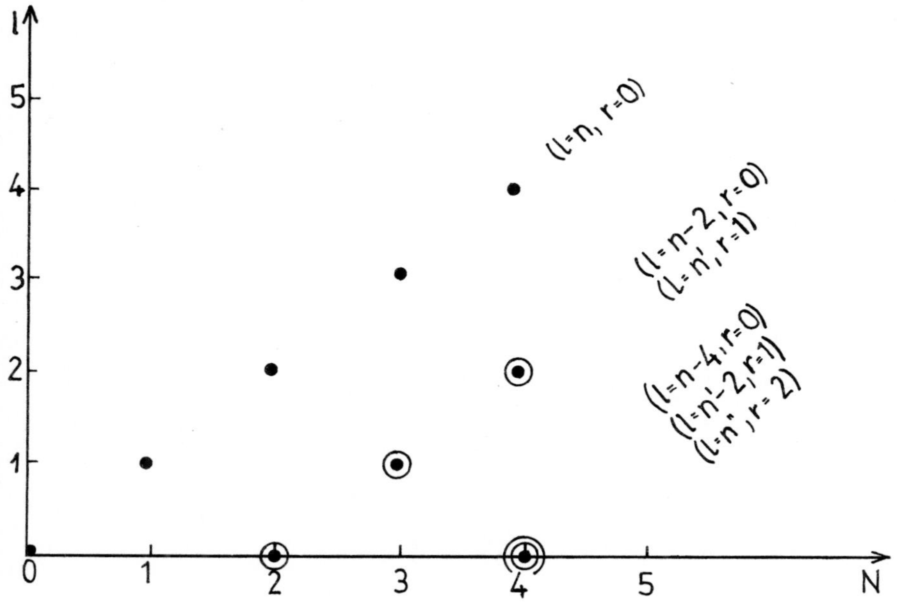

$M^2(N)$

Fig. 4

$$N = n + 2r$$
$$n' = n - 2$$
$$n'' = n' - 2$$

5. The Fermion-antifermion B S-equation.

Along the lines developed for the scalar model we shall attack now the actual problem, the solution of the B S equation for a fermion-antifermion system:[60,61]

$$(\gamma P_1 - m)\ \chi(q,P)\ (\gamma P_2 + m) = i\int K(q,q';P)\ \chi(q',P)\,dq'$$
$$P_1 = \frac{P}{2} + q; \qquad P_2 = \frac{P}{2} - q\ . \tag{5.1}$$

Again we shall restrict ourselves to convolution type kernels without derivative couplings. The general spin structure is:

$$K(q-q') = \lambda^1 K^1 1 \times 1 + \lambda^2 K^2 \gamma_\mu \times \gamma_\mu + \lambda^3 K^3 \sigma_{\mu\nu} \times \sigma_{\mu\nu} +$$
$$+\lambda^4 K^4 \gamma_5\gamma_\mu \times \gamma_5\gamma_\mu + \lambda^5 K^5 \gamma_5 \times \gamma_5 \tag{5.2}$$

It is rather convenient to rearrange the kernel by introducing projection operators P:

$$K = \lambda^S K^S P^S + \lambda^V K^V P^V + \lambda^T K^T P^T + \lambda^A K^A P^A + \lambda^P K^P P^P \tag{5.3}$$

with

$$\begin{pmatrix} P^S \\ P^V \\ P^T \\ P^A \\ P^P \end{pmatrix} = \frac{1}{32} \begin{pmatrix} 2 & 2 & 1 & -2 & 2 \\ 8 & -4 & 0 & -4 & -8 \\ 12 & 0 & -2 & 0 & 12 \\ 8 & 4 & 0 & 4 & -8 \\ 2 & -2 & 1 & 2 & 2 \end{pmatrix} \begin{pmatrix} 1 & \times 1 \\ \gamma_\mu & \times \gamma_\mu \\ \sigma_{\mu\nu} & \times \sigma_{\mu\nu} \\ \gamma_5\gamma_\mu & \times \gamma_5\gamma_\mu \\ \gamma_5 & \times \gamma_5 \end{pmatrix} \tag{5.4}$$

From the discussion of the scalar model we have learned that it is advisable to start from the Wick rotated B S equation for boundstate mass zero and to use the O(4)-symmetry of this equation by making an appropriate group theoretical decomposition of the amplitude.

5.1. $O^{\Pi,C}(4)$-symmetry.

The Wick rotated B S equation for M = O:

$$(\gamma q - im)\ \chi(q)\ (\gamma q - im) = \int K(q,q')\chi(q')dq' \qquad (5.5)$$

is invariant under the transformations of the group O(4) $(x_{\mu'} = O_{\mu}^{\nu}\ x_{\nu})$ extended by three-space reflections Π and charge conjugation C.

O(4) may be represented as the direct product of two-dimensional spinor rotations:[62]

$$O(4) = SU(2) \times SU(2)//Z_2$$
$$O_{\mu}^{\nu} = O_{\mu}^{\nu}(R_+,R_-); \qquad R_{\pm} \in SU(2) \qquad (5.6)$$

The irreducible representations of O(4), extended by Π and C, are characterized by two angular momenta j^+, j^-, an inner charge conjugation parity C' and, for $j^+ = j^-$, an inner O(4) parity π':

$$(j^+,j^-)^{\pi'C'} \qquad \text{for} \quad j^+ = j^-$$

$$(j^+,j^-)^{oC'} \qquad \text{for} \quad j^+ \neq j^- \qquad (5.7)$$

$$[(j^+,j^-)^o = (j^+,j^-) \oplus (j^-,j^+);\ j^+ > j^-] .$$

A standard base is the "product base"

$$\left| \begin{matrix} j^+, j^- \\ m^+, m^- \end{matrix} \right\rangle . \qquad (5.8)$$

The base in which the angular momentum j^2, $\vec{j} = \vec{j}^+ + \vec{j}^-$, is diagonal, is obtained by recoupling:

$$\left| \begin{matrix} j^+, j^- \\ j \ , j_3 \end{matrix} \right\rangle = \sum_{m^+, m^-} \left| \begin{matrix} j^+ j^- \\ m^+ m^- \end{matrix} \right\rangle (j^+m^+; j^-m^- | jj_3) . \qquad (5.9)$$

The O(4) spherical functions $Y^n_{\ell,m}(\Omega^4)$ correspond to the representation $(\frac{n}{2},\frac{n}{2})^+$. The spin of two Dirac particles belongs to the representation

$$(\tfrac{1}{2},0)^o \times (\tfrac{1}{2},0)^o .$$

As an example which shows how the techniques developed for the rotation group[63] can be applied directly in O(4) problems, we consider the reduction of the direct product of two O(4) representations:

$$\left| \begin{matrix} j_1^+ \ j_1^- \\ m_1^+ \ m_1^- \end{matrix} \right\rangle \left| \begin{matrix} j_2^+ \ j_2^- \\ m_2^+ \ m_2^- \end{matrix} \right\rangle = \qquad (5.10)$$

$$= \sum_{\substack{j^+, j^- \\ m^+, m^-}} \left| \begin{matrix} j^+ j^- \\ m^+ m^- \end{matrix} \right\rangle (j_1^+m_1^+; j_2^+m_2^+ | j^+m^+) \cdot (j_1^-m_1^-; j_2^-m_2^- | j^-m^-)$$

or in angular momentum base:

$$\left|\begin{matrix} j_1^+ \; j_1^- \\ \ell_1, m_1 \end{matrix}\right\rangle \left|\begin{matrix} j_2^+ \; j_2^- \\ \ell_2, m_2 \end{matrix}\right\rangle = \sum_{\substack{j^+, j^- \\ \ell, m}} \left|\begin{matrix} j^+ \; j^- \\ \ell, \; m \end{matrix}\right\rangle \left\langle \begin{matrix} j_1^+ \; j_1^- \\ \ell_1, m_1 \end{matrix} ; \begin{matrix} j_2^+ \; j_2^- \\ \ell_2, m_2 \end{matrix} \middle| \begin{matrix} j_2^+ \; j_2^- \\ \ell, \; m \end{matrix} \right\rangle$$

$$\left\langle \begin{matrix} j_1^+ \; j_1^- \\ \ell_1, m_1 \end{matrix} ; \begin{matrix} j_2^+ \; j_2^- \\ \ell_2, m_2 \end{matrix} \middle| \begin{matrix} j^+ \; j^- \\ \ell, \; m \end{matrix} \right\rangle = \sqrt{(2j^+ +1)(2j^- +1)(2\ell_1+1)(2\ell_2+1)} \, . \qquad (5.11)$$

$$\cdot (\ell_1 m_1 ; \ell_2 m_2 | \ell m) \begin{Bmatrix} j_1^+ & j_1^- & \ell_1 \\ j_2^+ & j_2^- & \ell_2 \\ j^+ & j^- & \ell \end{Bmatrix}$$

The fermion-antifermion B S amplitudes transform under $O^{\Pi, C}(4)$ as:

$$[U(R_+, R_-)\chi] \; (q) = \begin{bmatrix} R_+ & O \\ O & R_- \end{bmatrix} \chi[O^{-1}(R_+, R_-)q] \begin{bmatrix} R_+ & O \\ O & R_- \end{bmatrix}^{-1}$$

$$[U(\pi)\chi](q) = \begin{bmatrix} O & 1 \\ 1 & O \end{bmatrix} \chi(q_4, -\vec{q}) \begin{bmatrix} O & 1 \\ 1 & O \end{bmatrix} \qquad (5.12)$$

$$(C\chi)(q) = \begin{bmatrix} -\varepsilon & O \\ O & \varepsilon \end{bmatrix} \chi^T(-q) \begin{bmatrix} \varepsilon & O \\ O & -\varepsilon \end{bmatrix} ; \qquad \varepsilon = \begin{bmatrix} O & -1 \\ 1 & O \end{bmatrix}$$

Therefore they may be expanded in a base which is spanned by the product:

$$(\tfrac{1}{2},0)^{o}\times(\tfrac{1}{2},0)^{o}\times\sum_n(\tfrac{n}{2}\tfrac{n}{2})^{+} =$$

$$= [(0,0)^{++}+(0,0)^{-+}+(\tfrac{1}{2},\tfrac{1}{2})^{++}+(\tfrac{1}{2},\tfrac{1}{2})^{--}+(1,0)^{o+}]\times\sum_n(\tfrac{n}{2}\tfrac{n}{2})^{+}$$

$$\equiv \quad [1 + \gamma_5 + \gamma_\mu + \gamma_5\gamma_\mu + \sigma_{\mu\nu}] \times \sum_n Y^n_{\ell,m} \tag{5.13}$$

(= "Spin" x "Orbit")

Here we have given the first step of the decomposition of this product according to the Clebsch Gordan series. The complete decomposition into irreducible representations[64] is given in Table II.

sect.	$(j^+j^-)^{\pi 'C'}$	parity	charge conjugat.	numb.of equiv.re-present.	{spin, orbit}	$Z^{\xi,n}$
I.	$(\frac{n}{2}\frac{n}{2})^{++}$	$(-1)^j$	$(-1)^n$	3	$\{1;n\}$ $\{\gamma_\mu;n-1\}$ $\{\gamma_\mu;n+1\}$	S_n V_{n-} V_{n+}
II.	$(\frac{n}{2}\frac{n}{2})^{+-}$	$(-1)^j$	$(-1)^{n+1}$	1	$\{\sigma_{\mu\nu};n\}$	T_n
III.	$(\frac{n}{2}\frac{n}{2})^{-+}$	$(-1)^{j+1}$	$(-1)^n$	1	$\{\gamma_5;n\}$	P_n
IV.	$(\frac{n}{2}\frac{n}{2})^{--}$	$(-1)^{j+1}$	$(-1)^{n+1}$	3	$\{\sigma_{\mu\nu};n\}$ $\{\gamma_5\gamma_\mu;n-1\}$ $\{\gamma_5\gamma_\mu;n+1\}$	T_n A_{n-} A_{n+}
V.	$(\frac{n+1}{2}\,\frac{n-1}{2})^{o-}$		$(-1)^{n+1}$	1	$\{\gamma_\mu;n\}$	V_n

VI.	$(\frac{n+1}{2}\ \frac{n-1}{2})\ 0+$	$(-1)^n$	3	$\{\gamma_5\gamma_\mu;n\}$ $\{\sigma_{\mu\nu};n-1\}$ $\{\sigma_{\mu\nu};n+1\}$	A_n T_{n-} T_{n+}

Table II.

$\xi = \{j^+\ j^-;\pi\text{'}C\text{'}\}$ η = degeneracy parameter

For illustration we give examples for base functions in Sectors I and III which are chosen as eigenfunctions of the O(4) "angular momentum":

$$S_n^I(\hat{q}) = \tfrac{1}{2}Y^n(\hat{q}) \tag{5.14}$$

$$V_{n\pm}^I(\hat{q}) = \sum_{\substack{S,\ell \\ S_3,m}} Y_{S,S_3}^{\frac{1}{2}\frac{1}{2}}\ Y_{\ell,m}^{\frac{n\pm1}{2},\frac{n\pm1}{2}}(\hat{q})\cdot \left[\begin{array}{cc|c} \frac{1}{2}\ \frac{1}{2} & \frac{n\pm1}{2}\ \frac{n\pm1}{2} & \frac{n}{2}\ \frac{n}{2} \\ S,S_3; & \ell,m & j,j_3 \end{array}\right]$$

$Y_{0,0}^{\frac{1}{2}\frac{1}{2}} = \frac{i\gamma_4}{2}$, $Y_{1,S_3}^{\frac{1}{2}\frac{1}{2}} = \frac{\vec{\gamma}_{S_3}}{2}$, the Clebsch-Gordan-coefficient is given (5.11)

$$P_n^{III}(\hat{q}) = \frac{\gamma_5}{2}\ Y^n(\hat{q})$$

5.2. The hyperradial equations.

An expansion of the B S amplitude

$$\chi(q) = \sum_{\xi,\eta} z^{\xi,\eta}(|q|)\cdot Z^{\xi,\eta}(\hat{q}) \tag{5.15}$$

will lead, after insertion into the B S equation (5.5), to 6 systems of equations for the hyperradial functions $Z^{\xi,\eta}(|q|)$.[65]

An example of a one-dimensional sector is

$$\chi^{III}(q) = P_n(|q|)\cdot P_n(\hat{q}) \tag{5.16}$$

which satisfies the hyperradial equation:

$$(-q^2 - m^2)\; P_n(q) = (\lambda^P\; K_n^P\; P_n)(q) \tag{5.17}$$

(The symbol on the right hand side is to be understood as

$$(K_n\; F)(q) = \int_0^\infty q'^3\, dq'\; K_n(q,q')F(q') \tag{5.18}$$

and results from the expansion

$$K(q-q') = \sum_{n,\ell,m} K_n(q,q')Y^{*n}_{\ell,m}(\hat{q}')Y^n_{\ell,m}(\hat{q}) \tag{5.19}$$

$$\int d^4q'\; K(q-q')[F(q')Y^n(\hat{q}')] = Y^n(\hat{q})[(K_n\; F)(q)].)$$

In Sector I one obtains a system of three coupled hyperradial equations: (5.20)

$$\begin{bmatrix} q^2-m^2, & -\,imq\sqrt{\frac{2n}{n+1}}, & imq\sqrt{\frac{2(n+2)}{n+1}} \\ -\,imq\sqrt{\frac{2n}{n+1}}, & \frac{-q^2}{n+1} - m^2, & -q^2\frac{\sqrt{n(n+2)}}{n+1} \\ imq\sqrt{\frac{2(n+2)}{n+1}}, & -\,q^2\,\frac{\sqrt{n(n+2)}}{n+1}, & \frac{q^2}{n+1}-m^2 \end{bmatrix} \begin{bmatrix} S^I_n(q) \\ V^I_{n-}(q) \\ V^I_{n+}(q) \end{bmatrix} = \begin{bmatrix} (\lambda^S K^S_n S^I_n)(q) \\ (\lambda^V K^V_{n-1} V^I_{n-})(q) \\ (\lambda^V K^V_{n+1} V^I_{n+})(q) \end{bmatrix}$$

5.3. Exact solutions.

The hyperradial equations may be solved by analytic methods for potentials of the type:

a) $V(R) = \frac{\lambda}{R^2}$, b) the square well, c) $V(R) = \alpha + \beta R^2$.

In particular the case $V(R) = \frac{\lambda}{R^2}$, the socalled Wick-Cutcosky problem, was studied extensively in the literature.[66] The first classical result is due to Goldstein for a potential $V(R) = \lambda\frac{\gamma_\mu x \gamma_\mu}{R^2}$, namely a solution of the type:

$$\chi_n^{P(III)}(q) = F(-\rho + 1, \rho + n + 2; n + 2; \frac{q^2}{m^2})\gamma_5 Y^n_{\ell,m}(\hat{q})$$

$$\rho = -\frac{1}{2}(n + 2) + [\frac{1}{4}(n + 1)^2 - \lambda^P]^{1/2} \qquad (5.21)$$

which belongs to the one-dimensional sector III.

The Wick rotated relativistic "Coulomb" potential has a strong singularity at R = 0, which leads to the appearence of the socalled "continuous" spectrum. This phenomenon has been discussed in some detail by Bastai, Bertocchi, Furlan and Tonin.[66] (For nonsingular potentials, like the square well and the oscillator, it will not occur.) An important contribution to this problem is due to Kummer,[66] who studied $V(R) = \frac{\gamma_\mu x \gamma_\mu}{R^2}$, $\frac{\gamma_5\gamma_\mu x \gamma_5\gamma_\mu}{R^2}$ and found solutions which belong to a discrete spectrum:

$$\chi_{n,N}^{A(VI)}(q) = \frac{m^2 - q^2}{(m^2+q^2)^{(\xi+3)/2}}F(-N+n+1, -\xi+N+1; -\xi+1; 1+\frac{q^2}{m^2}).$$

$$\cdot A_n(\hat{q}) \tag{5.22}$$

$$N \geq n + 1, \qquad N \in N$$

$$\xi = \sqrt{8\,\lambda^A_{N,n} + 1} > 2\,N + 1\,,$$

$$\sqrt{8\,\lambda^A_{N,n} + 1} - \sqrt{4\,\lambda^A_{N,n} + (n+1)^2} = 2\,N - n$$

With respect to our decomposition into radial equations, this is a solution in sector VI, with $K^T = 0$: In this case two equations become algebraic ones and can be used to eliminate two amplitudes; the resulting single equation is not of the Goldstein type. Solutions of the same type exist in sectors I and IV.

The square well has been analyzed recently by Keam.[65b]

In connection with the relativistic quark model with strong binding we are interested in the harmonic oscillator. Exact solutions in all sectors, in particular also solutions of the three-coupled radial equations, were found.[60] An elegant representation of these solutions is obtained with help of annihilation and creation operators.[67] For this, we make the ansatz:

$$\chi(q) = \chi_o - \frac{i}{2m}(\not{q}\,\chi_o + \chi_o\,\not{q}) \equiv \chi_o + \chi_1 \tag{5.23}$$

and remark that if

χ_o is of S, V, T, A, P type, then
χ_1 is of V, S, A, T, O type.

This reflects the wellknown fact that the equal mass boundstate mass zero equation decomposes into a V + S, T + A and P sector. Insertion of the ansatz (5.23) into the B S equation (5.5) leads to the two equations for χ_o and χ_1 separately,

$$\begin{aligned}(- q^2 - m^2 - K)\ \chi_o &= 0\\ (- q^2 - m^2 + K)\ \chi_1 &= 0\end{aligned} \tag{5.24}$$

in which the spin-dependence only appears in the connection of χ_o and χ_1:

$$\chi_1 = - \frac{i}{2m} \{\not q,\ \chi_o\}_+ \quad . \tag{5.25}$$

We discuss the solution of Eqs. (5.24) and (5.25) in the case of the S + V sector. In this case, the relation (5.25) takes the form

$$\chi_o = \gamma_\mu\ \chi_o^\mu,\quad \chi_1 = - \frac{i}{m} q_\mu\ \chi_o^\mu \tag{5.26}$$

and, in our projection base for the kernel, equations (5.24) become

$$\begin{aligned}(- q^2 - m^2 - \lambda^V\ K^V)\ \chi_o^\mu &= 0\\ - \frac{i}{m}(- q^2 - m^2 + \lambda^S\ K^S)\ q_\mu\ \chi_o^\mu &= 0 \quad .\end{aligned} \tag{5.27}$$

This simplification of the Fermion-antifermion B S equation is independent of the special form of K^i. It is particularly useful for heavy quarks or the oscillator potential. In the latter case, if

$$\lambda^V K^V = \alpha^V - \beta^V \Box_q \tag{5.28}$$

$$\lambda^S K^S = \alpha^S - \beta^S \Box_q \ ; \qquad \beta^V = - \beta^S = \beta$$

χ is a solution of the B S equation, if χ_o and χ_1 are solutions of the same oscillator equation, in general for different eigenvalues α^V and α^S. In order to describe the relation between χ_o and χ_1 in a simple form, we introduce the wellknown annihilation and creation operators ($|\beta| = 1$, i.e. making the substitution $q \to \frac{q}{\sqrt{\beta}}$):

$$a_\mu^- = \frac{1}{\sqrt{2}}(q_\mu + \partial_\mu), \ a_\mu^+ = \frac{1}{\sqrt{2}}(q_\mu - \partial_\mu); \ [a_\mu^-, a_{\mu'}^+] = \delta_{\mu\mu'}$$

$$q_\mu = \frac{1}{\sqrt{2}}(a_\mu^- + a_\mu^+), \ \partial_\mu = \frac{1}{\sqrt{2}}(a_\mu^- - a_\mu^+) \tag{5.29}$$

$$q^2 - \Box = 2(a_\mu^+ a_\mu^- + 2) \ .$$

Now, if χ_o^μ is an oscillator solution, then $q_\mu \chi_o^\mu$ is also an oscillator function, if

$$q_\mu \chi_o^\mu = \frac{1}{\sqrt{2}} a_\mu^+ \chi_o^\mu \text{ or } \frac{1}{\sqrt{2}} a_\mu^- \chi_o^\mu \text{ or } 0 \ . \tag{5.30}$$

This condition is satisfied for χ_o^μ of the following form

$$\chi_o^{\mu} = (a_\nu^{\pm}\, a_\nu^{\pm}\, a_\mu^{\mp} - a_\mu^{\pm}\, a_\nu^{\pm}\, a_\nu^{\mp})\,\bar{\chi}$$

or (5.31)

$$\chi_o^{\mu} = \varepsilon^{\mu\nu\rho\sigma}\, a_\nu^{+}\, a_\rho^{-}\, c_\sigma\, \bar{\chi}$$

where $\bar{\chi}$ denotes an arbitrary eigenfunction of the harmonic oscillator. Therefore, insertion of χ_o^{μ} into Eq.(5.26) gives an exact solution of the Fermion-antifermion B S equation with harmonic interaction.

In a similar way one may construct solutions for the P and T + A sectors from arbitrary solutions of the scalar oscillator equation. We give a list of these expressions in Table III. By specialization of $\bar{\chi}$ to a solution which belongs to a fixed O(4) quantum number n

$$\bar{\chi}_{n,r}(q) = q^n\, L_r^{n+1}(q^2)\, e^{-\frac{q^2}{2}}\, Y^n(\hat{q}) \qquad (5.32)$$

we achieve that the solutions transform according to the irreducible representations of $O^{\Pi,C}(4)$, listed in Table II. It turns out that the different forms of the solutions correspond in a unique way to the sectors I - VI of the $O^{\Pi,C}(4)$ irreducible representations.

6. Mass spectrum and wave functions of mesons in the relativistic quark model.

In the nonrelativistic quark model the low mass mesons have the quantum numbers:[48]

Dirac-sector	Solutions of the BS equation	Sector
SΘV	$(\gamma_\mu - \frac{i}{m\sqrt{2}} a^+_\mu)(a^-_\nu a^-_\nu a^+_\mu - a^-_\mu a^-_\nu a^+_\nu)\bar{\chi}$ $(\gamma_\mu - \frac{i}{m\sqrt{2}} a^-_\mu)(a^+_\nu a^+_\nu a^-_\mu - a^+_\mu a^+_\nu a^-_\nu)\bar{\chi}$	I
	$\varepsilon^{\mu\nu\rho\sigma} \gamma_\mu a^+_\nu a^-_\rho c_\sigma \bar{\chi}$	$V^a: c^{(a)}_\sigma =$ $= \varepsilon^{4\sigma\alpha\beta} a^+_\alpha a^-_\beta$ $V^b: c^{(b)}_\sigma = \delta_{4\sigma}$
TΘA	$(\gamma_5\gamma_\nu a^-_\nu + \frac{1}{m\sqrt{2}} \gamma_5\sigma_{\mu\nu} a^+_\mu a^-_\nu)\ \bar{\chi}$ $(\gamma_5\gamma_\nu a^+_\nu + \frac{1}{m\sqrt{2}} \gamma_5\sigma_{\mu\nu} a^-_\mu a^+_\nu)\ \bar{\chi}$	IV
	$(\sigma_{\mu\nu} a^-_\nu + \frac{1}{m\sqrt{2}} \gamma_5\gamma_\alpha\ \varepsilon^{\alpha\beta\mu\nu}\ a^+_\beta a^-_\nu) \cdot$ $\cdot\ \varepsilon^{\mu\rho\sigma\kappa}\ a^+_\rho a^-_\sigma c_\kappa\ \bar{\chi}$ $(\sigma_{\mu\nu} a^+_\nu + \frac{1}{m\sqrt{2}} \gamma_5\gamma_\alpha\ \varepsilon^{\alpha\beta\mu\nu} a^-_\beta a^+_\nu)\ \cdot$ $\cdot\ \varepsilon^{\mu\rho\sigma\kappa}\ a^+_\rho a^-_\sigma c_\kappa\ \bar{\chi}$	$VI^a: c_\kappa^{(a)} =$ $= \varepsilon^{4\kappa\gamma\delta}\ a^+_\gamma a^-_\delta$ $VI^b: c_\kappa^{(b)} = \delta_{4\kappa}$
	$\sigma_{\mu\nu}\ a^+_\mu a^-_\nu\ \ \bar{\chi}$	II
P	$\gamma_5\ \bar{\chi}$	III

$(j^+j^-)^{\Pi'C'}$	Eigenvalues α^o	α^1
$(\frac{n}{2}\ \frac{n}{2})^{++}$	$\alpha^V=-m^2-2(N+1)\sqrt{\beta^V}$ $\alpha^V=-m^2-2(N+3)\sqrt{\beta^V}$	$\alpha^S=m^2+2(N+2)\sqrt{\beta^V}$ $\alpha^S=m^2+2(N+2)\sqrt{\beta^V}$ $\beta^S = -\beta^V$
$(\frac{n+1}{2}\ \frac{n-1}{2})^{O^-}$	$\alpha^V=-m^2-2(N+2)\sqrt{\beta^V}$	
$(\frac{n}{2}\ \frac{n}{2})^{--}$	$\alpha^A=-m^2-2(N+1)\sqrt{\beta^A}$ $\alpha^A=-m^2-2(N+3)\sqrt{\beta^A}$	$\alpha^T=m^2+2(N+2)\sqrt{\beta^A}$ $\alpha^T=m^2+2(N+2)\sqrt{\beta^A}$ $\beta^T = -\beta^A$
$(\frac{n+1}{2}\ \frac{n-1}{2})^{O^+}$	$\alpha^T=-m^2-2(N+1)\sqrt{\beta^T}$ $\alpha^T=-m^2-2(N+3)\sqrt{\beta^T}$	$\alpha^A=m^2+2(N+2)\sqrt{\beta^T}$ $\alpha^A=m^2+2(N+2)\sqrt{\beta^T}$ $\beta^A = -\beta^T$
$(\frac{n}{2}\ \frac{n}{2})^{+-}$	$\alpha^T=-m^2-2(N+2)\sqrt{\beta^T}$	
$(\frac{n}{2}\ \frac{n}{2})^{-+}$	$\alpha^P=-m^2-2(N+2)\sqrt{\beta^P}$	[N=n+2r]

T A B L E I I I

Exact solutions of the Fermion-antifermion BS equation for harmonic interaction.[67]

$j^{\Pi,C}$		orbital angular momentum $\ell = 0$	$\ell = 1$	$\ell = 2$
quark spin	$S = 0$	0^{-+}	1^{+-}	.....
			2^{++}	.....
	$S = 1$	1^{--}	1^{++}	.....
			0^{++}	.....

(6.1)

For spin independent forces, quark spin-singlet and -triplet states with the same orbital angular momentum will be degenerate (SU(6)-symmetry). We shall discuss now, how this structure can be relativistically generalized by simple B S models.

At this stage a remark on the SU(3) part of the amplitudes should be made: We can complete the amplitudes multiplying them by the familiar SU(3) $|q\bar{q}\rangle$-amplitudes of the mesons. This means that we treat the quarks in an SU(3)-symmetric fashion; in spite of the fact that the quark model offers successful schemes for the treatment of SU(3) breaking. Since we did not treat the B S equation for the case $m_{q_1} \neq m_{\bar{q}_2}$, the results will apply strictly speaking only to mesons composed of non-strange quarks, that is to mesons with isospin one.

The solutions for boundstate mass M = O will serve as a starting point for our quark model considerations. In the last section we have shown that the exact solutions for a harmonic interaction are of the form

$$\chi(q) = \chi_o(q) - \frac{i}{2m}\{\not{q},\chi_o(q)\}_+ \equiv \chi_o(q) + \chi_1(q). \quad (5.23)$$

In our application we shall have:

$$\frac{<q^2>}{m^2} \sim \frac{\sqrt{\beta}}{m^2} << 1 \ .$$

Therefore, $\chi_1(q)$ is a small component of the amplitude. This property may be used to obtain approximate solutions in case the potential parameters do not satisfy the eigenvalue conditions exactly. In the subsequent discussion we shall consider M = O solutions of the type

$$\chi(q) = \chi_o(q) - \frac{i}{2m} \{\not{q}, \chi_o(q)\}_+ + O(\frac{1}{m^2}) \qquad (6.2)$$

with

$$\chi_o(q) \sim q^{n_b} L_r^{n_b+1} (\frac{q^2}{\sqrt{\beta}}) \cdot e^{-\frac{q^2}{2\sqrt{\beta}}} \cdot Z^{\xi,\eta(n_b)}(\hat{q}) \ . \qquad (6.3)$$

To these belong the potential parameters:

$$\begin{aligned} \alpha^o &= - m^2 - 2(n_b + 2r + 2) \sqrt{\beta} \\ \alpha^1 &\approx + m^2 \ . \end{aligned} \qquad (6.4)$$

In Table IV the spectrum belonging to the different solutions (6.3.4) is shown. Quantum numbers in the same column belong to the same eigenvalue α, if the β's of the different interaction types are equal. Natural ($\Pi = (-1)^j$) and unnatural ($\Pi = (-1)^{j+1}$) parity states were separated, since they will not be mixed by O(4) breaking for $M \neq 0$.

Like in the scalar model, the field theoretical normalization condition (3.10), which is in our case

$$\frac{-1}{(2\pi)^4} \text{ Trace } \int d^4q \bar{\chi}_{r'}^{B}(q;P)\left(\frac{\not P}{2} - i\not q - m\right)\chi_r^B(q;P)\left(\frac{\not P}{2} + i\not q + m\right) =$$

$$= \left(\lambda \frac{\partial P^2}{\partial \lambda}\right)_{P^2=M_B^2} \cdot \delta_{rr'} \qquad (6.5)$$

$$\bar{\chi}(q_4,\vec{q};P) = \gamma_4 \ \chi^+(-q_4,\vec{q};P) \ \gamma_4,$$

yields selection rules, if we omit states with negative physical norm ("ghosts"). (In order to obtain them, we anticipate $\lambda \frac{\partial P^2}{\partial \lambda} < 0$, Table VII).

A property of the solutions in Table IV is that states belonging to different sectors, but which have the same Dirac structure and the same orbital quantum number, are degenerate. Therefore we may rearrange the degenerate states, thereby obtaining a structure which is akin to the nonrelativistic quark model. The result is shown in Table V ("negative norm" solutions have been left out), where we have grouped together the quarkspin-singlet states and the -triplet states.

Let us now discuss possible relativistic models. From the table we see that the interactions

A) P + V - S $\qquad (\alpha^P \approx \alpha^V \approx -\alpha^S \approx -m^2;\ \alpha^A = \alpha^T = 0)$ (6.6)

B) P + T - A

C) P + T - A - V + S

D) A - T + V - S

E) A - T + V - S - P

give one quarkspin-singlet and one quarkspin-triplet,

TABLE IV

$\chi_o(\hat{q})$	Π	C	K	$j^{\Pi,C}$	sign(Norm)
I. S_n	$(-1)^j$	$(-1)^n$	$\alpha^S\underset{\sim}{\approx}-\alpha^V\underset{\sim}{\approx}-m^2$	$(0^{++}\ 1^{--}\ldots)$* 0^{+-}	$(-1)^{n-j+1}$
V_{n-} V_{n+}			$\alpha^V\underset{\sim}{\approx}-\alpha^S\underset{\sim}{\approx}-m^2$	$1^{--}\ 2^{++}$ $(0^{+-}\ 1^{-+}\ )$ $0^{++}\ \ldots\ldots$ $0^{++}\ 1^{--}\ldots$ (0^{+-})	$(-1)^{n-j}$
II. T_n	$(-1)^j$	$(-1)^{n+1}$	$\alpha^T\underset{\sim}{\approx}-m^2$	$(1^{-+}\ 2^{+-})$ $0^{++}\ 1^{--}\ldots$ (0^{+-})	$(-1)^{n-j+1}$
V. V^a_n	$(-1)^j$	$(-1)^{n+1}$	$\alpha^V\underset{\sim}{\approx}-m^2$	$(1^{-+}\ 2^{+-})$ $1^{--}\ldots$	$(-1)^{n-j+1}$
VI. A^a_n	$(-1)^j$	$(-1)^n$	$\alpha^A\underset{\sim}{\approx}-\alpha^T\underset{\sim}{\approx}-m^2$	$(1^{--}\ 2^{++})$ 1^{-+}	$(-1)^{n-j+1}$
T^a_{n-} T^a_{n+}			$\alpha^T\underset{\sim}{\approx}-\alpha^A\underset{\sim}{\approx}-m^2$	$1^{--}\ 2^{++}$ $(1^{-+}\ )\ldots\ldots$ 1^{--}	$(-1)^{n-j}$
III. P_n	$(-1)^{j+1}$	$(-1)^n$	$\alpha^P\underset{\sim}{\approx}-m^2$	$0^{-+}\ 1^{+-}\ \ldots\ldots$ $(0^{--}\ )$	$(-1)^{n-j}$
IV. T_n	$(-1)^{j+1}$	$(-1)^{n+1}$	$\alpha^T\underset{\sim}{\approx}-\alpha^A\underset{\sim}{\approx}-m^2$	$1^{++}\ 2^{--}$ $(0^{-+}\ 1^{+-}\ ).$ 0^{--}	$(-1)^{n-j}$
A_{n-} A_{n+}			$\alpha^A\underset{\sim}{\approx}-\alpha^T\underset{\sim}{\approx}-m^2$	$(1^{++}\ 2^{--}\ )$ $0^{-+}\ 1^{+-}\ldots\ldots$ $(0^{--}\ )$ $(0^{--}\ 1^{++}\ )$ 0^{-+}	$(-1)^{n-j+1}$
V. V^b_n	$(-1)^{j+1}$	$(-1)^{n+1}$	$\alpha^V\underset{\sim}{\approx}-m^2$	$1^{++}\ 2^{--}\ldots$ $(1^{+-}\ )$	$(-1)^{n-j}$
VI. A^b_m	$(-1)^{j+1}$	$(-1)^n$	$\alpha^A\underset{\sim}{\approx}-\alpha^T\underset{\sim}{\approx}-m^2$	$1^{+-}\ 2^{-+}$ $(1^{++}\ )$	$(-1)^{n-j}$
T^b_{n-} T^b_{n+}			$\alpha^T\underset{\sim}{\approx}-\alpha^A\underset{\sim}{\approx}-m^2$	$(1^{+-}\ 2^{-+}\ )$ $1^{++}\ldots\ldots$ $(1^{+-}\ )$	$(-1)^{n-j+1}$

*Solutions in brackets belong to negative norm.

$\chi_o(\hat{q})$		Π	C	K	$j^{\Pi C}$	"on shell"
$\frac{\gamma_5}{2} Y^{n'}_{j,j_3}$	$n=0,1,\ldots$ $j=n,n-2,\ldots$	$(-1)^{j+1}$	$(-1)^n$	$\alpha^P_{\approx}-m^2$	0^{-+} 1^{+-} 2^{-+} 0^{-+}	
$\frac{\gamma_5\gamma_4}{2} Y^{n}_{j,j_3}$	$n=0,1,\ldots$ $j=n,n-2,\ldots$	$(-1)^{j+1}$	$(-1)^n$	$\alpha^A_{\approx}-\alpha^T_{\approx}$ $-m^2$	0^{-+} 1^{+-} 2^{-+} 0^{-+}	
$\frac{\vec{\gamma}s_3}{2} Y^{n_b}_{j-1,\ell_3}(1s_3;j-1\ell_3\|jj_3)$	$n_b=0,1,\ldots$ $j=n_b+1,n_b-1,\ldots$	$(-1)^j$			1^{--} $2^{++}\ldots.$	
$\frac{\vec{\gamma}s_3}{2} Y^{n_b}_{j,\ell_3}(1s_3;j\ell_3\|jj_3)$	$n_b=1,2,\ldots$ $j=n_b,n_b-2,\ldots$	$(-1)^{j+1}$	$(-1)^{n_b+1}$	$\alpha^V_{\approx}-\alpha^S_{\approx}$ $-m^2$	1^{++} 2^{--}	
$\frac{\vec{\gamma}s_3}{2} Y^{n_b}_{j+1,\ell_3}(1s_3;j+1\ell_3\|jj_3)$	$n_b=1,2,\ldots$ $j=n_b-1,n_b-3,\ldots$	$(-1)^j$			0^{++} 1^{--}	
$\frac{\sigma_4 s_3}{2} Y^{n_b}_{j-1,\ell_3}(1s_3;j-1\ell_3\|jj_3)$	$n_b=0,1,\ldots$ $j=n_b+1,n_b-1,\ldots$	$(-1)^j$			1^{--} $2^{++}\ldots.$	
$\frac{\sigma_4 s_3}{2} Y^{n_b}_{j,\ell_3}(1s_3;j\ell_3\|jj_3)$	$n_b=1,2,\ldots$ $j=n_b,n_b-2,\ldots$	$(-1)^{j+1}$	$(-1)^{n_b+1}$	$\alpha^T_{\approx}-\alpha^A_{\approx}$ $-m^2$	1^{++} 2^{--}	
$\frac{\sigma_4 s_3}{2} Y^{n_b}_{j+1,\ell_3}(1s_3;j+1\ell_3\|jj_3)$	$n_b=1,2,\ldots$ $j=n_b-1,n_b-3,\ldots$	$(-1)^j$			0^{++} 1^{--}	

$\frac{1}{2}Y^{n}_{j,j_3}$	$n=1,2,\ldots$ $j=n-1,n-3,\ldots$	$(-1)^{j}$	$(-1)^{n}$	$\alpha^{S}{}_{\approx}-\alpha^{V}{}_{2}$ ${}_{\approx}-m^{2}$	0^{+-} 1^{-+}	$=0$
$\frac{\gamma_4}{2}Y^{n}_{j,j_3}$	$n=1,2,\ldots$ $j=n-1,n-3,\ldots$	$(-1)^{j}$	$(-1)^{n+1}$	$\alpha^{V}{}_{\approx}-\alpha^{S}{}_{2}$ ${}_{\approx}-m^{2}$	0^{++} 1^{--}	$=0$
$\frac{\gamma_5\vec{\gamma}s_3}{2}\,Y^{n_b}_{j-1,\ell_3}(1s_3;j-1\ell_3\|jj_3)$	$n_b=1,2,\ldots$ $j^{b}=n_b,n_b-2,\ldots$	$(-1)^{j+1}$			1^{+-} 2^{-+}	
$\frac{\gamma_5\vec{\gamma}s_3}{2}\,Y^{n_b}_{j,\ell_3}(1s_3;j\ell_3\|jj_3)$	$n_b=2,3,\ldots$ $j^{b}=n_b-1,n_b-3,\ldots$	$(-1)^{j}$	$(-1)^{n_b}$	$\alpha^{A}{}_{\approx}-\alpha^{T}{}_{2}$ ${}_{\approx}-m^{2}$	1^{-+}	$=0$
$\frac{\gamma_5\vec{\gamma}s_3}{2}\,Y^{n_b}_{j+1,\ell_3}(1s_3;j+1\ell_3\|jj_3)$	$n_b=2,3,\ldots$ $j^{b}=n_b-2,n_b-4,\ldots$	$(-1)^{j+1}$			0^{-+}	
$\frac{\varepsilon^{4k\ell s_3}}{4}\sigma_{k\ell}Y^{n_b}_{j-1,\ell_3}(1s_3;j-1\ell_3\|jj_3)$	$n_b=1,2,\ldots$ $j^{b}=n_b,n_b-2,\ldots$	$(-1)^{j+1}$			1^{++} 2^{--}	
$\frac{\varepsilon^{4k\ell s_3}}{4}\sigma_{k\ell}Y^{n_b}_{j,\ell_3}(1s_3;j\ell_3\|jj_3)$	$n_b=2,3,\ldots$ $j^{b}=n_b-1,\ldots$	$(-1)^{j}$	$(-1)^{n_b+1}$	$\alpha^{T}{}_{\approx}-\alpha^{A}{}_{2}$ ${}_{\approx}-m^{2}$	1^{--}	$=0$
$\frac{\varepsilon^{4k\ell s_3}}{4}\sigma_{k\ell}Y^{n_b}_{j+1,\ell_3}(1s_3;j+1\ell_3\|jj_3)$	$n_b=2,3,\ldots$ $j^{b}=n_b-2,\ldots$	$(-1)^{j+1}$			0^{--}	

T a b l e   V

which is analogous to the nonrelativistic model (6.1). The models have the property that, besides the wanted states, there occur further solutions, which have no analogon in the nonrelativistic quark model (6.1). To these belong

i) the "exotics of second kind"[68], i.e.: 0^{--}; 0^{+-}, 1^{-+}, 2^{+-} ...

ii) 0^{-+}, 1^{--}, ... states with anomalous orbital angular momentum in comparison to (6.1).

Thirring has shown[69] that exotics will not occur, if the fields $\psi_\alpha(x_1)$, $\bar{\psi}_\alpha(x_2)$ are taken at equal time. In agreement with that we find

$$\int \chi_o(q)\, dq_4 = 0 \tag{6.7}$$

for the exotics in our models. Furthermore, these amplitudes vanish for both quarks on shell (in the c.m.s.):

$$(1+\gamma_4)\chi_o(1-\gamma_4)\Big|_{q_4=0,\,\vec{q}^2=\frac{M^2}{4}-m^2} = 0. \tag{6.8}$$

That would mean that these states are not coupled to free quarks. However, we cannot use this criterium offhand, because our approximations are not valid for the quarks on-shell.

Now we start with the consideration of the B S equation for boundstate mass $M \neq 0$:

$$(\gamma q - im + \frac{iM}{2}\gamma_4)\chi(q,M)(\gamma q - im - \frac{iM}{2}\gamma_4) = \int K(q-q')\chi(q',M)dq'. \tag{6.9}$$

Because of the conditions

$$\frac{M^2}{4m^2} << 1, \quad \frac{<q^2>}{m^2} << 1 \tag{6.10}$$

we neglect the O(4)-breaking spin orbit term

$$(\gamma q-im)\ \chi(q,M)(\gamma q-im) - K\chi \approx - \frac{M^2}{4}\gamma_4\chi\ \gamma_4 - \frac{Mm}{2}(\gamma_4\chi-\chi\gamma_4)\ . \tag{6.11}$$

Following our procedure for the boundstate mass zero case (Eq.6.2) we make the ansatz

$$\chi(q,M) = \chi_o(q,M) - \frac{i}{2m}\{\not{q},\chi_o(q,M)\}_+ + \chi_{br}(q,M) + O\left(\frac{1}{m^2}\right) \tag{6.12}$$

leading to the equations:

$$(-q^2 - m^2 - K)\chi_o = - \frac{M^2}{4}\gamma_4\ \chi_o\ \gamma_4 - \frac{Mm}{2}(\gamma_4\ \chi_{br} - \chi_{br}\ \gamma_4) \tag{6.13a}$$

$$(-q^2 - m^2 - K)\chi_{br} = - \frac{M^2}{4}\gamma_4\ \chi_{br}\ \gamma_4 - \frac{Mm}{2}(\gamma_4\ \chi_o - \chi_o\ \gamma_4)\ . \tag{6.13b}$$

Under the condition that the potential does not compensate the quark mass term in Eq.(6.13b) we can approximate it by the algebraic equation:

$$(-m^2 - \alpha^{br})\ \chi_{br} = - \frac{Mm}{2}\ [\gamma_4,\chi_o]_- \quad . \tag{6.14}$$

Insertion of χ_{br} into Eq.(6.13a) yields

$$(-q^2-m^2-\kappa)\chi_o + \frac{M^2\ m^2}{4(m^2+\alpha^{br})}\ [2\chi_o - \frac{(m^2-\alpha^{br})}{m^2}\gamma_4\chi_o\gamma_4] = 0 \qquad (6.15)$$

from which all spin-dependence gets **eliminated** if $\gamma_4\ \chi_o\ \gamma_4 = \rho\chi_o$, $\rho = \pm 1$. According to Table V, $\rho = -1$ ($\rho = +1$) corresponds to "normal" ("exotic") states.

In this order of approximation the **dynamics** is described by the simple equation:[70]

$$(-q^2 - m^2 + \frac{M^2}{4} \cdot \frac{2m^2 - \rho(m^2 - \alpha^{br})}{m^2 + \alpha^{br}} - \kappa)\ \chi_o = 0 \qquad (6.16)$$

to which corresponds the solution of the B S equation (6.9):

$$\chi(q,P) = \chi_o(q,P) - \frac{i}{2m}\{\not{q},\chi_o(q,P)\}_+ + \frac{m}{2(m^2+\alpha^{br})}[\not{P},\chi_o(q,P)]_- \qquad (6.17)$$

We recall the conditions of the validity of Eq.(6.16):

i) strong binding $(\frac{M^2}{4m^2} \ll 1)$

ii) in smooth potentials $(\frac{\langle q^2\rangle}{m^2} \ll 1)$

iii) the Dirac structure of the interaction kernel satisfies the condition leading to Eq.(6.14).

Let us now treat our five quark models (6.6) according to these methods. In all of them the interaction does not compensate the quark mass in the equation for χ_{br}. The α^{br} are listed in Table VI.

Model	Singlets				Triplets			
	α^{br}	κ	α^o	β^o	α^{br}	κ	α^o	β^o
P+V-S	O	$\frac{4}{3}$	α^P	β^P	O	$\frac{4}{3}$	α^V	β^V
P+T-A	$+m^2$	4	α^P	β^P	O	$\frac{4}{3}$	α^T	β^T
P+T-A-V+S	$+m^2$	4	α^P	β^P	$+m^2$	4	α^T	β^T
A-T+V-S	O	$\frac{4}{3}$	α^A	β^A	$+m^2$	4	α^V	β^V
A-T+V-S-P	$+m^2$	4	α^A	β^A	$+m^2$	4	α^V	β^V

Table VI.

Since the normal states are eigenstates of $\gamma_4\ \chi_o\ \gamma_4$ with $\rho = -1$, we may solve the eigenvalue equation (6.16) with this specification and get for harmonic forces the mass spectrum

$$M^2 = \kappa(\alpha^o + m^2 + 2(n_b + 2\,r + 2)\ \sqrt{\beta^o})$$
$$n_b = 0, 1, 2, \ldots; \quad r = 0, 1, 2, \ldots \qquad (6.18)$$

with κ according to Table VI.

In this approximation the models give linear Regge trajectories. If we require degeneracy of singlet and triplet masses (analogous to the nonrelativistic SU(6) symmetry), then the models B and D lead to amplitudes with a range differing by a factor 3 from singlets to triplets. We, therefore, shall not discuss them further. To the models A, C and D correspond in this symmetry limit the interactions (compare (5.2) - (5.4)):

P+V-S model: $K = -\frac{1}{4}(\gamma_5 x \gamma_5 + \gamma_\mu x \gamma_\mu - 1x1)(\alpha+\beta R^2)$ (eigenvector of the Fierz-transformation)

P+T-A-V+S model: $K = \gamma_5 \times \gamma_5 (\alpha + \beta R^2)$ (6.19)

A-T+V-S-P model: $K = -\gamma_5 \times \gamma_5 (\alpha + \beta R^2)$

We calculated the B S amplitudes in the centre of mass frame in the Wick rotated form. Lorentz covariance is restored,[71] if we analytically continue q_4, $-iq_4 \to q_o$, and apply the Lorentzboost $L^{-1}(P)$:

$$L^{-1}(P) \begin{bmatrix} q_o \\ \vec{q} \end{bmatrix} = \begin{bmatrix} \frac{q \cdot P}{M} \\ \vec{q} - \frac{(qP+q_oM)}{M(M+P_o)}\vec{P} \end{bmatrix} \equiv \begin{bmatrix} \frac{q \cdot P}{M} \\ \frac{1}{M}\sqrt{(q \cdot P)^2 - q^2P^2}\,.\vec{e}_q(P) \end{bmatrix} \tag{6.20}$$

$$L^{-1}(P) \begin{bmatrix} \gamma_o \\ \vec{\gamma} \end{bmatrix} = \begin{bmatrix} \frac{\not{P}}{M} \\ \vec{\gamma} - \frac{(\not{P}+M\gamma_o)}{M(M+P_o)}\vec{P} \end{bmatrix} \equiv \begin{bmatrix} \frac{\not{P}}{M} \\ \vec{\gamma}(P) \end{bmatrix}$$

This leads to the substitutions:

$$F(q^2, iq_o) \to F(q^2, i\,\frac{q \cdot P}{M})$$

$$Y\left(\frac{\vec{q}}{|\vec{q}|}\right) \to Y[\vec{e}_q(P)] \tag{6.21}$$

$$\gamma_o \to \frac{\not{P}}{M}$$

$$\vec{\gamma} \to \vec{\gamma}(P) \quad .$$

The analytic continuation of the radial part, the Gauss function, to real values of q_o would lead to amplitudes increasing exponentially. This is a shortcoming of our oscillator approximation of the Wick-rotated potential. The wavefunctions are reliable only in the neighbourhood of Euclidean points:

$$q_E^2 = q_4^2 + \vec{q}^{\,2} \quad > \quad 0 \quad .$$

The normalization of the B S amplitudes is determined by the field theoretical condition (6.5) with

$$\lambda \frac{\partial P^2}{\partial \lambda} = - \quad \kappa\, m^2 + O(m^o) \quad .$$

For further use we give a complete compilation of the B S amplitudes for all three models in Table VII.[60,70]

Before we determine the free parameters by comparison with experiment, we repeat some general features of the spectrum common to all models:

a) Solution with $j = n$, $\ell = n_b$ $(j = \ell + 1, \ell, \ell - 1)$ **respectively** correspond to the quarkspin-singlet, -triplet states resp. of the nonrelativistic quark model.

b) There are solutions with $\ell < n_b$. They are characteristic of the B S treatment and are related to oscillations in time ("abnormal solutions").

c) There are positive norm solutions which "vanish on-shell".

d) For all modes there are hyperradial excitations, analogous to radial excitations in the nonrelativistic case.

For the determination of the interaction parameters we use the masses[72] of the following isospin triplets:

$j^{\Pi,C}$ =		0^{-+}	1^{+-}	2^{-+}	
n =		0	1	2	
r =	0	$\pi(140)$	B(1235)	$\pi'(1640)$	
M^2 =		0.02	1.53	2.67	GeV^2

$2\kappa\sqrt{\beta}$ = 1.40GeV^2, $\alpha = (-\,m^2 - \frac{2.8}{\kappa})$ GeV^2

$j^{\Pi,C}$ =		1^{--}	2^{++}	3^{--}	(6.22)
n_b =		0	1	2	
r =	0				
		$\rho(765)$	$A_2(1310)$	g(1660)	
M^2 =		0.59	1.72	2.76	GeV^2

$2\kappa\sqrt{\beta}$ = 1.11 GeV^2, $\alpha = (-\,m^2 - \frac{1.6}{\kappa})$ GeV^2

It is seen that from the spectrum we obtain the combination $\alpha + m^2$ only, which property is typical for strong binding in smooth potentials. Also, the leading terms of the B S amplitudes do not depend on the quark mass. Finally, because singlets and triplets are determined by

MODEL A

$$\chi(q,P) = \chi_o(q,P) + \frac{1}{2m}[\not{P},\chi_o] - \frac{i}{2m}\{q,\chi_o\}$$

$$M^2 = \frac{4}{3}[\alpha^P+m^2+2(n+2r+2)\sqrt{\beta^P}]$$

$$\chi^{n,r(q,P)}_{o_{j,j_3}} = \gamma_5 \cdot Y^n_{j,j_3} \cdot R_{n,r}(q_E^2,\beta^P)$$

$$M^2 = \frac{4}{3}[\alpha^V + m^2 + 2(n_b+2r+2)\sqrt{\beta^V}]$$

$$\chi^{\ell}_{o\; n_b,r\; j,j_3}(q,P) =$$

$$= \sum_{s_3,\ell_3} \vec{\gamma}_{s_3}(P)\cdot Y^{n_b}_{\ell,\ell_3}\cdot(1s_3;\ell\ell_3|jj_3)\cdot R_{n_b,r}(q_E^2,\beta^V);$$

$$M^2 = 4[\alpha^V + m^2 + 2(n_b+2r+2)\sqrt{\beta^V}]$$

$$\Pi = (-1)^j \qquad C = (-1)^j$$

$$\chi_{o\; n,r\; j,j_3}(q,P) = \frac{\not{P}}{M} Y^n_{j,j_3} \cdot\sqrt{3}\cdot R_{n,r}(q_E^2,\beta^V)$$

MODEL C

$$\chi(q,P) = \chi_o(q,P) + \frac{1}{4m}[\not P,\chi_o] - \frac{i}{2m}\{\not q,\chi_o\}$$

$$\Pi = (-1)^{j+1} \qquad C = (-1)^j$$

$$M^2 = 4[\alpha^P+m^2+2(n+2r+2)\sqrt{\beta^P}]$$

$$\chi_{o\,\substack{n,r\\ j,j_3}}(q,P) = \gamma_5\cdot Y^n_{j,j_3}\cdot\sqrt{3}\cdot R_{n,r}(q_E^2,\beta^P)$$

$$M^2 = 4[\alpha^T + m^2 + 2(n_b+2r+2)\sqrt{\beta^T}]$$

$$\chi^{\ell}_{o\,\substack{n_b,r\\ j,j_3}}(q,P) =$$

$$= \frac{\not P}{M}\sum_{s_3\ell_3}\vec{\gamma}_{s_3}(P)\cdot Y^{n_b}_{\ell,\ell_3}(1s_3;\ell\ell_3|jj_3)\cdot\sqrt{3}\cdot R_{n_b,r}(q_E^2,\beta^T);$$

$$\Pi = (-1)^j \qquad C = (-1)^j$$
$$\ell = j-1$$

$$\Pi = (-1)^{j+1} \qquad C = (-1)^{j+1}$$
$$\ell = j$$

$$\Pi = (-1)^j \qquad C = (-1)^j$$
$$\ell = j+1$$

$$M^2 = 4[\alpha^S + m^2\ 2(n+2r+2)\sqrt{\beta^S}]$$

$$\Pi = (-1)^j \qquad C = (-1)^{j+1}$$

$$\chi_{o\ \substack{n,r \\ j,j_3}}(q,P) = Y^n_{j,j_3} \cdot \sqrt{3} \cdot R_{n,r}(q_E^2, \beta^S)$$

$$M^2 = 4[\alpha^T + m^2 + 2(n_b+2r+2)\sqrt{\beta^T}]$$

$$\chi^{\ell}_{o\ \substack{n_b,r \\ j,j_3}}(q,P) =$$

$$= \frac{P_\mu}{2M}\varepsilon^{\mu\nu\rho\sigma}\ \gamma_\nu\gamma_\rho \sum_{s_3,\ell_3} \delta_{\sigma,s_3}(P)\cdot Y^{n_b}_{\ell,\ell_3}\cdot(1s_3;\ell\ell_3|jj_3)\cdot$$

$$\cdot\ \sqrt{3}\ \cdot\ R_{n_b,r}(q_E^2, \beta^T)$$

$$\Pi = (-1)^{j+1} \qquad C = (-1)^{j+1} \qquad \ell = j-1$$

$n_b = 1,2,\ldots$
$r = 0,1,\ldots$
$j = n_b, n_b-2,\ldots$

$$\Pi = (-1)^{j} \qquad C = (-1)^{j} \qquad \ell = j$$

$n_b = 2,3,\ldots$
$r = 0,1,\ldots$
$j = n_b-1, n_b-3,\ldots$

$$\Pi = (-1)^{j+1} \qquad C = (-1)^{j+1} \qquad \ell = j+1$$

$n_b = 2,3,\ldots$
$r = 0,1,\ldots$
$j = n_b-2, n_b-4,\ldots$

M O D E L   E

$$\chi(q,P) = \chi_o(q,P) + \frac{1}{4m}[\not{P},\chi_o] - \frac{i}{2m}\{\not{q},\chi_o\}$$

$$M^2 = 4[\alpha^A + m^2 + 2(n+2r+2)\sqrt{\beta^A}]$$

$n = 0,1,\ldots.$

$r = 0,1,\ldots.$

$j = n,n-2,\ldots$

$$\chi_{o\ \begin{smallmatrix} n,r \\ j,j_3 \end{smallmatrix}}(q,P) = \gamma_5\ \frac{\not{P}}{M}\cdot Y^n_{j,j_3}\cdot\sqrt{3}\cdot R_{n,r}(q_E^2,\beta^A)$$

$$M^2 = 4[\alpha^V + m^2 + 2(n_b+2r+2)\sqrt{\beta^V}]$$

$$\chi^{\ell}_{o\ \begin{smallmatrix} n_b,r \\ j,j_3 \end{smallmatrix}}(q,P) =$$

$$= \sum_{s_3,\ell_3} \vec{\gamma}_{s_3}(P)\ Y^{n_b}_{\ell,\ell_3}\cdot(1s_3;\ell\ell_3|jj_3)\cdot\sqrt{3}\cdot R_{n_b,r}(q_E^2,\beta^V)$$

$n_b = 0,1,\ldots$
$r = 0,1,\ldots$
$j = n_b+1,\ n_b-1,\ldots$

$n_b = 1,2,\ldots$
$r = 0,1,\ldots$
$j = n_b,\ n_b-2,\ldots$

$n_b = 1,2,\ldots$
$r = 0,1,\ldots$
$j = n_b-1,\ n_b-3,\ldots$

$$M^2 = 4[\alpha^V + m^2 + 2(n+2r+2)\sqrt{\beta^V}]$$

$\Pi = (-1)^j \qquad C = (-1)^j \qquad n = 1,2,\ldots$

$r = 0,1,\ldots$

$$\chi_{o\ \, n,r \atop j,j_3}(q,P) = \frac{\not{P}}{M}\cdot Y^n_{j,j_3} \cdot \sqrt{3}\cdot R_{n,r}(q_E^2,\beta^V) \qquad j = n-1, n-3,\ldots$$

$$M^2 = 4[\alpha^A + m^2 + 2(n_b+2r+2)\sqrt{\beta^A}]$$

$$\chi^{\ell}_{o\ \, n_b,r \atop j,j_3}(q,P) =$$

$$= \gamma_5 \cdot \sum_{s_3 \ell_3} \vec{\gamma}_{s_3}(P)\, Y^{n_b}_{\ell,\ell_3}\, (1s_3;\ell\ell_3|jj_3)\cdot\sqrt{3}\cdot R_{n_b,r}(q_E^2,\beta^A)$$

$\Pi = (-1)^{j+1} \qquad C = (-1)^j \qquad \ell = j-1$

$\Pi = (-1)^j \qquad C = (-1)^{j+1} \qquad \ell = j$

$\Pi = (-1)^{j+1} \qquad C = (-1)^j \qquad \ell = j+1$

TABLE VII

$$R_{n,r}(q_E^2,\beta) = (2\pi)^2 \sqrt{\frac{2}{3\beta}} \cdot \sqrt{\frac{r!}{(n+r+1)!}} \left(\frac{\sqrt{q_E^2}}{4\sqrt{\beta}}\right)^n \cdot L_r^{n+1}\left(\frac{q_E^2}{\sqrt{\beta}}\right)\cdot e^{\frac{-q_E^2}{2\sqrt{\beta}}};$$

$q_E^2 = q_4^2 + \vec{q}^{\,2} > 0$. The hyperspherical functions are:

$$Y^n_{\ell,\ell_3} = (-1)^{\frac{n-\ell}{2}}\, Y_{\ell,\ell_3}(\vec{e}_q(P))\, C^n_\ell\left(\frac{i(qP)}{\sqrt{P^2\cdot q_E^2}}\right) ;$$

$$C^n_\ell(\cos\beta) = N^n_\ell \sin^\ell\beta\cdot C^{\ell+1}_{n-\ell}(\cos\beta),\quad N^n_\ell = \ell!\sqrt{\frac{2^{2\ell+1}(n+1)(n-\ell)!}{\pi\cdot(n+\ell+1)!}}$$

different interaction parameters, we can, in principle, account for SU(6) breaking.

Although the models differ in the Dirac-structure of the amplitudes there seems to be no preference at this stage.

C. PHENOMENOLOGICAL APPLICATIONS

7. Current matrix elements.

A good field for testing our general ideas on relativistic quark dynamics by application to phenomenological problems are leptonic decays of vector mesons and the total hadronic cross-section of e^+e^--annihilation.[73] These processes are described by some simple current matrix elements, which may be expressed immediately with help of the B S amplitudes of the vector mesons. We restrict ourselves to electromagnetic processes.

7.1. Leptonic vector meson decays and higher mass vector mesons.

We start with the assumption that the electromagnetic current might be expressed by the quark fields in the form (Sect.3.3):

$$j_\mu^{e.m.}(x) = [\bar{\psi}(x), \gamma_\mu Q\psi(x)]; \quad Q = \frac{1}{3}\begin{pmatrix} 2 & & 0 \\ & -1 & \\ 0 & & -1 \end{pmatrix} \qquad (7.1)$$

suppressing the renormalization constant.

The decay of the neutral vector mesons $V \to e^+e^-$ is determined by the current matrix element

$$\langle 0|j_\mu^{e.m.}(0)|{}^{M}_{P}\,{}^{1^{--}}_{S_3}\rangle = (2\pi)^{-3/2}\frac{M_V^2}{2\gamma_V}\varepsilon_\mu^{S_3}$$

$$\Gamma_{V\to e^+e^-} = \frac{\alpha^2}{12}\cdot\frac{4\pi}{\gamma_V^2}\cdot M_V \quad . \tag{7.2}$$

Inserting the expression for the current into the decay matrix element, γ_V gets related to the B S amplitude:

$$\langle 0|j_\mu^{e.m.}(0)|M,1^{--}\rangle = \langle 0|T(\bar\psi(0)Q\gamma_\mu\psi(0))|M,1^{--}\rangle = \tag{7.3}$$

$$= (2\pi)^{-11/2} T_r\ \gamma_\mu\cdot Q\int\chi(q,P)d^4q.$$

Besides the groundstate vector mesons $\rho(765)$ etc., which belong to the quantum number $n_b = \ell = 0$, $r = 0$, our model contains a whole series of higher mass 1^{--}-states. For example at a mass $M \approx 1.65$ GeV there are three 1^{--}-iso-triplet states - as members of nonets - with $N = n_b + 2r = 2$:

$$n_b = 0 \quad \ell = 0 \quad r = 1$$
$$n_b = 2 \quad \ell = 2 \quad r = 0$$
$$n_b = 2 \quad \ell = 0 \quad r = 0 \ .$$

Using the B S amplitudes of the three models given in Table VII and inserting them in Eq.(7.3) we obtain

$1/\gamma_{V_r}$	A	C	E
$n_b=\ell=0$ $r=0,1,\ldots$	$(-1)^r \frac{8}{\pi} \cdot \frac{\sqrt{r+1}\,\sqrt{\beta^V}}{M_{V_r}^2 \cdot \sqrt{3}} \langle Q\rangle_V$	$(-1)^r \frac{4}{\pi} \cdot \frac{\sqrt{r+1}\,\sqrt{\beta^T}}{m\, M_{V_r}} \langle Q_V\rangle$	$(-1)^r \frac{8}{\pi} \cdot \frac{\sqrt{r+1}\,\sqrt{\beta^V}}{M_{V_r}^2} \langle Q_V\rangle$
$n_b\neq 0$ $\ell=0,2$ $r=0,1,\ldots$	0	0	0

$\langle Q\rangle_V$ denotes the coupling to the quark antiquark charge, it is $\langle Q\rangle_V = \frac{1}{\sqrt{2}}, \frac{1}{3\sqrt{2}}, \frac{-1}{3}$ for ρ^0, ω, ϕ with ideal mixing.

(7.4)

Therefore, for the local current (7.1) only one of the N+1 states, namely the hyperradial excited state, is coupled to the photon. This corresponds to the intuitive picture of the non-relativistic quark model that the annihilation rate of a $q\bar{q}$ pair is proportional to the probability density at zero distance.[74]

Inspite of having neglected renormalization effects we give a short comparison with experiment. Using the interaction parameters of Eq.(6.22) we get

model	A	E	exp.[75]	
$\frac{\gamma^2_{\rho^0}}{4\pi} =$	0.15	0.45	0.64 ±0.07	(7.5)

In model C γ_ρ depends on the quark mass. The experimental value would lead to $m_q \sim 460$ MeV. In the following we shall disregard it.

It is particularly gratifying for our model that recently there was found a strong experimental indication for the existence of a ρ' with a mass around 1600 MeV in e^+e^- annihilation[76] and photoproduction[77] of $2\pi^+2\pi^-$.

For the ratio of the ρ',ρ coupling we obtain

$$\frac{\gamma^2_{\rho'}(1600)}{\gamma^2_\rho\ (765)} = \frac{M^4_{\rho'}}{2M^4_\rho} \approx 10 \qquad \text{(models A, E)} \tag{7.6}$$

$$= 6.6 \pm 2.0 \qquad \text{(exp.result}^{78,75}\text{)}.$$

This ratio is independent of renormalization effects.

We wish to point out that in our scheme there are 3 linear independent 1^{--} states around 1600 MeV, but only one of them couples to the photon. Of course we cannot say that the corresponding linear combination represents a physical particle. Without considering the decay channels and spin-orbit splitting we cannot obtain the complete mass matrix, whose eigenvectors are the physical states. In general we expect the state $|1^{--}, n_b = \ell = 0, r = 1\rangle$ to be a linear combination of the physical vector mesons. This would lead to interesting interference and mass shift effects in the different final states of the e^+e^- induced reactions in this energy range. Precise experiments could therefore give information on the additional vector meson related to the relativistic degree of freedom.

Finally we want to mention that the coupling of a series of vector mesons to the e.m. current described by our model gives a dynamical explanation of the "generalized vector meson dominance model" of e.m. interactions.[79]

7.2. The total hadronic cross section. A resonance model dual to pointlike structure.

The total cross section for the reaction $e^+e^- \to$ hadrons is directly related to the current two-point function (quark fourpoint function):[80]

$$\langle 0|j_\mu^{e.m.}(x)\, j_\nu^{e.m.}(0)|0\rangle = (2\pi)^{-3}\int d^4P\cdot\theta(P_0)\cdot \cdot\, e^{-iPx}\cdot\rho_V(P^2)\,\left[g_{\mu\nu} - \frac{P_\mu P_\nu}{P^2}\right] \tag{7.7}$$

$$\sigma_{tot}(e^+e^- \to \text{hadrons}) = \frac{16\,\pi^3\,\alpha^2}{s^2}\cdot\rho_V(s)$$

The discussion of this Green's function is a particular simple and good example for illustrating dual dynamics in the relativistic quark model, as introduced in Sect.3.2.

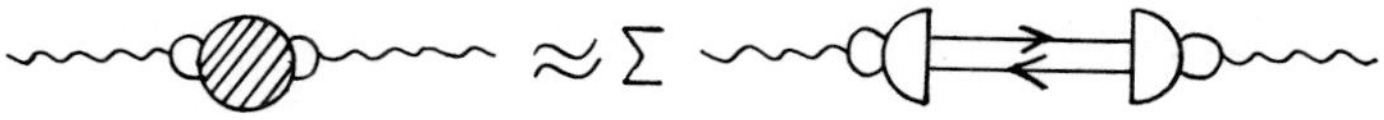 (7.8)

vector mesons

Analytically we get in zero width approximation:

$$\rho_V(s) = \sum_r \frac{16\beta^V}{3\pi^2}(r+1)\,\delta(s-M_r^2) \sum_{i=\rho,\omega,\phi} |\langle Q\rangle_i|^2 \begin{bmatrix} \times\ 1 \text{ model A} \\ \times\ 3 \text{ model E} \end{bmatrix} \tag{7.9}$$

Since the vector mesons are unstable we have to modify the tree-graph approximation by taking into account the finite widths. The effect is obtainable by smoothening the δ-functions. Using the mass formula Eq.(6.18) we obtain

$$\sigma_{tot} = \frac{2\pi\alpha^2}{s}\left(1 + \frac{1.6\ \text{GeV}^2}{s}\right) \begin{bmatrix} \times\ 1 \text{ model A} \\ \times\ \frac{1}{3} \text{ model E} \end{bmatrix} \quad . \tag{7.10}$$

Let us compare this result with the "pointlike" cross section in the quark parton model:[81]

$$\sigma_{tot} = \frac{4\pi\alpha^2}{3s} \quad \Sigma Q_i^2 = \frac{8\pi\alpha^2}{9s} \quad . \tag{7.11}$$

We see that our boundstate model gives, in the average, the asymptotic behaviour generally attributed to the light cone (scaling) singularity of the currents of free quark fields.[82] We consider this as an interesting result, especially because it gives some hint on the meaning of "current quarks" and "constituent quarks" in the frame work of relativistic QUARFT with strong binding.[46] This relation between resonance structure and the 1/s behaviour of σ_{tot} is reminiscent of the relation of resonance structure and the scaling limit of the structure functions in deep inelastic electron scattering (Bloom-Gilman duality[38]). Therefore our QUARFT leads to a dual model for $\sigma_{tot}(e^+e^- \to \text{hadrons})$. Unfortunately, in our model the axial current Green's functions are not "saturated" in the same way by axial vector mesons. Therefore our B S model, at this stage, does not fulfil the requirement of current algebra on the level of the Weinberg spectral sum rules.[83]

8. Hadronic vertices.

For further applications the hadronic vertices are of particular importance (see Sect.3.2). We shall use our B S amplitudes of Table VII for the calculation of the coupling constants according to Eq.(3.11), that is in triangle approximation.

8.1. The evaluation of mesonic coupling constants.

The matrix element M for the mesonic decay $M \to M_1 + M_2$ is related to the six-point function:

$$\left[\prod_{i=1}^{3}((p_i-\bar{p}_i)^2 - M_i^2)\right] \eta(p_1p_2p_3;\bar{p}_1\bar{p}_2\bar{p}_3)\Big|_{\substack{(p_i-\bar{p}_i)^2=M_i^2 \\ (p_i-\bar{p}_i)_o<0}} =$$

$$= \frac{i}{\sqrt{2\pi}} \left[\prod_{i=1}^{3} \bar{\chi}_{\left[{M_i \atop P_i}\right]} (p_i \bar{p}_i) \right] M_{\left[{M_i \atop P_i}\right]} \tag{8.1}$$

and in triangle approximation[41] it is given by the graph

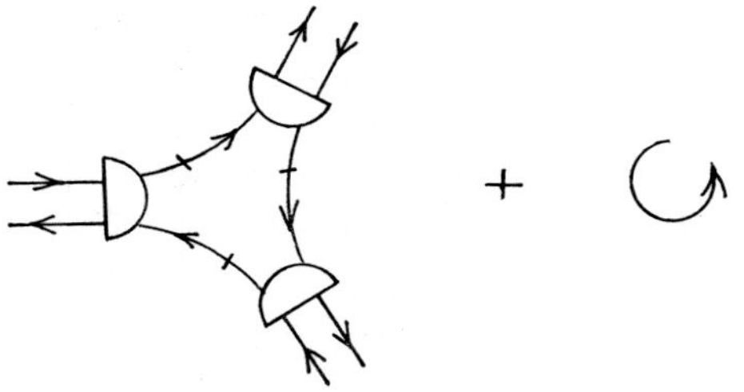

(8.2)

For the inner lines we insert free propagators in accordance with our treatment of the B S equation. The meson-quark-antiquark vertices are given by the amputated B S amplitudes, that is the vertex functions. If we express them in terms of the B S amplitudes, we obtain:

$$M_L = [\mathrm{Trace}(SU(3))] \{\mathrm{Trace}(2\pi)^{-9/2} i \int d^4k \chi^{(3)}(k,P)(\not{k}-\tfrac{\not{P}}{2}-m) \cdot$$
$$\cdot \chi^{(1)}(k+\frac{P_2}{2},P_1)(\not{k}-\frac{\not{P}_1-\not{P}_2}{2}-m) \cdot$$
$$\cdot \chi^{(2)}(k-\frac{P_1}{2},P_2)(\not{k}+\frac{\not{P}}{2}-m)\} \quad . \tag{8.3}$$

In the vertex (8.2) we consider all meson lines as outgoing: $P_1 + P_2 + P = 0$.

For the amplitudes we used the factorized form

$$\chi^{SU(3)}(q,P) = T^{SU(3)} \cdot \chi(q,P) \quad , \tag{8.4}$$

where $T^{SU(3)}$ denotes the familiar $3 \times \bar{3}$ wave functions of the mesons. Then

$$\text{Trace } (SU(3)) = \text{Tr}[T^3 (T^{1^T} T^{2^T} + \eta_1 \eta_2 \eta_3 T^{2^T} T^{1^T})]$$

η_i = charge conjugation parity of particle i

$$\eta_1 \eta_2 \eta_3 = +1 : \text{D-coupling}$$
$$= -1 : \text{F-coupling.}$$

In order to give an impression of the problems which may arise in such a calculation based on a dynamical model, we consider as an example the decay $\rho \to 2\pi$ with the amplitudes of P + V - S model, which are (Table VII):

$$\chi_L^V(q,P) = \frac{4\pi}{\sqrt{3\beta^V}} [\not{\varepsilon}(1 - \frac{\not{P}}{m}) + \frac{(\varepsilon q)}{m}] \cdot e^{\frac{-q_E^2}{2\sqrt{\beta^V}}} + O[\frac{1}{m^2}]$$

$$\chi_L^{PS}(q,P) = \frac{4\pi}{\sqrt{3\beta^P}} [\gamma_5(1 - \frac{\not{P}}{m})] \cdot e^{\frac{-q_E^2}{2\sqrt{\beta^P}}} + O[\frac{1}{m^2}] \quad . \tag{8.5}$$

For the calculation of the integral (8.3) we deform, as usual, the path of the dk_o integration into the Euclidean region and then insert our approximate solution (8.5).

Evaluation of the traces leads to

$$M = \sqrt{2}\ (2\pi)^{-9/2} [\frac{4\pi}{\sqrt{3\beta}}]^3\ 2m^2(\varepsilon, P_1 - P_2) e^{\frac{\Sigma M^2}{24\sqrt{\beta}}} \int d^4k\ e^{\frac{-3k^2}{2\sqrt{\beta}}} \tag{8.6}$$

and by integrating over the Euclidean Gauss function we get:

$$M_{\rho\to 2\pi} = \frac{32\sqrt{\pi}}{27\sqrt{3}} \frac{m^2}{\sqrt{\beta}} e^{\frac{\Sigma M^2}{24\sqrt{\beta}}} (\varepsilon, P_1 - P_2) \tag{8.7}$$

If we take the experimental width $\Gamma_\rho \sim 125$ MeV and $\sqrt{\beta} \sim 0.4$ GeV^2, as determined by the slope of the Regge trajectories, we obtain $m_q \sim 800$ MeV. Such a value of the quark mass does not correspond to our general scheme of heavy quarks. Apparently the dynamical determination of the meson coupling constants from field theoretically normalized B S amplitudes via triangle graphs leads to some problems. Of course, one may adopt the standpoint that this dynamical approximation scheme is incorrect and use the amplitudes only with the aim of having a covariant formulation of a phenomenological model[84] without absolute normalization (similar to the $\tilde{U}(12)$ formulation of the quark model[85,86]). However, we shall take the dynamical aspect seriously and search for a model consistent with heavy quarks.

8.2. Quark mass dependence of hadronic vertices and saturation of superstrong forces.

In our example (8.7) the strength of the meson couplings increases with the quark mass. The reason for this is that in our model the quark mass measures the strength of the superstrong interaction which is responsible for the strong binding of the quarks. Since the mesonic interaction in the triangle graph approximation is derived from the superstrong interaction, it is not surprising that, in general, it will increase with it. Thus in a scalar quark model the triangle graph is proportional to

m^3, but in our spinor models the leading power of the B S amplitude into a Dirac part and an orbital part allows the determination of the leading power in m by simply examining the Dirac traces. For the four different types of couplings, namely

quark-spin - triplet - singlet - singlet ≡ T - S - S
- triplet - triplet - singlet ≡ T - T - S
- triplet - triplet - triplet ≡ T - T - T
- singlet - singlet - singlet ≡ S - S - S

we give the results for the different models in Table VIII.

Leading terms of B S amplitudes			T-S-S	T-T-S	T-T-T	S-S-S
	S	T				
A)	γ_5	$\not\varepsilon$	m^2	m	m^2	m^{-1}
C)	γ_5	$\not\varepsilon\frac{\not P}{M}$	m	m^3	m^3	m^{-1}
E)	$\gamma_5\frac{\not P}{M}$	$\not\varepsilon$	m^2	m^2	m^2	m^2

Table VIII

It is seen that the leading power in m of the meson couplings is different in the various models and is sometimes less than in the scalar model. We interpret this as a "saturation" effect of the superstrong forces resulting from their spin dependence. Therefore it is natural to ask for a model in which the superstrong forces are completely saturated, i.e. in which the mesonic forces are independent of the quark mass in leading order. (We remark that saturation of superstrong forces has also to be assumed in order to explain the absence of other boundstates than those of $q\bar{q}$ and qqq.[87] It is not known whether these two kinds of saturation properties are related to each other.)

A candidate which may show complete saturation is model E ($-\gamma_5 \times \gamma_5$ interaction) because the "formal" leading terms are the same for all channels. Indeed, it turns out that the leading term is not proportional to m^2 but to m^o. In order to see this, we rewrite the B S amplitudes of this model in the form:

$$\begin{aligned} \chi_L(q,P) &= \chi_o + \frac{1}{2m}\{\not{q},\chi_o\} + \frac{1}{4m}[\not{P},\chi_o] \\ &= \frac{1}{2m}\,[\bar{S}^{-1}(\tfrac{P}{2}+q)\chi_o - \chi_o\, S^{-1}(\tfrac{P}{2}-q)] \end{aligned} \tag{8.8}$$

where $S^{-1}(k) = \not{k} - m$

$\bar{S}^{-1}(k) = \not{k} + m$

are the inverse free propagators. Insertion into Eq. (8.3) leads to

$$M \sim \frac{1}{8m^3}\mathrm{Tr}\{2\chi_o^{(3)}\chi_o^{(1)}\chi_o^{(2)}D^{-1}(k-\tfrac{P}{2})D^{-1}(k-\frac{P_1-P_2}{2})D^{-1}(k+\tfrac{P}{2})\}+$$

$$+\frac{1}{8m^2}\mathrm{Tr}\{\chi_o^{(3)}\chi_o^{(1)}(\not{k}-\frac{\not{P}_1-\not{P}_2}{2})\chi_o^{(2)}[2D^{-1}(k-\tfrac{P}{2})$$

$$D^{-1}(k+\frac{P}{2}) - D^{-1}(k+\frac{P}{2})D^{-1}(k-\frac{P_1-P_2}{2}) -$$

$$- D^{-1}(k-\frac{P_1-P_2}{2})D^{-1}(k-\frac{P}{2})] \qquad (8.9)$$

$$+ 2 \text{ cyclic terms}\}$$

where

$$D^{-1}(k) = \bar{S}^{-1}(k) \quad S^{-1}(k) = k^2 - m^2 \quad .$$

The first term vanishes for all three types of meson couplings, because χ_o is always composed of an odd number of Dirac matrices. In the second term the bracket vanishes in the order m^4, therefore the complete matrix element is of order m^o.

In order to discuss correctly the meson decays in this saturating model, we need the B S amplitudes, including terms $1/m^3$. Proceeding in a way similar to the first order calculation which was presented in Sect.6, and which is analogous to the Foldy-Wouthuysen transformation, we get for the B S amplitude:

$$\chi_L(q,P) = \frac{1}{2m}\{\bar{S}^{-1}(\frac{P}{2}+q)(\chi_o+\frac{\chi_2}{m^2}) - (\chi_o+\frac{\chi_2}{m^2})\, S^{-1}(\frac{P}{2}-q)\}$$

$$+ \frac{(Pq)}{4m^3}\{\bar{S}^{-1}(\frac{P}{2}+q)\cdot\chi_o + \chi_o\cdot S^{-1}(\frac{P}{2}-q)\} \qquad (8.10)$$

$$+ \frac{\varepsilon_1}{8m^3}[\not{P},\chi_o] + \frac{\varepsilon_2}{4m^3}\{\not{q},\chi_o\} + \frac{\beta}{2m^3}\{\not{\beta},\chi_o\} \quad .$$

In this expression, χ_o and χ_2 are both either of vector or axial-vector type. The contribution in order m^o is of the form:

$$M = [\text{Trace SU(3)}]\cdot[-(2\pi)^{-9/2} i \int d^4k \; \mathrm{Tr}\{\Delta\chi^{(3)}(k,P) \; \chi_o^{(1)}(k+\frac{P_2}{2},P_1)\cdot\chi_o^{(2)}(k-\frac{P_1}{2},P_2) + 2 \text{ cyclic terms}\}] \tag{8.11}$$

where $\Delta\chi$ is given by:

$$\Delta\chi = \frac{\varepsilon_1}{8}[\not{P},\chi_o] + \frac{\varepsilon_2}{4}\{\not{q},\chi_o\} + \frac{\beta}{2}\{\not{\gamma},\chi_o\} \tag{8.12}$$

$\varepsilon_1 = -\alpha_V - \alpha_T \qquad \varepsilon_2 = -\alpha_V - \alpha_S$ for V-type

$\varepsilon_1 = -\alpha_A - \alpha_P \qquad \varepsilon_2 = -\alpha_A - \alpha_T$ for A-type.

These calculations show that there exists a spin dependence of the superstrong forces which leads to saturation. This is the main result of this section. Equations (8.11) and (8.12) describe the meson vertices resulting from this model.

The small components $\Delta\chi$ of the amplitudes depend on the fine structure of the interaction. We have not yet performed a phenomenological analysis based on these equations. But we can see already that the dependence of the decay matrix element on the small components of the B S amplitudes has important phenomenological consequences: There are couplings which are absent in the nonrelativistic quark model.[88,89] They will show up most distinctly in decays where more than one coupling is present, for instance in the decay $B \to \pi + \omega$, $A_1 \to \pi + \rho$.

9. Baryons, meson scattering and so on.

In this final section we report on some research which is in an even more preliminary stage.

9.1. The three-quark problem.

In the spirit of the dynamical assumptions, discussed in Sect.4 for $q\bar{q}$-binding in mesons, R. Meyer studied the relativistic three quark boundstate problem for scalar quarks.[36] We already described the structure of the B S equation in Sect.3.1. in graphical form. The B S amplitude $\chi(p_1 p_2 p_3)$:

$$\chi_P(p_1 p_2 p_3)\,\delta(p_1+p_2+p_3-P) = \frac{1}{(2\pi)^4}\iint dx_1 dx_2 dx_3$$

$$e^{i(p_1x_1+p_2x_2+p_3x_3)}\cdot\langle 0|T\psi(x_1)\psi(x_2)\psi(x_3)\left|\begin{matrix} M & j \\ P & j_3\end{matrix}\right.\rangle \tag{9.1}$$

is a solution of the B S equation

$$(p_1-m^2)(p_2^2-m^2)(p_3^2-m^2)\chi_P(p_1p_2p_3) = \iint dp_1' dp_2' dp_3'\,\delta(p_1'+p_2'+p_3'-P)\cdot$$

$$\cdot\{\sum_{\gamma=1}^{3}(p_\gamma^2-m^2)\langle p_\alpha p_\beta|K_2|p_\alpha' p_\beta'\rangle\delta(p_\gamma-p_\gamma') + \langle p_1p_2p_3|K_3|p_1'p_2'p_3'\rangle\}\chi_P(p_1p_2p_3) \tag{9.2}$$

where $p_\alpha p_\beta p_\gamma$ denote the momenta in cyclic permutation. For the inverse propagator we already assumed an effective one-particle form.

We shall make a phenomenological ansatz of the convolution type for the two-quark- and three-quark-irreducible interaction kernels K_2 and K_3 respectively.

First we introduce relative coordinates:

$$P = P_1 + P_2 + P_3; \; k_3 = \frac{1}{2}(p_1 - p_2), \; q_3 = \frac{1}{3}(p_1 + p_2 - 2\, p_3). \tag{9.3}$$

The choice of Jacobian coordinates distinguishes the subsystem (12). The relative coordinates, obtained from Eq.(9.3) by cyclic permutation, are linearly related:

$$\begin{bmatrix} k_3 \\ q_3 \end{bmatrix} = \begin{bmatrix} -\frac{1}{2} & -\frac{3}{4} \\ 1 & -\frac{1}{2} \end{bmatrix} \cdot \begin{bmatrix} k_1 \\ q_1 \end{bmatrix} = \begin{bmatrix} -\frac{1}{2} & \frac{3}{4} \\ -1 & -\frac{1}{2} \end{bmatrix} \cdot \begin{bmatrix} k_2 \\ q_2 \end{bmatrix} \tag{9.4}$$

The possibility of Wick rotation for three-particle B S equations has not been proven in general. But there are some arguments that for small bound state masses and heavy quarks the Wick rotation $k_{30} \to ik_{34}$, $q_{30} \to iq_{34}$ (k_{34}, q_{34} real) may be performed.[90] Because of the linear relation (9.4), this means also a Wick rotation for the other relative coordinates.

Following the arguments of Sect.4 we approximate in the Wick rotated form of Eq.(9.2) the two and three quark interaction kernels by harmonic forces

$$K_2^{12} = (\alpha + \frac{1}{2} m^4 \omega^2 \Box_{k_3}) \quad \text{etc.}$$

$$K_3 = m^2 A + \frac{1}{2} m^6 B [\frac{3}{2} \Box_{k_3} + 2 \Box_{q_3}] \tag{9.5}$$

and get after further approximations ($\langle q^2 \rangle / m^2 \ll 1$) an eightdimensional oscillator equation for the bound state masses M^2:

$$[6(2 k_3^2 + \frac{3}{2} q_3^2) - \frac{b^2}{12}(\frac{1}{2}\Box_{k_3} + \frac{2}{3}\Box_{q_3})]\chi(p_3,q_3) = (M^2 - a)\chi(p_3,q_3). \tag{9.6}$$

From this equation we get immediately the following results:

The mass spectrum is given by:

$$M^2 = a + 4 b + b \sum_{i=1}^{8} n_i . \tag{9.7}$$

It corresponds to linear "Regge trajectories". Due to oscillations in relative energy there is much higher degeneracy than in the non-relativistic oscillator quark model, even if some of the excited levels have to be eliminated because they belong to negative norm.

The numerical values are the result of a fit to the low lying states:

$$a = 3 m^2 \frac{m^4 + A - 3\alpha}{m^4 - \alpha} \approx - 3.3 \quad \text{to} \quad -2.9 \text{ GeV}^2$$

$$b = 3\, m^3 \sqrt{6\, \frac{\omega^2 - B}{m^4 - \alpha}} \approx 1\ \mathrm{GeV}^2 \quad .$$

The ground state is not degenerate, the radius of its B S amplitude

$$\chi_o(k_3, q_3) = e^{-\frac{6}{b}(k_3^2 + \frac{3}{4} q_3^2)} \tag{9.9}$$

is determined by the Regge slope.

In this model too, the large quark mass is compensated by the depth of the four-dimensional potential. If we consider the relativistic bound-state model for more than three quarks, using only the two and three quark interactions already introduced here, then the compensation of the quark mass by the potential does not always take place. It was shown by R. Meyer that it is possible to explain the absence of low lying many-quark bound states by a suitable choice of the potential parameters. This model is a relativistic generalization of a non-relativistic saturation model proposed by L.I. Schiff.[87]

The sequence of the baryon states in the non-relativistic quark model, $[\underline{56}, 0^+]$, $[\underline{70}, 1^-]$, $[\underline{56}, 2^+]$,... in SU(6) x O(3) notation, seems to indicate that not all oscillator excitations exist in nature. This suggests to examine another approach to the three quark problem, in which the low lying baryons are considered as the excitations of an effective two-body system, the quark diquark system.[91] In particular the SU(6) x O(3) assignment [21, 0] to the diquark bound state could explain the baryon spectrum in the non-relativistic quark model. The starting point for such a relativistic model form the relativistic Faddeev equations, as discussed by R. Meyer. We refer for further information to the original literature.[92,36]

The outstanding problem of the relativistic three-quark system is the spin and statistics puzzle for quarks, appearing in the non-relativistic quark model of the baryons.[93] In order to study this problem one clearly has to treat the relativistic three-Fermi-quark bound-states. In the case of the mesons we have found that the relativistic B S amplitudes might look quite different from a non-relativistic wave function. Therefore we are not convinced that the baryon spectrum brings unavoidable difficulties for a Fermi-quark model. But the problem is not yet solved.

9.2. Resonance excitation scattering.

"Resonance excitation scattering" (in zero width approximation) of scalar mesons in a scalar quark model has been analyzed by G.Berendt and E.Weimar.[94] For the calculation of the scattering amplitude

$$A(p_1p_2;p_3p_4) \sim \sum_{\text{Mes}}$$ 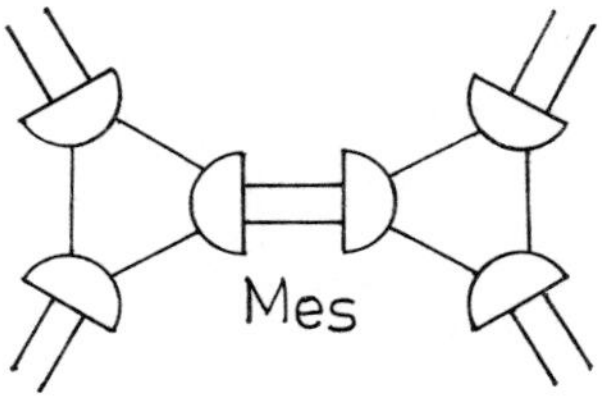 (9.10)

they determine the meson vertices with help of Wick rotated B S amplitudes in oscillator approximation (Sect. 4.2.). The sum in (9.10) may be calculated by a technique, useful for all problems in which SU(n)-symmetric oscillator wave functions are involved.[95] Their main result is that the model does not satisfy the requirement of planar duality.[96] Hence, there remains an essential difference between our field theoretical oscillator approximations and the infinite number of oscillators which must be excited in order to give the Veneziano amplitude![16]

9.3. Conclusion and outlook.

In these lectures we tried to develop a possible scheme of hadron dynamics based on quark fields and general field theory. Motivated by phenomenological considerations we persued in this general framework the following dynamical patterns:

i) heavy quarks with a superstrong, nonsingular, interaction;

ii) dual dynamics for hadron processes in tree graph approximation with internal lines corresponding to all mesons and baryons appearing in the relativistic quark model, and vertices calculated in quark triangle approximation;

iii) saturation of the superstrong quark interaction leading to hadronic vertices of conventional strength and thereby introducing a hierarchy in the interactions.

The application of this scheme to simple problems (mass

spectrum, leptonic decays, e^+e^- -annihilation, hadronic vertices) showed its compatibility with our experimental knowledge. It is clear that these dynamical patterns should be tested more detailed and in more complicated processes in order to find out their predictive power, or the necessity of modifications. We mention a few of these equations:

A detailed phenomenological analysis of meson decays should verify the predictions of our model of hadronic vertices; it would check the assumption of triangle approximation and saturation.

Our dynamical patterns lead to a dual scaling behaviour for $\sigma_{tot}(e^+e^- \longrightarrow \text{hadrons})$. It would be interesting if this mechanism could be extended to deep inelastic electron scattering in order to explain the observed scaling behaviour and duality of the structure functions.

In our model the postulate of current algebra is not yet incorporated. It has to be examined to what extent experimental facts or theoretical arguments impose modifications.

The extension of the Fermi-quark model to baryons is important for two reasons: Theoretically the spin-statistics puzzle is a crucial problem for a field theoretical quark model. From the practical point of view there are many experimental data on baryons relevant for the check of the quark model.

In the quark model mesons and baryons are composed of the same constituents. Relations between total cross sections or magnetic moments of mesons and baryons derived by the additivity assumption in the nonrelativistic quark model support this basic feature. In the frame work of QUARFT generalized crossing relations of n-point functions connect meson and baryon properties. It would be very interesting to derive from these general properties with help of our dynamical assumptions new relations between mesons and baryons.

Even a very liberal cut-off on speculations imposes now an end to this outlook.

REFERENCES

1. Chapter A is partly based on Lectures by H. Joos, Quark Theory of Elementary Particles, Instituto de Fisica Teorica, Sao Paulo (1969).

2. Textbooks on general Quantum Field Theory, for example, L. Klein (Editor), "Dispersion Relations and the Abstract Approach to Field Theory" (New York, Gordon and Breach Publishers Inc. 1961).
R. Jost, "The General Theory of Quantized Fields", Providence: American Mathematical Soc. (1965).
R. Jost (Editor), Teoria quantistica locale, New York and London, Academic Press (1969).

3. R. Haag, Phys. Rev. 112, (1958) 669.
W. Zimmermann, Nuovo Cim. X, (1958) 597.
D. Ruelle, Helv. Phys. Acta 35 (1962) 147.

4. W. Zimmermann, Nuovo Cim. 13 (1959) 503, ibid. 16 (1960) 690.

5. For a review on this work see:
R. Haag, Brandeis Lectures 1970.

6. W. Thirring, Nuclear Phys. 10, 97 (1959).

7. S. Sakata, Progr. Theor. Phys. 16, 686 (1956).

8. J. Wess, Nuovo Cim. 15, 52 (1960).

9. M. Gell-Mann, Physics Lett. 8, 214 (1964).
G. Zweig, CERN preprints TH 401, 412 (1964) (unpublished).

10. H. J. Lipkin, Proc. of the Lund Int. Conf. on Elementary Particles, p. 51 (Lund 1969).

11. G. Preparata, Massive Quarks and Deep Inelastic Phenomena, Universita di Roma (Preprint).

12. O. W. Greenberg, Phys. Rev. Lett. 13, 598 (1964).
For arguments from weak interactions see:
T. Goto, O. Hara, and S. Ishida, Progr. Theor. Phys. 43, 849 (1970).
J. G. Körner, Nuclear Phys. B25, 282 (1970).

13. M. Y. Han and Y. Nambu, Phys. Rev. 139B, 1006 (1965).

14. M. Gell-Mann, in "Elementary Particle Physics", p.733 (P.Urban ed., Springer, Wien-New York, 1972).

15. We thank Prof. K. Symanzik for a hint in this direction.

16. Dynamics based on generalized Veneziano formulas seem to correspond to infinite component fields. For ref. see:
V. Alessandrini, D. Amati, M. Le Bellac, D. Olive, Phys. Rep. 1c, No. 6 (1971).
R. J. Rivers, Nuovo Cim. 11A, 178 (1972).
For dynamical quark n-point functions based on the Veneziano formula see:
M. Bando, S. Machida, H. Nakkagawe, and K. Yamawaki, Progr. Theor. Phys. 47, 626 (1972) and literature quoted there.

17. On a more kinematical level, phenomenological questions in the framework of QUARFT were discussed by:
T. Gudehus, DESY 68/11.
C. H. Llewellyn-Smith, Ann. Phys. (N.Y.) 53, 521 (1969).

18. M. Gell-Mann, Physica 1, 63 (1964).

19. a) J. V. Allaby et al., Nuovo Cim. 64A, 75 (1969).
b) Yu. M. Antipov et al., Physics Lett. 29B, 245 (1969).
c) M. Bott-Bodenhausen et al., Physics Lett. 40B, 693 (1972).

20. R. Hagedorn, Nuovo Cim. Suppl. 6, 311 (1968).
V. M. Maksimenko et al., Sov. Phys. - JETP Lett. 3, 214 (1966).

21. T. Massam, The Quark Hunters' Progress, CERN 68-24.
Y. S. Kim, N. Kwak, Fields and Quanta 3, 1 (1972).

22. J. J. de Swart, Phys. Rev. Lett. 18, 618 (1967).

23. F. Low, Comments Nucl. and Part. Phys. 1, 52, 85 (1967).
H. Fritzsch, M. Gell-Mann, Proc. of the XVI Int. Conference on High Energy Physics, Chicago-Batavia, Vol. 2, 135 (1972).

24. K. Johnson, Phys. Rev. D6, 1101 (1972).

25. Similar ideas were proposed by:
H. Suura, Physics Lett. 42B, 237 (1972).

26. Heavy quarks were introduced in a non-relativistic framework by:
G. Morpurgo, Physics 2, 95 (1965).

27. G. Källen, Helv. Phys. Acta, 25, 417 (1952).

H. Lehmann, Nuovo Cim. 11, 342 (1954).
S. S. Schweber, Relativistic Quantum Field Theory, p. 659 (Harper and Row, New York 1961).

28. We thank G. Preparata for a clarifying discussion on this point.

29. We omit technical details. In view of the fact, that all hadrons may be generated by e^+e^- annihilation, this assumption might be even justified in physics. Compare also: R. F. Dashen, D.H. Sharp, Phys. Rev. 165, 1857 (1968).

30. J. M. G. Fell, Transact. Am. Math. Soc. 94, 365 (1960).
R. Haag, D. Kastler, Journ. Math. Phys. 5, 848 (1964).
S. Doplicher, R. Haag, J. E. Roberts, Commun. Math. Phys. 23, 199 (1971).

31. K. Symanzik, Journ. Math. Phys. 1, 249 (1960).
for references see:
K. Symanzik, "Many-Particle Structure of Green's Functions", Symposia on Theoretical Physics, Vol.3, 121 (1967), (Plenum Press, 1967).

32. J. G. Taylor, Phys. Rev. 150, 1321 (1966).
D. Z. Freedman, C. Lovelace and J. M. Namyslowski, Nuovo Cim. 43A, 258 (1966).

33. This form is due to K. Symanzik (unpublished).

34. H. A. Bethe, E. E. Salpeter, Phys. Rev. 84, 1232 (1951).
M. Gell-Mann, F. Low, Phys. Rev. 84, 350 (1951).

35. S. Mandelstam, Proc. Roy. Soc. A233, 248 (1955).
R. E. Cutkosky, M. Leon, Phys. Rev. 135B, 1445 (1964).

36. R. Meyer, Dissertation, Hamburg (1972); internal report DESY-T-72/10.

37. R. Dolen, D. Horn, and C. Schmid, Phys. Rev. Lett. 19, 402 (1967), Phys. Rev. 166, 1772 (1968). Application to meson-meson scattering: C. Lovelace, Physics Lett. 28B, 264 (1968).

38. E. D. Bloom, F. J. Gilman, Phys. Rev. Lett. 25, 1140 (1970), Phys. Rev. D4, 2901 (1971).

39. M. Böhm, H. Joos, M. Krammer, DESY 72/62 (1972). A. Bramon, E. Etim, M. Greco, Physics Lett. 41B, 609 (1972).

40. A. Salam, R. Delbourgo, J. Strathdee, Proc. Roy. Soc. (London) A284, 146 (1965).

41. T. Gudehus, DESY-Report 68/11 (1968). M. Böhm, T. Gudehus, Nuovo Cim. 57A, 578 (1968). H. Suura, B.-L. Young, Nuovo Cim. 11A, 101 (1972).

42. see f.i.: G. Morpurgo in "Theory and Phenomenology in Particle Physics" part A, p. 84 (Editor A. Zichichi, Academic Press New York, London (1969)). J. J. J. Kokkedee, "The Quark Model" (W. A. Benjamin, Inc. New York (1969)).

43. H. Harari, Phys. Rev. Lett. 22, 562 (1969). J. Rosner, Phys. Rev. Lett. 22, 689 (1969).

44. Y. Takahashi, Nuovo Cim. 6, 370 (1957).

45. F. Gutbrod told us about the importance of considering modifications of S'(q) according to (3.14) in phenomenological bound state models of e.m. interactions. (Compare F. Gutbrod, DESY 72/74 (1972).

46. M. Gell-Mann, Ref. 14, for the relation between constituent quarks and current quarks.

47. J. J. J. Kokkedee, l.c.;

O.W. Greenberg, Proc. of the Lund Int. Conf. on Elementary Particles, p. 385 (Lund 1969).

48. R. H. Dalitz, "Symmetries and the Strong Interactions", Proc. of the XIIIth International Conference on High Energy Physics, Berkeley (1966) 215. R. H. Dalitz, "Mesonic Resonance States", Meson Spectroscopy, p. 497 (C. Baltay, A. H. Rosenfeld ed., New York (1968)).

49. Ref. 34.
An extensive review of the theory of the B S equation has been given by: N. Nakanishi, Progr. Theor. Phys. Suppl. 43, 1 (1969).

50. M. Böhm, H. Joos, M. Krammer, Nuovo Cim. 7A, 21 (1972), and in "Concepts in Hadron Physics", p. 407 (Springer-Verlag, Wien, New York, 1971).

51. M. K. Sundaresan, P. J. S. Watson, Ann. Phys. (N.Y.) 59, 375 (1970).
G. Preparata, in "Subnuclear Phenomena", p. 240 (1969 International School of Physics E. Majorana, Erice).

52. G. C. Wick, Phys. Rev. 96, 1124 (1954).

53. M. Gourdin, Nuovo Cim. 7, 338 (1958).

54. A. Erdelyi, ed., "Higher Transcendental Functions", Vol. 2 (McGraw-Hill, New York, 1953).

55. E. Zur Linden, H. Mitter, Nuovo Cim. 61B, 389 (1969).

56. A. Pagnamenta, Nuovo Cim. 53A, 30 (1968).

57. See Ref. 51 and
P. Becher, Diplomarbeit, Würzburg (1972).

58. M. Ciafaloni, P. Menotti, Phys. Rev. 140B, 929 (1965).

59. R. P. Feynman, M. Kislinger, F. Ravndal, Phys. Rev. D3, 2706 (1971).

60. M. Böhm, H. Joos, M. Krammer, Nuclear Phys. B51, 397 (1973).

61. M. K. Sundaresan and P. J. S. Watson, Ref. 51.

62. B. L. v. d. Waerden, "Die gruppentheoretische Methode in der Quantenmechanik", p. 78 ... (Springer, Berlin 1932).

63. See f.i.:
A. R. Edmonds, "Angular Momentum in Quantum Mechanics" (Princeton Univ. Press, 1957).

64. S. Mandelstam, Proc. Roy. Soc. A237, 496 (1956).

65. H. M. Lipinski, D. R. Snider, Phys. Rev. 176, 2055 (1968).
R. F. Keam, a) Journ. Math. Phys. 9, 1462 (1968).
b) ibid. 10, 594 (1969).
c) ibid. 11, 394 (1970).
d) ibid. 12, 515 (1971).
R. Delbourgo, A. Salam, J. Strathdee, Nuovo Cim. 50A, 193 (1967).
P. Breitenlohner, MPI-Preprint (1970).
P. Narayanaswamy, A. Pagnamenta, Nuovo Cim. 53A, 635 (1968).
K. Ladânyi, Preprint, Tübingen (1972).

66. J. S. Goldstein, Phys. Rev. 91, 1516 (1953).
S. N. Biswas, H. S. Green, Nuclear Phys. 2, 177 (1956)
A. Bastai, L. Bertocchi, G. Furlan, M. Tonin, Nuovo Cim. 30, 1532 (1963).
W. Kummer, Nuovo Cim. 31, 219 (1964); ibid. 34, 1840 (1964).
K. Seto, Progr. Theor. Phys. 42, 1394 (1969).

H. Ito, Progr. Theor. Phys. 43, 1035 (1970).
N. Nakanishi, Journ. Math. Phys. 12, 1578 (1971).

67. D. Zum Winkel, Diplomarbeit, Hamburg (in preparation).

68. H. Harari, Proc. of the 14th International Conference on High-Energy Physics, Vienna (1968), p. 195.
H. J. Lipkin, Ref. 10.

69. W. Thirring, in "Subnuclear Phenomena" (l.c.) p.200.

70. M. Krammer, internal report DESY-T 73/1.

71. H. Joos, Fortschritte der Physik 10, 65 (1962).

72. Review of particle properties, Particle Data Group, Physics Lett. 39B, 1 (1972).

73. M. Böhm, H. Joos, M. Krammer, DESY 72/62 (1972).
T. Kobayashi, Progr. Theor. Phys. 48, 335 (1972).

74. H. Pietschmann, W. Thirring, University of Vienna, Scientific Note No. 32 (1965).
R. van Royen, V. F. Weisskopf, Nuovo Cim. 50, 617 (1967); ibid. 51, 583 (1967).

75. J. Lefrancois, Proc. of the 1971 Int. Symposium on Electron and Photon Interactions at High Energies, p. 51 (Cornell Univ.).

76. G. Barbarino et al., Lett. Nuovo Cim. 3, 689 (1972).

77. G. Smadja et al., in "Experimental Meson Spectroscopy 1972" (Third Philadelphia Conference) p. 349.

78. F. Ceradini et al., Physics Lett. 43B, 341 (1973).

79. J. J. Sakurai, D. Schildknecht, Physics Lett. 40B, 121 (1972).
A. Bramon, E. Etim, M. Greco, Physics Lett. 41B, 609 (1972).

80. N. Cabibbo, R. Gatto, Phys. Rev. 124, 1577 (1961).

81. R. P. Feynman, Proc. of the Third Topical Conference on High Energy Collisions of Hadrons, Stony Brook. (Gordon and Breach, New York, 1969).

82. H. Fritzsch, M. Gell-Mann, Proc. of the Coral Gables Conference on Fundamental Interactions at High Energy, Vol. 2, p. 1 (Gordon and Breach, New York, London, Paris, 1971).

83. S. Weinberg, Phys. Rev. Lett. 18, 507 (1967).

84. Electromagnetic meson decays in the P+V-S model were discussed recently by D. Flamm and J. Sanchez; Lett. Nuovo Cim. 6, 129 (1973).

85. R. Delbourgo, M. A. Rashid, A. Salam, J. Strathdee, Proc. of the Seminar on High Energy Physics and Elementary Particles (Trieste 1965) p.455.

86. For a review on relativistic generalizations of SU(6) we refer to:
H. Ruegg, W. Rühl, T.S. Santhanam, Helv. Phys. Acta 40, 9 (1967).

87. L. I. Schiff, Phys. Rev. Lett. 17, 612 (1966).
O. W. Greenberg, Phys. Rev. 150, 1177 (1966).
G. Morpurgo, Physics Lett. 20, 684 (1966).

88. G. Morpurgo, Ref. 42.
H. J. Lipkin, Phys. Rev. 159, 1303 (1967).

89. Such modified couplings are suggested by experiment:
M. Afzal et al., Nuovo Cim. 15A, 61 (1973).
This paper contains references to earlier experiments.

90. J. Nuttal, Phys. Rev. 160, 1459 (1967).

91. D. B. Lichtenberg, Phys. Rev. 178, 2197 (1969).
A. N. Mitra, Nuovo Cim. 56A, 1164 (1968).

92. For a review of three-particle scattering we refer to:
W. Sandhas, in "Elementary Particle Physics" (l.c.) p. 57.
J. G. Taylor, Ref. 32.

93. O. W. Greenberg, Ref. 12.

94. G. Berendt, E. Weimar, private communication.

95. G. Berendt, E. Weimar, Lett. Nuovo Cim. 5, 613 (1972).

96. L. Susskind, Phys. Rev. Lett. 23, 545 (1969).
D. Geffen (private communication).

Acta Physica Austriaca, Suppl.XI, 117–138 (1973)

PARTONS IN THE LIGHT OF THE CONE*

BY

H. LEUTWYLER
Inst. f. theoretische Physik, Universität Bern, Switzerland and
California Institute of Technology, Pasadena California

1. INTRODUCTION

These lecture notes are meant as an introduction to some concepts of Light Cone Physics developped in the last few years. Like Regge Pole theory this branch of physics is still essentially a kinematical science. To my knowledge there are still no consistent dynamical models which lead to the type of scaling behaviour discussed in the following. One of the most important problems in this field is undoubtedly the construction of such a model which shows that the assumptions about the behaviour of currents on the light cone are consistent with the basic principles of field theory[1].

According to Quantum Electrodynamics - the one and only one - theory in particle physics - photons and leptons are point particles. They do not seem to have degrees of freedom other than those of translation and spin, which are properly accounted for by

*Lecture given at XII. Internationale Universitätswochen für Kernphysik, Schladming, February 5 - 17, 1973.

Maxwells and Diracs equations. We do not have nearly comparable information about the structure of hadrons. In fact, the many hadron-hadron collisions investigated thus far did not shed much light on the structure of these particles. What these experiments did lead to, however, is a periodic system for the classification of valencies - SU(3) - and some indications about the level structure of the excited states - Regge trajectories - of a hadron.

Electron-hadron scattering has been a more effective tool in the study of the structure of hadrons, basically because the structure of one of the collision partners, viz. the electron, is known. The electron serves as a probe to test out the structure of the hadron which it hits.

2. ELASTIC ELECTRON SCATTERING

The dominant contribution to electron-hadron scattering arises through one-photon exchange. For an unpolarized target of arbitrary spin the elastic cross section involves two unknown functions, the form factors $G_E(q^2)$ and $G_M(q^2)$ which depend only on the mass of the virtual photon:[2]

$$\frac{d\sigma}{d\Omega} = \frac{\alpha^2}{4E^2\sin^4\frac{\Theta}{2}} \frac{\cos^2\frac{\Theta}{2}}{1+\frac{2E}{M}\sin^2\frac{\Theta}{2}} \left[\frac{4M^2G_E^2 - q^2G_M^2}{4M^2-q^2} + \frac{(-q^2)}{2M^2} + g^2\frac{\Theta}{2}G_M{}^2\right] . \quad (1)$$

The functions G_E and G_M describe the distribution of electric charge and magnetic moment within the target respectively and may intuitively be viewed as Fourier transforms of these distributions

$$G_{E,M}\,(-\vec{q}^{\,2}) = \int d^3 \mathbf{x}\; e^{i\vec{q}\cdot\vec{x}}\; \rho_{E,M}(\vec{x}) \quad . \tag{2}$$

For scattering on protons the data are well fitted by

$$\begin{aligned} G_E(q^2) &= \left(1 - \frac{q^2}{k^2}\right)^{-2} \\ G_M(q^2) &= 2.79\; G_E(q^2) \quad . \end{aligned} \tag{3}$$

These form factors correspond to a charge distribution of the form

$$\rho_E(\vec{x}) = \frac{1}{\pi b^3} \exp\left\{-\frac{2|\vec{x}|}{b}\right\} \quad . \tag{4}$$

It is amusing to note that this distribution is exactly the same as the distribution of the electron charge in the ground state of the hydrogen atom:

$$|\psi_o(\vec{x})|^2 = \frac{1}{\pi a^3} \exp\left\{-\frac{2|\vec{x}|}{a}\right\} \tag{5}$$

except for the scale ($b \simeq 0.47$ Fermi, $a \simeq 0.53$ Å).

Note, however, that the interpretation of the form factors as Fourier transforms of the corresponding distributions in x-space cannot be taken literally. This is best seen if one works out the form factors corresponding to point particles. The form factors for minimal coupling of the electromagnetic current to point particles of spin 0, $\frac{1}{2}$ and 1 are given in Table I.

TABLE I

Spin	G_E^{point}	G_M^{point}
0	$(1-q^2/4M^2)^{1/2}$	0
$\frac{1}{2}$	1	1
1	$(1-q^2/4M^2)^{1/2}$	$\{\frac{2}{3}(1-q^2/4M^2)\}^{1/2}$

Clearly, except for the case of spin $\frac{1}{2}$, the Fourier transforms of these form factors are not concentrated at $\vec{x} = 0$, despite the fact that they describe point particles.

3. INELASTIC ELECTRON SCATTERING

Let us first consider inelastic electron scattering on a nucleus of mass number A and charge Ze:

$$e + (A,Z) \rightarrow e + \ldots$$

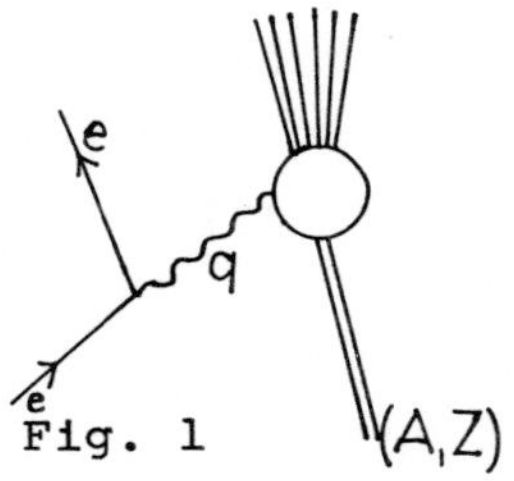

Fig. 1

We look at the total cross section for fixed electron variables E (electron energy before the collision), E' (electron energy after the collision) and scattering angle Θ of the electron in the Laboratory frame. For an arbitrary unpolarized target this cross section may again be expressed in terms of two unknown functions:[2]

$$\frac{d^2\sigma}{d\Omega dE'} = \frac{\alpha^2}{2ME^2\sin^2\frac{\Theta}{2}} \{q^2V_1 + (\nu^2+2EE'\cos^2\frac{\Theta}{2})V_2\} \quad . \tag{6}$$

In addition to q^2 - the $(\text{mass})^2$ of the virtual photon - these functions now depend on a second variable, the energy ν of the photon (energy loss $\nu = E - E'$ of the electron). For elastic scattering the energy loss is determined by the scattering angle:

$$\nu = \frac{-q^2}{2M_{target}} \qquad \text{elastic scattering .}$$

For inelastic scattering the energy loss is larger:

$$\nu > \frac{-q^2}{2M_{target}} \qquad \text{inelastic scattering.}$$

A plot of the accessible region in q^2 versus ν therefore looks as follows:

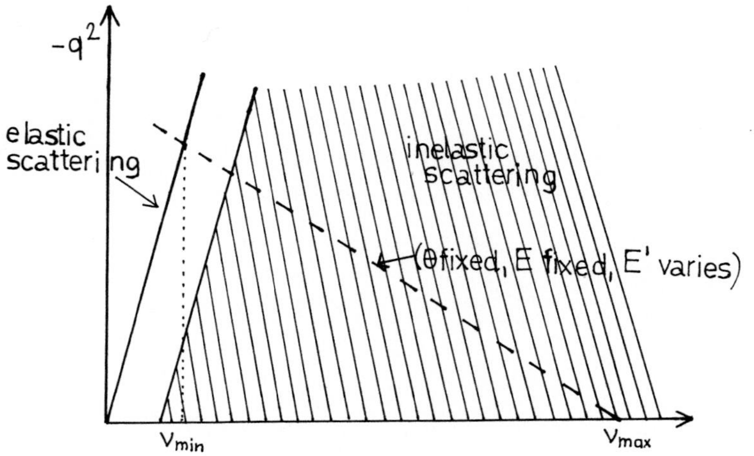

Fig. 2

The data for inelastic scattering on a nucleus qualitatively look as follws[2] (values of E and Θ fixed corresponding to the dashed line in Figure 2):

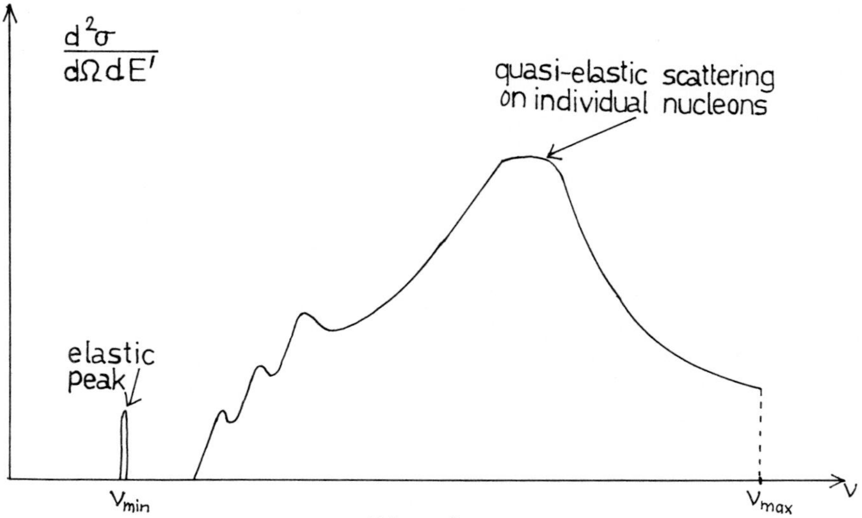

Fig. 3

The broad peak arises from quasi-elastic scattering on the constituents of the nucleus, i.e. on the individual nucleons. Indeed the corresponding values of q^2 and ν roughly satisfy

$$\frac{-q^2}{2\nu} \simeq M_{\text{nucleon}}$$

as it is appropriate for elastic scattering on particles of mass M_{nucleon}.

If the <u>proton</u> was composed of weakly bound objects of mass μ we would expect to see an analogous quasi-elastic peak in the electron-proton cross section at $-q^2/2\nu \simeq \mu$.

The data on electron-proton scattering seem to confirm the <u>scaling law</u> conjectured by Bjorken:

$$V_2(\nu,q^2) \simeq \frac{1}{2\nu^2}\,\frac{F_2(\xi)}{\xi} \; ; \qquad \xi = \frac{-q^2}{2\nu M} \tag{7}$$

in the deep inelastic region, i.e. if both ν and $(-q^2)$ are large. Fig. 4 shows a plot of the proton scaling function $F_2(\xi)$.

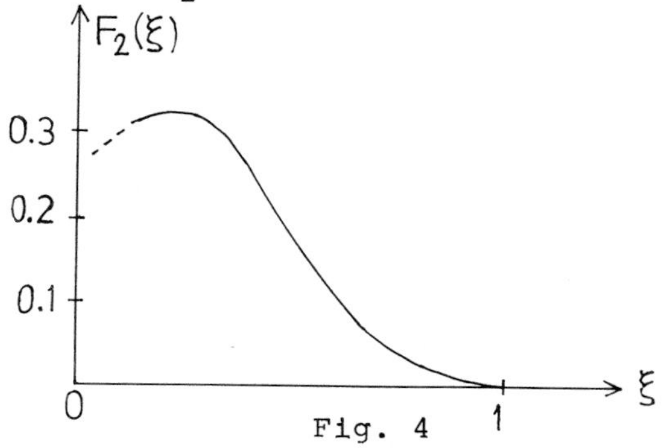

Fig. 4

The function V_1 is more difficult to measure. In the following we make the essentially untested assumption that this function scales in the same manner as V_2:

$$V_1(\nu,q^2) \to \frac{1}{2\nu^2} H_1(\xi) \quad . \tag{8}$$

(This hypothesis requires the ratio R of longitudinal to transverse cross sections to tend to zero like ν^{-1} in the deep inelastic region.)

4. PRIMITIVE PARTON MODEL

The scaling law may be understood in the following simple picture which provides us with a zero'th approximation to the parton model. Suppose that the proton consists of pointlike constituents of charge Z_1e, Z_2e,... and assume that the mass of these particles is continuously distributed with a probability distribution $\rho_i(x)$. (The quantity $\rho_i(x)dx$ is the probability to find the parton of charge Z_ie in the mass interval $x \le \mu \le x + dx$.) We neglect binding effects as well as the relative motion of the partons and take them to be at rest in the Laboratory. In this model, the total mass of the remaining partons is simply 1 - x, the proton mass being normalized to 1. The cross section for scattering of electrons on protons is given by

$$\frac{d\sigma}{d\Omega} = \sum_i \int dx \; \rho_i(x) \; \left(\frac{d\sigma}{d\Omega}\right)^{\text{point}} \tag{9}$$

where $(d\sigma/d\Omega)^{\text{point}}$ denotes the elastic cross section for scattering on a pointlike particle of mass x and charge $Z_i e$. This cross section depends on the spin of the partons. Suppose the partons have spin $\frac{1}{2}$. In this case the parton form factors G_E and G_M that appear in the Rosenbluth formula (1) for the elastic cross section are both equal to 1 (times Z_i) and we get

$$\left(\frac{d\sigma}{d\Omega}\right)^{\text{point}}_{\text{spin } \frac{1}{2}} = \frac{\alpha^2 Z_i^2}{4E^2 \sin^2\frac{\Theta}{2}} \frac{1}{1+\frac{2E}{x}\sin^2\frac{\Theta}{2}} \left[\cot g^2\frac{\Theta}{2} - \frac{q^2}{2x^2}\right] . \tag{10}$$

It is a simple matter to work out the resulting inelastic cross section. One finds that it indeed obeys the scaling law

$$V_2(\nu,q^2) = \frac{1}{\nu(-q^2)} \quad F_2\left(\frac{-q^2}{2\nu}\right) \tag{11}$$

with

$$I_2(x) = \sum_i Z_i^2 \; x \; \rho_i(x) . \tag{12}$$

As was to be expected from our previous discussion, partons of mass x only contribute along the line $-q^2 = 2x\nu$. The function V_1 vanishes in this model.

The probability interpretation of the function $\rho_i(x)$ immediately implies the sum rules

$$\int_0^1 \frac{dx}{x} F_2(x) = \sum_i Z_i^2$$

$$\int_0^1 dx\, F_2(x) = \langle x \rangle \sum_i Z_i^2 \simeq \frac{1}{\langle N \rangle} \sum_i Z_i^2 = \langle Z_i^2 \rangle .$$

Experimentally, one finds

$$\int_0^1 \frac{dx}{x} F_2(x) > 0.8$$

$$\int_0^1 dx\, F_2(x) = 0.187 \pm 0.009$$

whereas, if the Partons are identified with Quarks, ($Z_1 = 2/3$, $Z_2 = 2/3$, $Z_3 = -1/3$), one would expect

$$\sum_i Z_i^2 = 1$$

$$\langle Z_i^2 \rangle = 0.33 \quad .$$

If the partons are taken to be scalar particles rather than particles of spin $\frac{1}{2}$, then the relevant form factors must be taken from the first line of Table I. One obtains again the scaling law (7), but now, instead of $V_1 = 0$, one finds

$$V_1 = \frac{\nu^2}{(-q^2)} V_2 \quad .$$

This expression for V_1 disagrees with the data - it is too large. (Note that it does not satisfy the scaling law (8)). For partons of spin 1 the situation is even worse: the corresponding fuction V_2 scales with the power ν^{-1} rather than according to (7). The data therefore seem to indicate that the charged constituents of the proton have spin $\frac{1}{2}$.

The model we have just described is of course false. If the constituents were indeed loosely bound they should easily be knocked out of the proton and we should therefore observe particles with fractional charge and with a continuous mass distribution. This obvious clash with observation can be avoided if one reinterprets the model in terms of a null-plane picture of the proton. We will describe this picture in section 6.

5. CURRENT CORRELATION FUNCTION

Electron scattering amounts to a measurement of certain matrix elements of the electromagnetic current operator. In particular, elastic electron scattering tests the matrix element of the electromagnetic current between two proton states,

$$<p's'| \; j_\mu(x) \; |p,s> \quad .$$

Inelastic electron scattering on unpolarized protons provides us instead with a measurement of the current correlation function:

$$<p|[j_\mu(x), \; j_\nu(o)]|p> = \frac{1}{2} \sum_s <p,s|[j_\mu(x), \; j_\nu(o)]|p,s>$$

where $|p,s>$ denotes a proton of momentum p^μ and spin direction s, normalized by

$$\langle p',s'|p,s\rangle = (2\pi)^3\ 2p^o\ \delta^3(\vec{p}'-\vec{p})\ \delta_{s's}\ .$$

The functions V_1 and V_2 which appear in the formula (6) for inelastic electron scattering are the Fourier transforms of the current correlation function[2]

$$\frac{1}{4\pi}\int dx e^{iqx}\langle p|[j_\mu(x),j_\nu(o)]|p\rangle = \{q_\mu q_\nu - g_{\mu\nu}q^2\}V_1(\nu,q^2)$$
$$+\{(p_\mu q_\nu + p_\nu q_\mu)\nu - g_{\mu\nu}\nu^2 - p_\mu p_\nu q^2\}V_2(\nu,q^2). \quad (13)$$

The variable ν stands for the scalar product pq - the proton mass is set equal to 1.

Unfortunately, it is not possible to directly invert this Fourier transform and to compute the correlation function in terms of the data on V_1 and V_2, because the data only cover the spacelike region $q^2 \leq 0$. To undo the Fourier transform one has to know the functions V_1, V_2 in the timelike region as well.

In principle, causality of the current correlation function suffices to fix the functions V_1, V_2 essentially uniquely from their spacelike values. In fact, causality even restricts the data in the spacelike region. (see ref.3).

In order to analyze the implications of scaling for the current correlation function it is convenient to formulate the Bjorken limit in terms of the four-vector q^μ rather than in terms of the invariants q^2, $\nu = pq$. It is a simple matter to verify that the limit $q^2 \to \infty$, $\nu \to \infty$ with q^2/ν held fixed is obtained by letting the vector q^μ tend to infinity along a null ray, i.e.[3]

$$q^\mu = k^\mu + En^\mu \qquad \begin{matrix} k,n \text{ fixed}, \; n^2 = 0 \\ E \to \infty \end{matrix} \quad .$$

In this limit, the phase exp iqx in (13) oscillates rapidly except on the plane $nx \simeq 0$. Since causality requires the current correlation function to vanish outside the light cone, $nx \simeq 0$ implies $x^2 \simeq 0$. The Bjorken limit therefore tests the behaviour of the correlation function in the vicinity of the light cone. This qualitative argument can easily be made more precise (see ref.3). The particular behaviour described by the scaling laws (7) and (8) requires that the Fourier transforms of V_1 and V_2

$$\tilde{V}_i(x) = \frac{1}{(2\pi)^3} \int dq \; e^{-iqx} \; V_i(q^2,pq)$$

are discontinuous at $x^2 = 0$:

$$\tilde{V}_i(x) = \frac{i}{4\pi} \varepsilon(x^0)\theta(x^2) \; h_i(px) + \begin{matrix} \text{terms which vanish as} \\ x^2 \to 0 \end{matrix} \quad . \tag{14}$$

The coefficients $h_1(px)$ and $h_2(px)$ are essentially Fourier transforms of the scaling functions $H_1(\xi)$ and $F_2(\xi)$ respectively:

$$\begin{aligned} px \; h_1(px) &= i \int_{-1}^{1} d\xi \; e^{i\xi px} \; H_1(\xi) \\ px \; h_2(px) &= i \int_{-1}^{1} d\xi \; e^{i\xi px} \; F_2(\xi)/\xi \quad . \end{aligned} \tag{15}$$

It is easy to work out the corresponding light cone singularity of the correlation function with the result

$$i\langle p|[j_\mu(x),j_\nu(o)]|p\rangle = J_\mu\partial_\nu D+J_\nu\partial_\mu D - g_{\mu\nu}\, J^\rho\partial_\rho D$$
$$+ \text{ less singular terms} \qquad (16)$$

where

$$D(x) = \frac{1}{2\pi}\,\varepsilon(x^o)\,\delta(x^2)$$
$$J_\mu(x) = x_\mu\, h_1(px) + 2p_\mu\, px\, h_2(px)\ . \qquad (17)$$

6. PARTONS ON A NULL-PLANE

The singularity structure (16) of the current correlation function is characteristic of a current belonging to a free spin $\frac{1}{2}$ field:

$$j_\mu(x) = :\bar{\psi}(x)\; Z^2\; \gamma_\mu\psi(x): \qquad (18)$$

(Z denotes the charge matrix of the field ψ). For free fields the leading light cone singularity of the current is easily worked out from the anticommutation rule

$$\{\psi(x),\bar{\psi}(y)\} = \gamma^\mu\;\partial_\mu D(x-y)+\ldots \qquad (19)$$

with the result

$$i[j_\mu(x),j_\nu(y)] = B_\mu\;\partial_\nu D + B_\nu\;\partial_\mu D - g_{\mu\nu}\;B^\rho\;\partial_\rho D$$
$$+ \varepsilon_{\mu\nu\rho\sigma}\;\tilde{B}^\rho\;\partial^\sigma\;D$$
$$+ \text{ less singular terms}\ . \qquad (20)$$

The operators B_μ and $\tilde{B}_\mu$ stand for the bilocal fields

$$B_\mu(x,y) = i\,\bar{\psi}(x)Z^2\,\gamma_\mu\psi(y) - i\,\bar{\psi}(y)Z^2\,\gamma_\mu\psi(x)$$

$$\tilde{B}_\mu(x,y) = \bar{\psi}(x)Z^2\gamma_\mu\gamma_5\,\psi(y) + \bar{\psi}(y)Z^2\gamma_\mu\gamma_5\,\psi(x)\,. \tag{21}$$

The current associated with a charged free scalar field or a field of spin 1 possesses a more singular structure on the light cone. If the data indeed confirm the scaling laws (7) and (8) then the corresponding light cone singularity (16) suggests the following simple picture.[4] It suggests that the electromagnetic current is indeed coupled to spin $\frac{1}{2}$ fields which are of course not free fields, but nevertheless have the canonical light cone singularity (19). The coefficient J_μ of the leading light cone singularity of the current commutator then measures the proton expectation value of the bilocal operator B_μ

$$J_\mu(x) = \frac{1}{2}\sum_s \langle p,s|B_\mu(x,0)|p,s\rangle \tag{22}$$

whereas the spin average of the expectation value of the axial bilocal field vanishes on account of parity conservation. This picture easily lends itself to a parton interpretation.

For this purpose we have to look more closely at the properties of the bilocal field $B_\mu(x,0)$ if x is on the light cone. This is most easily done by introducing a null-plane reference frame defined by the coordinates

$$x^\pm = \frac{1}{\sqrt{2}}(x^0 \pm x^3)$$

and by looking at the field $\psi(x)$ on the null plane $x^- = 0$. There is an essential difference between the null plane restriction of a field and its restriction to a spacelike plane, which is best understood by looking at the corresponding stability groups. (This is the set of all Lorentz transformations which map the given plane onto itself). In the case of a null plane, say $x^- = 0$, the three generators of translations within the plane are

$$\underset{\sim}{P} = (P_+, P_1, P_2)$$

where

$$P_\pm = \frac{1}{\sqrt{2}} (P_o \pm P_3) .$$

One of these three generators, viz. P_+ has positive spectrum, since the energy of a physical state always exceeds the absolute value of the third component of its momentum[5]. (Of course, the translations within a spacelike plane, say $x^o = 0$, are P_1, P_2, P_3 - none of them is positive.)

A further essential difference between the two cases emerges if we look at the various components of $\psi(x)$. On a null plane, not all components are on the same footing[5]. There are "good" components $\psi_+(x) = G\,\psi(x)$ obtained through the projection matrix G

$$G = \frac{1}{2}\ \gamma_-\ \gamma_+$$

$$\gamma_\pm = \frac{1}{\sqrt{2}} (\gamma_o \pm \gamma_3)$$

and bad components, defined by $\psi_-(x) = (1-G)\ \psi(x)$. The reason why such a distinction arises is again best seen by looking at the stability group of the plane. If Λ belongs to the stability group, then the Lorentz transformation $\psi \to U(\Lambda)\psi U(\Lambda)^+$ transforms the good components among themselves, but takes the bad components into a mixture of good and bad ones (i.e. the representation is reducible, but does not decompose). The difference between good and bad components shows up drastically in the anticommutation relation of the field. Whereas the good components of a free field satisfy the simple null plane commutation rule ($x^- = y^-$):

$$\{\psi_+(x),\ \bar{\psi}_+(y)\} = \tfrac{1}{2}\gamma_-\delta(x^+-y^+)\delta(x'-y')\delta(x^2-y^2) \qquad (23)$$

the leading singularity (19) implies that the anticommutator of the bad components explodes as x tends to the null plane passing through y. In other words, the bad components cannot be restricted to a null plane.

Let us focus on the good components of the parton field $\psi(x)$ and let us assume that the anticommuation relation (23) is valid even for interacting partons. (This assumption is slightly stronger than the requirement that the leading singularity be given by (19).) We define the annihilation operators of the interacting field $\psi(x)$ by

$$\psi_+(x) = (2\pi)^{-3/2}2^{-1/4}\int_0^\infty dk_+ \int_{-\infty}^\infty dk^1 dk^2 \{e^{-i\underset{\sim}{k}\underset{\sim}{x}}a(\underset{\sim}{k}) + e^{i\underset{\sim}{k}\underset{\sim}{x}}b(\underset{\sim}{k})^+\} \qquad (24)$$

where $\underset{\sim}{k}$, $\underset{\sim}{x}$ stand for the triplets (k_+, k_1, k_2), (x^+, x^1, x^2) respectively and the scalar product is $\underset{\sim}{k}\underset{\sim}{x} = k_+x^+ + k_1x^1 + k_2x^2$. The spectrum condition mentioned above guarantees that $a(\underset{\sim}{k})$ and $b(\underset{\sim}{k})$ are indeed annihilation operators: they annihilate all physical states whose momentum component P_+ is less than k_+ and it is therefore appropriate to interpret them as annihilation operators for partons and antipartons respectively. Note that only three components of the parton momentum are specified, the "energy" k_- is unspecified as we are working at fixed "time" x^-. The annihilation operators still carry a Dirac spin index and satisfy the anticommutation rule

$$\{a(\underset{\sim}{k}),\ a(\underset{\sim}{k}')^+\} = G\ \delta^3(\underset{\sim}{k} - \underset{\sim}{k}') \ . \tag{25}$$

Let us now look at the proton matrix element of the good component of the bilocal field

$$<p|\bar{\psi}(x)\gamma_+\psi(o)|p> = <p|\bar{\psi}_+(x)\gamma_+\psi_+(o)|p> \qquad .$$

(We consider one particular charge component of the parton field, say partons with charge $Z_1 e$ and antipartons with charge $-Z_1 e$). Since we are not interested in the disconnected part of the matrix element, we order the operators such that the creation operators stand to the left of the annihilation operators. The forward matrix element of the terms involving a^+b^+ and ab vanishes, because these operators raise or lower the value of p_+. Only the matrix elements

$$\begin{aligned} <p|a^+(\underset{\sim}{k})a(\underset{\sim}{k}')|p> &= \delta^3(\underset{\sim}{k}-\underset{\sim}{k}')(2\pi)^3 \cdot 2p_+\ n(\underset{\sim}{k},p) \\ <p|b^+(\underset{\sim}{k})b(\underset{\sim}{k}')|p> &= \delta^3(\underset{\sim}{k}-\underset{\sim}{k}')(2\pi)^3 \cdot 2p_+\ \bar{n}(\underset{\sim}{k},p) \end{aligned} \tag{26}$$

contribute and we get

$$<p|\bar{\psi}(x)\gamma_+\psi(o)|p> = 2p_+\int d^3\underset{\sim}{k}\{e^{i\underset{\sim}{k}\underset{\sim}{x}}n(\underset{\sim}{k},p)-e^{-i\underset{\sim}{k}\underset{\sim}{x}}\bar{n}(\underset{\sim}{k},p)\}. \quad (27)$$

The normalization factors in (26) have been chosen such that the quantity $n(\underset{\sim}{k},p)d^3\underset{\sim}{k}$ gives us the probability to find a parton whose momentum $\underset{\sim}{k}$ lies in the interval $d^3\underset{\sim}{k}$ (plus anything with momentum $\underset{\sim}{p} - \underset{\sim}{k}$). The probability to find a parton whose momentum component k_+ is larger than the momentum p_+ of the proton vahishes. This was to be expected in a picture that describes the proton as composed of several partons,

$$\underset{\sim}{p} = \underset{\sim}{k}^{(1)} + \underset{\sim}{k}^{(2)} + \ldots$$

where the momentum of each one of the partons has a positive longitudinal component k_+.

We are now in a position to express the scaling function $F_2(\xi)$ in terms of the parton probability densities n and $\bar{n}$. For this purpose we evaluate the matrix element

$$J_+(x) = i<p|\bar{\psi}(x)Z^2\gamma_+\psi(o)-\bar{\psi}(o)Z^2\gamma_+\psi(x)|p>$$

$$= 2ip_+\sum_i Z_i^{\,2}\int d^3\underset{\sim}{k}\ \{e^{i\underset{\sim}{k}\underset{\sim}{x}}[n_i(\underset{\sim}{k},p)+\bar{n}_i(\underset{\sim}{k},p)] - h.c.\}$$

for a point x on the light cone ($x^1 = x^2 = 0$) and compare the result with the good component of the relation (17)

$$J_+(x) = 2p_+ \, px \, h_2 \, (px)$$
$$= 2ip_+ \int_{-1}^{1} d\xi \; e^{i\xi px} \; F_2(\xi)/\xi \quad .$$

This leads to

$$\frac{1}{k_+} F_2(\frac{k_+}{p_+}) = \sum_i Z_i^{\;2} \int_{-\infty}^{\infty} dk_1 dk_2 \{n_i(\underset{\sim}{k},p) + \bar{n}_i(\underset{\sim}{k},p)\} \; . \tag{28}$$

The quantity $F_2(\xi)\ d\xi/\xi$ is therefore to be interpreted as the sum over all partons of $Z_i^{\;2}$ times the probability to find a parton of charge $Z_i e$ (or its antiparticle) in the interval $d\xi$ of the longitudinal momentum fraction, the transverse momentum being arbitrary. This interpretation is of course only consistent if the densities n_i and $\bar{n}_i$ vanish for $k_+ > p_+$ $(\xi > 1)$ - which is the case as we have shown above. (The fact that F_2 is a function only of the momentum fraction k_+/p_+ follows from Lorentz invariance). The only difference to the primitive model discussed earlier turns out to be that the mass fraction must be replaced by the fraction of longitudinal momentum which the interacting parton carries - the assumption that the partons are all at rest within the proton and that there is no binding may be dropped.

At this point it is interesting to observe that the densities $\int d^2kn$ and $\int d^2k\bar{n}$ as defined by (27) must be positive, even if one does not assume that there exists an underlying field $\psi(x)$, but instead only assumes that the commutators of vector and axial currents have the structure (20). (To derive this result[6] one looks at the positivity conditions for the correlation function of the current $\alpha V_\mu(x) + \beta A_\mu(x)$.)

Note that the scaling function $H_1(\xi)$ does not have an analogous interpretation in terms of probability densities, because it relates to the bad components of the parton field.

Finally, I should emphasize an important point which I have skipped in the above discussion. It concerns the behaviour of the scaling function at $\xi = 0$, i.e. partons with very small longitudinal momentum ("wee" partons). The scaling function F_2 does probably not vanish at $\xi = 0$. This means that the probability densities n and $\bar{n}$ are singular at $k_+ = 0$ in which case the integral $\int d^3\underset{\sim}{k}\, n(\underset{\sim}{k},p)$ is divergent (infinitely many wee's within the proton). This indicates that the splitting of the field $\psi(x)$ into creation and annihilation parts is well-defined for $k_+ \neq 0$, but the operators $a(\underset{\sim}{k})$ and $b(\underset{\sim}{k})$ are not defined on test functions $f(\underset{\sim}{k})$ that do not vanish at $k_+ = 0$. One should therefore look for a way to treat the point $k_+ = 0$ of the spectrum more carefully.

Lack of time prevents me from describing the extension of the above discussion to the vector and axial vector currents associated with weak interactions of hadrons, to matrix elements of the current commutator between vacuum and mesons, to the disconnected piece of the current commutator (i.e. its vacuum expectation value) to the decay $\pi^0 \to 2\gamma$ etc. I hope that these lecture notes provide sufficient incentive for the reader to look at some of the literature on the subject[7].

FOOTNOTES AND REFERENCES

1. The model proposed by K. Symanzik, DESY preprint 72/68, provides an example of a dynamical scheme which leads to scaling. It is not clear to me, however, whether this model is consistent with general principles of field theory. An investigation of the scaling properties of this model has been carried out by Parisi (to be published).

2. F. J. Gilman, Proceedings of the 4th International Symposium on Electron and Photon Interactions, Liverpool, 1969.

3. H. Leutwyler, Proceedings of the Kaiserslautern Summer School, Lecture Notes in Physics, Springer, 1972. H. Leutwyler and P. Otterson, Proceedings of the Frascati Meeting on Outlook for Broken Conformal Symmetry, May 1972 (to be published).

4. H. Fritzsch and M. Gell-Mann, Proceedings of the 1971 Conference on Duality and Symmetry in Hadron Physics (Weizmann Science Press).

5. F. Rohrlich, Acta Physica Austriaca, Suppl. VIII, 277 (1971).
   H. Leutwyler, Springer Tracts in Modern Physics, 50, 29 (1969).
   J. R. Klauder, H. Leutwyler and L. Streit, Nuovo Cim. 66A, 536 (1970).

6. I am indebted to H. Fritzsch for this remark.

7. For an up-to-date review which contains many references to the original literature the reader is referred to Y. Frishman, Proceedings of the XVI International Conference on High Energy Physics, Chicago, 1972.

Acta Physica Austriaca, Suppl.XI, 139–166 (1973)

CHIRAL INVARIANCE AND LOW-ENERGY PION-PION SCATTERING*

BY
H. LEHMANN
Universität Hamburg

INTRODUCTION

The foundations of chiral invariance and its relation to current algebra are well-known since a number of years. Several excellent review articles or lecture notes[1] are available. I recommend them to those among the audience who wish to study the subject in a systematic manner.

It is my belief that interesting aspects of chiral invariance still remain to be analyzed. In these lectures I shall describe recent work on its consequences for low-energy pion-pion scattering. In the first lecture I give a formulation of chiral invariance in the context of general quantum field theory and derive from it the Adler-Weinberg theorem for $\pi\pi$-scattering. The second lecture deals with an extension of this theorem to higher pion momenta and its application to the scattering phase shifts in an effective range approximation. Finally, in the third lecture I discuss attempts

*Lectures given at XII. Internationale Universitätswochen für Kernphysik, Schladming, February 5 - 17, 1973.

to obtain further information on $\pi\pi$-scattering by making dynamical assumptions beyond chiral invariance and treating a Lagrangian field-theoretic model for the interaction of pions and nucleons in the one-loop approximation.

Let me first remind you of the basic ideas of chiral invariance as the limiting case of current algebra and PCAC. It is taken for granted that local operators $\vec{V}_\mu(x)$ and $\vec{A}_\mu(x)$ exist for the hadronic interactions. While $\vec{V}_\mu$ is the isovector current which expresses the conservation of isospin, i.e. $\partial^\mu \vec{V}_\mu = 0$, the existence of $\vec{A}_\mu$ is suggested by the theory of weak interactions. The decay of charged pions implies

$$\langle 0|A_{\mu,i}(x)|\pi_j\rangle = i\delta_{ij}\, F_\pi \pi_\mu \frac{e^{-i\pi\cdot x}}{(2\pi)^{3/2}} , \qquad (1)$$

where $F_\pi = 92$ MeV is the pion decay constant. It follows directly that

$$\langle 0|\partial^\mu A_{\mu,i}(x)|\pi_j\rangle = \delta_{ij}\, F_\pi\, m_\pi^2 \frac{e^{-i\pi\cdot x}}{(2\pi)^{3/2}} . \qquad (2)$$

In the chiral-invariant limit we study a hypothetical world where $m_\pi = 0$, $F_\pi \neq 0$. The chiral group SU(2) x SU(2) is introduced by Gell-Mann's idea that the currents

$$\vec{W}_\mu^{\pm} = 1/2(\vec{V}_\mu \pm \vec{A}_\mu) \qquad (3)$$

generate independent rotations so that for infinitesimal

transformations $\delta\vec{\omega}^{\pm}$

$$\delta\vec{W}_{\mu}^{\pm} = \delta\vec{\omega}^{\pm} \times \vec{W}_{\mu}^{\pm} \quad . \tag{4}$$

The "charges"

$$"\vec{Q}^{\pm}(x_o) = \int d^3x \; \vec{W}_o^{\pm}(x)"$$

should satisfy the current algebra commutation relations for equal times; the "charges" being time-dependent if $\partial^{\mu} \vec{A}_{\mu} \neq 0$. Even in the symmetric case with $m_{\pi} = 0$ and $\partial^{\mu} \vec{A}_{\mu} = 0$ these relations are not valid as operator equations and we shall not use them. It is important to realize that, since $\langle 0|\vec{A}_{\mu}|\pi\rangle \neq 0$, the axial currents do not leave the space of one-particle states invariant and chiral invariance is a Goldstone symmetry (or spontaneously broken symmetry). It does not lead to multiplets of particles which correspond to representations of SU(2) x SU(2). Instead, the known consequences of chiral invariance are low-energy theorems.

Chiral-invariant Lagrangian field theories were formulated by Gursey[2] and by Gell-Mann and Levy[3] in 1960. Their relation to current algebra was pointed out by Weinberg[4] in 1967, who discovered the connection between low-energy theorems and tree diagrams. The view that results, for reactions which involve pions, obtained in the chiral-invariant limit with $m_{\pi} = 0$ are approximately valid in the real world has been expressed by various authors, very clearly by Dashen and Weinstein[5] in 1969. It offers a systematic way of understanding the success of current algebra theorems, which are off-shell statements involving the limit $m_{\pi} \rightarrow 0$, when applied to actual processes.

I. CHIRAL INVARIANCE IN GENERAL QUANTUM FIELD THEORY. DERIVATION OF LOW-ENERGY THEOREMS.

1)

It is convenient first to formulate a set of conditions which express chiral invariance, on a modest mathematical level, in the framework of general quantum field theory. We take the perturbation treatment of Lagrangian field theories as a guide. In particular, we infer from it that the chiral invariant interaction of massless pions does not lead to infrared problems and scattering matrix elements are finite. We take the incoming states of pions and other (massive) hadrons as a complete basis and denote the incoming fields by $\vec{\pi}_{in}$ and ψ_{in} where the latter symbol stands for all particles other than pions.

Our system is to be invariant under infinitesimal vector and axial rotations $\delta\vec{\omega}_V$ and $\delta\vec{\omega}_A$. Accordingly, we require the existence of local and locally conserved vector operators $\vec{V}_\mu(x)$ and $\vec{A}_\mu(x)$, which are vectors in isospin space, and of vector charges $\vec{Q}_V$ associated with $\vec{V}_\mu$. The following conditions are to be satisfied:

$$\partial^\mu \vec{V}_\mu = 0 , \quad \partial_\mu \vec{A}^\mu = 0 \tag{5}$$

$$\vec{A}_\mu(x) \xrightarrow[(x_o \to \mp\infty)]{} \vec{A}_{\mu,in}(x) = - F_\mu \partial_\mu \vec{\pi}_{\substack{in \\ out}}(x) \tag{6}$$

$$[Q_{vi}, Q_{vj}] = i\varepsilon_{ijk} Q_{vk} \tag{7}$$

$$\delta_A \vec{V}_\mu = \delta\vec{\omega}_A \times \vec{A}_\mu , \quad \delta_A \vec{A}_\mu = \delta\vec{\omega}_A \times \vec{V}_\mu \tag{8}$$

$$\delta_A \vec{\pi}_{in}(x) = - F_\pi \delta\vec{\omega}_A , \quad \delta_A \psi_{in}(x) = 0. \tag{9}$$

In (6) a weak limit is to be taken, the condition is equivalent to eq.(1). (8) states the behaviour of $\vec{V}_\mu$ and $\vec{A}_\mu$ under axial transformations and corresponds to eq.(4); it replaces the customary condition on the commutator of two axial currents which we want to avoid since, due to $\langle 0|\vec{A}_\mu|\pi\rangle \neq 0$, axial charge operators $\vec{Q}_A$ do not exist and a precise formulation of axial commutators is complicated. On a formal level the relation $[Q_{Ai}, Q_{Aj}] = i\varepsilon_{ijk} Q_{Vk}$ follows from (7) and (8).

To obtain (9) we remark first that we apply the axial rotations $\delta\vec{\omega}_A$ to operators with the standard properties $\langle 0|\vec{\pi}_{in}|0\rangle = 0$, $\langle 0|\vec{V}_\mu|\pi\rangle = 0$. Therefore, it follows from (8) that $\delta_A \vec{A}_\mu$ has no one-pion component or $\delta_A(\partial_\mu \vec{\pi}_{in}) = 0$ in view of (6). Hence $\delta_A \vec{\pi}_{in}$ is a constant, $\delta_A \vec{\pi}_{in} = C\cdot\delta\vec{\omega}_A$, and we determine C from $\vec{V}_{\mu,2\pi}$, the two-pion component of $\vec{V}_\mu$ in its incoming fields expansion. $\vec{V}_{\mu,2\pi}$ is related to the pion electromagnetic form factor $\tilde{F}(q^2)$ by

$$\vec{V}_{\mu,2\pi}(x) = \int d^4\xi\, F(x-\xi) :\vec{\pi}_{in}(\xi) \times \partial_\mu \vec{\pi}_{in}(\xi):, \qquad (9')$$

$$\tilde{F}(q^2) = \int d^4x\, e^{iq\cdot x}\, F(x) \;, \qquad \tilde{F}(0) = 1 \;.$$

According to (8) we have

$$\delta_A \vec{V}_{\mu,2\pi}(x) = C\int d^4\xi F(x-\xi)\, \delta\vec{\omega}_A \times \partial_\mu \vec{\pi}_{in}(\xi)$$

$$= C\; \delta\vec{\omega}_A \times \partial_\mu \vec{\pi}_{in}(x) \;.$$

This is linear in $\vec{\pi}_{in}$ and therefore related to the one-pion component of $\vec{A}_\mu$:

$$\delta_A \vec{V}_{\mu,2\pi} = \delta\vec{\omega}_A \times \vec{A}_{\mu,in} = - F_\pi \delta\vec{\omega}_A \times \partial_\mu \vec{\pi}_{in},$$

$$C = - F_\pi \quad .$$

So far, we have only assumed the existence of currents $\vec{V}_\mu$ and $\vec{A}_\mu$. We add now the requirement that an interpolating pion field $\vec{\pi}(x)$ exists which is local and relatively local with respect to the currents. It follows from (6) that

$$<0|\vec{A}_\mu|\pi> = - F_\pi <0|\partial_\mu\vec{\pi}|\pi> \quad .$$

Therefore, $\vec{A}_\mu$ has the structure

$$\vec{A}_\mu = - F_\pi \partial_\mu \vec{\pi} + \vec{B}_\mu \; ; \; <0|\vec{B}_\mu|\pi> = 0, \tag{10}$$

with $\vec{B}_\mu$ defined by (10) as a local field. Taking the divergence we obtain

$$\Box \vec{\pi} = F_\pi^{-1} \partial_\mu \vec{B}^\mu, \tag{10'}$$

i.e. the "pion current" $\vec{j} = \Box\vec{\pi}$ is necessarily the divergence of an axial vector field. However, the decomposition of $\vec{A}_\mu$ in (10) will not be unique since it depends on the choice of a pion field $\vec{\pi}(x)$. The addition of a gradient to $\vec{B}_\mu$,

$$\vec{B}_\mu \to \vec{B}_\mu + \partial_\mu \vec{C}$$

is equivalent to a change of the pion field

$$\vec{\pi} \to \vec{\pi} + \vec{C}$$

and does not affect $\vec{A}_\mu$ or the scattering matrix (due to $\langle 0|\vec{C}|\pi\rangle = 0$).

For the operator $\vec{j}_\mu = \Box\vec{A}_\mu$ eq.(10) implies

$$\Box\vec{A}_\mu = \vec{j}_\mu = \partial^\nu(\partial_\nu\vec{B}_\mu - \partial_\mu\vec{B}_\nu). \qquad (11)$$

Since, according to (6)

$$\vec{A}_\mu(x) = - F_\pi \partial_\mu \vec{\pi}_{in}(x) - \int d^4\xi D_{ret}(x-\xi)\vec{j}_\mu(\xi), \qquad (12)$$

the axial current, with $\partial^\mu \vec{A}_\mu = 0$, can be expressed in terms of $\vec{B}_\mu$ which has no condition on its divergence.

It is not known to me whether the existence of a local pion field $\vec{\pi}(x)$ can be derived from the conditions (5) to (9). Clearly, this is equivalent to the question whether eq. (11) has for given $\vec{A}_\mu$ a solution $\vec{B}_\mu$ which is local and relatively local to $\vec{A}_\mu$ and satisfies $\langle 0|\vec{B}_\mu|\pi\rangle = 0$.

2)

We derive now the Adler-Weinberg low-energy theorem for $\pi\pi$-scattering. We are concerned, of course, with the chiral-invariant case of massless pions which has been treated by Dashen and Weinstein.[5]

Since $\vec{A}_\mu$ is asymptotically related to $\partial_\mu \vec{\pi}_{in}$, the scattering amplitude is given by the three-pion component of $\vec{j}_\mu$ restricted to the mass-shell. We express $\vec{j}_\mu$ in terms of $\vec{B}_\mu$ as in (11) and write the general form of $\vec{B}_{\mu,3\pi}$ as it follows from Lorentz- and isospin invariance.

$$\vec{B}_{\mu,3\pi}(x)=\int d^4x_1 d^4x_2 d^4x_3 b_\mu(x-x_1,x-x_2,x-x_3)$$

$$:\vec{\pi}_{in}(x_1)(\vec{\pi}_{in}(x_2)\vec{\pi}_{in}(x_3)): \qquad (13)$$

with $b_\mu(\xi_1,\xi_2,\xi_3)$ symmetric in ξ_2,ξ_3.

In momentum space we have

$$\tilde{b}_\mu(q_1,q_2,q_3)= -\frac{i}{2}[q_{2\mu}\tilde{b}(q_1,q_2,q_3)+q_{3\mu}\tilde{b}(q_1,q_3,q_2)] \ . \qquad (14)$$

In (14) we have used the possibility of adding a gradient $\partial_\mu \vec{C}$ to $\vec{B}_\mu$ (or $(q_1 + q_2 + q_3)_\mu \ \tilde{C}(q_1\ q_2\ q_3)$ to $i\tilde{b}_\mu(q_1,\ q_2,\ q_3)$) to eliminate a term proportional to q_{1_μ}. Therefore, without loss of generality we obtain from (13) and (14)

$$\vec{B}_{\mu,3\pi}(x) = \int d^4x_1 d^4x_2 d^4x_3 b(x-x_1,x-x_2,x-x_3)$$

$$:\vec{\pi}_{in}(x_1)(\partial_\mu\vec{\pi}_{in}(x_2)\vec{\pi}_{in}(x_3)) : \qquad (13')$$

The change $\delta_A \vec{B}_{\mu,3\pi}$ under an axial transformation follows by transforming the incoming fields according to (9). Eq.(11) relates this to $\delta_A \vec{A}_{\mu,3\pi}$. The result is

$$\delta_A \vec{A}_{\mu,3\pi}(x) = -\frac{F_\pi}{2}\int d^4x_1 d^4x_2 d^4x_3 b(x-x_1,x-x_2,x-x_3)$$

$$:\{\vec{\pi}_{in}(x_1)(\partial_\mu\vec{\pi}_{in}(x_2)\delta\vec{\omega}_A)-\partial_\mu\vec{\pi}_{in}(x_1)(\vec{\pi}_{in}(x_2)\delta\vec{\omega}_A)+$$

$$+\delta\vec{\omega}_A[(\partial_\mu\vec{\pi}_{in}(x_2)\vec{\pi}_{in}(x_3))-(\vec{\pi}_{in}(x_2)\partial_\mu\vec{\pi}_{in}(x_3))]\} :$$

$$(14)$$

Eq. (8) states that

$$\delta_A \vec{A}_{\mu,3\pi}(x) = \delta\vec{\omega}_A x\ \vec{V}_{\mu,2\pi}(x) = \int d^4\xi\ F(x-\xi)\cdot$$

$$:[\vec{\pi}_{in}(\xi)(\partial_\mu \vec{\pi}_{in}(\xi)\delta\vec{\omega}_A) - \partial_\mu \vec{\pi}_{in}(\xi)(\vec{\pi}_{in}(\xi)\delta\vec{\omega}_A)]: \tag{15}$$

where (9') has been used.

Equating (14) and (15), we obtain the following conditions:

$$\tilde{b}(0,q_2,q_3) = \tilde{b}(0,q_3,q_2);\ \tilde{b}(q_1,q_2,0) = -\frac{2}{F_\pi}\tilde{F}(q_1+q_2). \tag{16}$$

This consequence of chiral invariance relates the 3π-component of $\vec{A}_\mu$ to the pion formfactor. Always, of course, we have $q_1^2 = q_2^2 = q_3^2 = 0$. For the scattering amplitude, where also $(q_1 + q_2 + q_3)^2 = 0$, the only information is

$$\tilde{b}(0,\ 0,\ 0) = -\ 2/F_\pi\quad . \tag{16'}$$

We define the scattering amplitude $T_{i1,\ldots,i4}(k_1,\ldots,k_4)$ for the process $k_1 + k_2 \rightarrow\ - k_3 - k_4$ by

$$_{in}\langle -k_3,-k_4|S-1|k_1,k_2\rangle_{in} = i(4\pi^2)^{-1}\delta(k_1+k_2+k_3+k_4)$$

$$T_{i_1,\ldots i_4}(k_1,\ldots,k_4). \tag{17}$$

Eq.(10') leads to

$$T_{i_1,\ldots i_4}(k_1,\ldots,k_4) = F_\pi^{-1}(2\pi)^{9/2}\langle -k_3|\partial_\mu B^\mu_{i_4}(0)|k_1,k_2\rangle_{in}. \tag{18}$$

For $\partial_\mu \vec{B}^\mu$ we obtain from (13'):

$$\partial_\mu \vec{B}^\mu_{3\pi}(x) = - \int d^4x_1 d^4x_2 d^4x_3 b(x-x_1,x-x_2,x-x_3)$$

$$:\partial_\mu \vec{\pi}_{in}(x_1)(\vec{\pi}_{in}(x_2)\partial^\mu \vec{\pi}_{in}(x_3): + \frac{1}{2}\Box\int d^4x_1 d^4x_2 d^4x_3 \cdot$$

$$b(x-x_1,x-x_2,x-x_3):\vec{\pi}_{in}(x_1)(\vec{\pi}_{in}(x_2)\vec{\pi}_{in}(x_3)): .$$

Only the first term contributes on the energy shell. With the usual isospin decomposition and Mandelstam variables,

$$T_{i_1,\ldots i_4}(k_1,\ldots k_4) = \delta_{i_1 i_2}\delta_{i_3\ i_4}A(s,t,u) + \delta_{i_1 i_3}\delta_{i_2\ i_4}$$

$$A(t,s,u) + \delta_{i_1 i_4}\delta_{i_2 i_3}\ A(u,t,s), \tag{19}$$

$$A(s,t,u) = A(s,u,t),$$

the result is

$$A(s,t,u) = F_\pi^{-1}[k_1\cdot k_3\tilde{b}(k_3,k_2,k_1)+k_2\cdot k_3\tilde{b}(k_3,k_1,k_2)] \ . \tag{20}$$

Since we know $\tilde{b}$ for zero momenta, the leading (quadratic) term for small pion momenta follows from (20) as

$$A(s,t,u) = - 2F_\pi^{-1}(k_1\cdot k_3+k_2\cdot k_3)+\ldots = \frac{s}{F_\pi^2} +\ldots \ . \tag{21}$$

This is the low-energy theorem. The remainder is of higher than second order in the pion momenta and we shall analyze it in the second lecture. It is easily checked that the quadratic term in (21) is identical with the result obtained by calculating the four-pion vertex with the interaction Lagrangian ($2f_\pi = F_\pi^{-1}$)

$$L_{4\pi} = - f_\pi^2 : (\partial_\mu \vec{\pi}_{in})^2 \vec{\pi}_{in}^2 : \qquad (22)$$

The consequences of chiral invariance for more general processes involving pions are well known. I quote the result obtained by Weinberg[6] and Dashen and Weinstein[5] for the purely pionic reactions

2 pions → 2 n pions (n = 1,2,3,...) .

The amplitudes vanish for zero pion momenta. The leading term is of second order in the momenta and can be calculated exactly from the tree-approximation to the Lagrangian

$$L_I = 1/2 \sum_{n=1}^{\infty} (-1)^n (n+1) f_\pi^{2n} : (\partial_\mu \vec{\pi}_{in})^2 (\vec{\pi}_{in}^2)^n : =$$

$$= 1/2 : (\partial_\mu \vec{\pi}_{in})^2 \, [(1+f_\pi^2 \vec{\pi}_{in}^2)^{-2} - 1] : \quad . \qquad (23)$$

The Lagrangian can be written in different ways if we redefine the pion field by a point transformation. For the low-energy theorems which follow from chiral invariance for reactions which involve other hadrons besides pions I refer to the quoted literature.

II. APPLICATION TO LOW-ENERGY PION-PION SCATTERING

It is of interest to see if the low-energy theorem for

$\pi\pi$-scattering stated in eq.(21) can be extended to higher powers in the momenta and to compare such results with experimental data. Following Weinberg's[7] calculation of the $\pi\pi$-scattering lengths, a number of authors[8] have tried this. I shall take the view that the pion-mass is not an important parameter in $\pi\pi$-scattering, except very near threshold, and analyze the problem in a chiral-invariant world with zero-mass pions.[9] This simplifies the task considerably. As already mentioned, the success of current algebra gives support to this point of view. Also, a phenomenological analysis by Brown and Goble[10] has led these authors to propose that the energy scale in $\pi\pi$-scattering is not set by the pion mass but by $4\pi\ F_\pi \approx 1.2$ GeV. I use the term "low-energy $\pi\pi$-scattering" as referring to C.M. energies $\sqrt{s} < 1$ GeV. Empirically, in this energy range only a few partial waves, the isospin I = 0 and I = 2 S-waves, the I = 1 P-wave and the I = 0 D-wave are essential. The first important inelastic threshold appears to be the $\pi\pi \to K\bar{K}$ threshold at $\sqrt{s} = 980$ MeV (for the I = 0 S-wave).

Let us return to the problem of extending the low-energy theorem of eq.(21) assuming only chiral-invariance of the hadronic interactions. It is convenient to use the C.M. momentum q as an expansion parameter. s,t,u are of order q^2. Eq.(21) gives the amplitude to order q^2. We shall see that the next terms are of order q^4 and $q^4\log q^2$ and are determined by unitarity and analyticity (ordinary dispersion relations) up to two arbitrary parameters. I state first the result:

$$A(s,t,u) = 4f_\pi^2 s - \frac{f_\pi^4}{6\pi^2}\{3\ s^2\log[\alpha_1(-s-i0) + \qquad (24)$$

$$+ t(t-u)\log[\alpha_2(-t-i0)]+u(u-t)\log[\alpha_2(-u-i0)]\} + O(q^6)$$

α_1 and α_2 cannot be determined from chiral invariance alone. As indicated, the remainder is of order q^6 (this includes terms $q^6\log^2 q^2$ and $q^6\log q^2$).

To obtain (24) we calculate first the leading of the imaginary parts of the isospin amplitudes $T^{(I)}$ ($I = 0,1,2$) which are, in terms of $A(s,t,u)$,

$$\begin{aligned} T^{(0)} &= 3\,A(s,t,u) + T^{(2)} \\ T^{(1)} &= A\,(t,s,u) - A(u,t,s) \\ T^{(2)} &= A\,(t,s,u) + A(u,t,s)\;. \end{aligned} \tag{25}$$

The behaviour of $\text{Im}\,T^{(I)}$ for $q^2 \to 0$ follows from the q^2-terms of (21) using elastic unitarity for the S- and P-waves which are the only partial waves of order q^2. The result is

$$\begin{aligned} \text{Im}\,T^{(0)} &= \frac{2}{\pi} f_\pi^4\, s^2 + O(q^6) \\ \text{Im}\,T^{(1)} &= \frac{1}{6\pi}\, f_\pi^4\, s(t-u) + O(q^6) \\ \text{Im}\,T^{(2)} &= \frac{1}{2\pi}\, f_\pi^4\, s^2 + O(q^6)\,. \end{aligned} \tag{26}$$

To justify the neglect of the contributions from inelastic (pion production) channels to (26) we note that these are of order q^8. This is due to the small momentum behaviour of the phase space of n zero-mass particles which is of order q^{2n-4}. Since the production amplitudes are themselves of order q^2 (eq.(23)), the first inelastic contribution (from $2\pi \to 4\pi$) to $\text{Im}\,T^{(I)}$ is $O(q^8)$. Inelastic channels which involve massive particles do not, of course, contribute for small q^2. We rewrite (26) in terms of Im A using (25). Indicating the s,t,u channels by $\theta(s)$ etc. we have

$$\text{Im } A(s,t,u) = \frac{f_\pi^4}{6\pi}[3s^2\theta(s)+t(t-u)\theta(t)+u(u-t)\theta(u)] \ . \quad (27)$$

Knowing the behaviour of the imaginary part for $q \to 0$ we try to obtain information on the real part from dispersion relations. For massless particles no proof of these relations is known; we assume their validity on the basis of perturbation theory. In a fixed t-relation A(s,t,u) is the boundary value of an analytic function with cuts for $s > 0$ and $u > 0$. Its behaviour at the point $s = u = 0$ is determined by the discontinuity near the beginning of the cuts, which is given by (27), apart from polynomials in s and u whose coefficients are functions of t. To determine the non-polynomial terms we note simply that the functions

$$-\frac{1}{2}\frac{f_\pi^4}{\pi^2}s^2\log[\alpha_1(-s-i0)], \quad -\frac{1}{6}\frac{f_\pi^4}{\pi^2}u(u-t)\log[\alpha_2(-u-i0)]$$

have the discontinuities required by (27). The t-dependence of the polynomials is obtained from the t-u symmetry of A(s,t,u). The remaining arbitrariness to order q^4 is then a quadratic form in s,t,u which is symmetric in t,u. This can be written as $c_1 s^2 + c_2 \cdot (t-u)^2$, and is already accounted for by the parameters α_1 and α_2 introduced in the logarithmic terms.

The q^2-expansion can be carried further. The q^6 terms of A(s,t,u) can be calculated with the result that two new undetermined parameters appear. The next order, which is q^8, would involve three more parameters. However, we shall neglect terms of order q^6.

If we expand the amplitudes into partial waves

$$T^{(I)} = 32\pi \sum_{\ell=0}^{\infty} (2\ell+1) P_\ell(\cos\theta) [\cot\delta_\ell^I - i]^{-1}, \qquad (28)$$

the information contained in (24) can be stated as follows: For the three S- and P-waves we obtain the first two terms (which are of order q^2 and q^4 for massless pions) in a low-energy expansion. For the two D-waves which are $O(q^4)$ we obtain the first term. Chiral symmetry reduces these eight parameters to the two unknowns α_1 and α_2.

To allow for large phases we go over to an effective range approximation by expanding $\cot\delta_\ell^I$ at $q^2 = 0$. The result is for the S- and P-waves:

$$\cot\delta_0^0 = \frac{4\pi F_\pi^2}{q^2} + \frac{25}{18\pi}\log(q^2/4\pi^2 F_\pi^2) - \frac{\eta}{\pi} \qquad (29)$$

$$\cot\delta_0^2 = -\frac{8\pi F_\pi^2}{q^2} + \frac{20}{9\pi}\log(q^2/4\pi^2 F_\pi^2) - \frac{8}{5\pi}(\xi+\eta) - \frac{7}{18\pi} \qquad (30)$$

$$\cot\delta_1^1 = \frac{24\pi F_\pi^2}{q^2} - \frac{6\xi}{\pi} \quad . \qquad (31)$$

In these equations we have replaced α_1 and α_2 by two dimensionless parameters ξ and η which are defined by

$$\beta_i = -\log[16\pi^2 F_\pi^2 \alpha_i] \qquad (32)$$

and

$$\xi = \frac{1}{2}(\beta_2-\beta_1-1/9), \qquad \eta = \frac{1}{36}(33\beta_1+17\beta_2+\frac{11}{3}) \quad .$$

Next, I discuss briefly the consequences of these relations for the phase shifts.

1. Independently of ξ and η it follows from (29), (30) and (31) that $\delta_0^0(q^2)$ and $\delta_1^1(q^2)$ are positive and monotonically increasing with q^2, $\delta_0^2(q^2)$ is negative and monotonically decreasing. This qualitative result is in agreement with experimental data.

2. Since there are only two free parameters, one combination of the S- and P-waves is independent of ϵ and η. In the effective range approximation this is

$$\cot\delta_0^0(q^2) + \frac{1}{6}\cot\delta_1^1(q^2) - \frac{5}{8}\cot\delta_0^2(q^2) = \frac{13\pi F_\pi^2}{q^2} + \frac{35}{144\pi} \qquad (33)$$

where the last term is negligible. As a first check we evaluate (33) at the ρ-resonance energy ($\sqrt{s} = m_\rho$) where $\cot\delta_1^1 = 0$. The resulting relation for the two S-waves is shown in Fig.1.

If we take $\delta_0^0(q^2) = 80^\circ \pm 10^\circ$ [11] then $\delta_0^2(q^2) = -13.5^\circ \pm 1^\circ$. From experiment this number appears to lie between -10° and -15°.

Eliminating δ_1^1 from (33) with the help of (31), and with ξ fixed by the ρ-mass, we obtain an equation for the two S-waves at all energies:

$$\cot\delta_0^0(q^2) - \frac{5}{8}\cot\delta_0^2(q^2) = \pi F_\pi^2(9/q^2 + 4/9\, q^2), \qquad (34)$$

where the last term of (32) has been neglected. If experimental data on δ_0^0 are inserted the values predicted for δ_0^2 are rather insensitive to errors ($\Delta\delta_0^0 = 10^0$ leads to $\Delta\delta_0^2 \approx 1^0$) and are illustrated in Fig.2. Accurate data for δ_0^2 would therefore allow a quantitative test of eq. (34) to be made.

3. For the P-wave, eq.(31) is a standard effective range relation. ξ is adjusted by the ρ-mass as $\xi = 4\pi^2 F_\pi^2/q_\rho^2$. (31) was first derived by Brown and Goble[12] from current algebra. For the width the Brown-Goble relation follows:

$$\Gamma_\rho = \frac{q_\rho^5}{3\pi F_\pi^2 m_\rho^2} = \frac{m_\rho^3}{96\pi F_\pi^2} \left[1 - \frac{4m_\pi^2}{m_\rho^2}\right]^{5/2} . \qquad (35)$$

The pion mass appears in (35) since we compare results for massless and massive pions at fixed C.M. momentum and use the finite mass connection between momenta and energies. In addition, we have applied a pion mass correction factor $[1 + m_\pi^2/q^2]^{1/2}$ in eq.(31) which gives the correct threshold behaviour. It leads to a factor $2q_\rho/m_\rho$ in Γ_ρ, a 7% effect.

4. For the I = 0 S-wave data up to energies $\sqrt{s} \approx 800$ MeV are reasonably well reproduced by fitting the parameter η. The influence of the $K\bar{K}$-threshold on this partial wave is of course not included in our approximation.

5. If we calculate the D-waves from (24) and extrapolate them in the scattering length approximation we obtain $\delta_2^0(\sqrt{s} = 800\text{ MeV}) \approx 6^0$, $\delta_2^2(\sqrt{s} = 800\text{ MeV}) \approx 1^0$.

6. Finally, a remark on pion mass corrections.

Their size depends of course on how chiral invariance is broken. For plausible breaking mechanisms such as pure mass-breaking or Weinberg's model the corrections can be estimated in the tree approximation and they change the large phases at most by a few degrees.

Summarizing these results it appears that chiral invariance combined with an effective range approach leads to a reasonable two-parameter description of the main features of $\pi\pi$-scattering below 800 MeV. Present data are not precise enough for a detailed quantitative comparison.

III. ONE-LOOP APPROXIMATION IN CHIRAL-INVARIANT FIELD THEORIES

In this lecture I discuss attempts[13,14] to obtain results for the two parameters α_1 and α_2 of eq.(24) from a perturbation treatment of a chiral-invariant Lagrangian field theory. A basis for the applicability of perturbation theory is provided by the Adler-Weinberg theorem which assures us that in the low-energy limit first-order perturbation theory gives the exact result. Therefore, higher-order corrections are small for small momenta and chiral-invariant Lagrangians offer at least a hope that meaningful results for low-energy hadronic interactions can be obtained in perturbation theory. This will, of course, depend on the relative importance of successive orders of perturbation theory. For our present problem, this reduces to the question whether reliable results can be expected if the parameters α_1 and α_2 are calculated in a one-loop approximation. For the direct $\pi\pi$-interaction of eq. (22) the one-loop contribution should, on dimensional grounds, give the exact result. For the pion-nucleon interaction, which will be important, the effective expansion parameter in a multi-loop expansion is not a small number and it is easily realized that vertex corrections to the one-loop approximation are essential. One may hope that their inclusion takes higher-order effects largely into account. However, it is obvious that with our present knowledge no real argument for the validity of the one-loop approximation can be given; it can only be justified by successes.

Choice of the Lagrangian

The next question we have to consider is the choice of a Lagrangian. The simplest possibility is to take only the self-interactions of pions (eq.(22)) into account. Its contribution to our parameters, which arises from the pion-loop diagram (Fig. 3) cannot be calculated by standard renormalization theory, due to the non-renormalizability of the Lagrangian. Arguments based on the concept of minimal singularity or the technique of superpropagators suggest that the direct $\pi\pi$-interaction is too weak to lead, in the isospin $I = 1$ channel, to the ρ-resonance. Moreover, we shall see that the pion-nucleon coupling gives large contributions of the correct order of magnitude. Therefore, we take a Lagrangian which contains pion and nucleon fields. Contributions from other hadrons, which we neglect, may be important.[15] We do not consider the linear σ-model[16] since it has an additional parameter (m_σ) and can therefore only give one relation between α_1 and α_2. For the chiral-invariant interaction of pions and nucleons we choose the Lagrangian which is minimal in the number of derivatives (and therefore least singular in a given order of perturbation theory). The classical Lagrangian can be written as[2]

$$L_{class} = \frac{1}{16f_\pi^2} \mathrm{Tr}\{\partial_\mu e^{2if_\pi\vec{\tau}\cdot\vec{\pi}} \partial^\mu e^{-2if_\pi\vec{\tau}\cdot\vec{\pi}}\} +$$

$$+ \bar{\psi} i\gamma\partial\psi - M_N\bar{\psi}\, e^{2if_\pi\gamma_5\vec{\tau}\vec{\pi}}\psi. \tag{36}$$

With this Lagrangian the axial constant $g_A = 1.25$ is not an independent parameter. Its deviation from one is to be accounted for by a vertex correction, i.e.

$$g_A = g_A\left(\frac{f_\pi^2\, M_N{}^2}{4\pi^2}\right) \text{ with } g_A(0) = 1 \quad ,$$

and not by a derivative coupling term which could be added in eq. (36).

For the one-loop approximation, which is of the order f_π^4, the quantized interaction Lagrangian which follows from eq. (36) is

$$L_I = L_{\pi\pi} - 2f_\pi\, M_N : \bar{\psi}_{in}\, i\gamma_5\vec{\tau}\vec{\pi}_{in}\psi_{in} : + 2f_\pi^2\, M_N :$$

$$: \bar{\psi}_{in}\psi_{in}\vec{\pi}_{in}^2 + O(f_\pi) . \tag{37}$$

Nucleon contributions.

First, we neglect the contribution from the direct $\pi\pi$-interaction $L_{\pi\pi}$. The πN-part of eq.(37) leads to the one-loop diagrams shown in Fig.4. It is essential to realize that the low-energy theorem allows us to calculate the contributions from these diagrams without ambiguities due to subtractions. In fact, the subtraction constants are uniquely determined since we know that the sum of all f_π^4-diagrams does not contribute terms independent of q^2 or proportional to q^2. A calculation of the q^4-terms of these diagrams gives the following result:

$$A_{(a)} = \frac{2}{5}\,\frac{f_\pi^4}{\pi^2}\,s^2,\quad A_{(b)} = -\frac{2}{15}\,\frac{f_\pi^4}{\pi^2}\,s^2,\quad A_{(c)} = -\frac{1}{15}\,\frac{f_\pi^4}{\pi^2}\,(9s^2-5t^2-5u^2)\,. \tag{38}$$

The total nucleon contribution is therefore

$$A_N = A_{(a)}+A_{(b)}+A_{(c)} = -\frac{1}{3}\,\frac{f_\pi^4}{\pi^2}(s^2-t^2-u^2) + O(q^6)\ . \tag{39}$$

The remainder is a correction of the order $\frac{q^2}{M_N^2}$ and is negligible for the momenta we are interested in.

Comparing eq.(39) with eqs.(24) and (32) we find that the nucleon one-loop contribution to the parameters ξ and η is

$$\xi_{N,1\ \mathrm{loop}} = \frac{2}{3},\quad \eta_{N,1\ \mathrm{loop}} = \frac{1}{6}\ . \tag{40}$$

We realize now that these one-loop results must be supplemented by a vertex correction. This is evident from the Goldberger-Treiman relation

$$2f_\pi\, M_N\, g_A = G_{\pi NN}\ . \tag{41}$$

It shows that for the three-vertex in the Lagrangian (37)

f_π must be multiplied by g_A to obtain the correct normalization. The renormalization of the four-vertex is more complicated as we shall see presently.

For a first numerical comparison we consider the isospin I = 1 amplitude which contains only the parameter ξ. The box-diagram (c) gives the main contribution

$$\xi_{(c)} = \frac{14}{15} \quad \rightarrow \quad \frac{14}{15} g_A^4 , \tag{42}$$

with vertex renormalization. If we neglect the contributions $(-\frac{4}{15})$ from the diagrams (a) and (b) and from the pion-loop and insert eq.(42) into eq.(31) we obtain (with $m_\pi = 0$) a value for the ρ-mass

$$m_{\rho(c)} \approx \sqrt{\frac{15}{14}} \frac{F_\pi \cdot 4\pi}{g_A^2} \approx \sqrt{\frac{15}{14}} \frac{1}{G^2_{\pi NN/4\pi}} \cdot \frac{M_N^2}{F_\pi} .$$

The two expressions are connected by the Goldberger-Treiman relation which is exact for zero-mass pions. This shows that the box-diagram alone provides an attractive I = 1 interaction of the correct magnitude, leading to $m_{\rho(c)} \approx 800$ MeV (corrections to the Goldberger-Treiman relation imply an uncertainty of ≈ 100 MeV).

Unfortunately, the neglect of the contribution from diagrams (a) and (b) is not really justified, since they might become important if vertex corrections are included. Let us therefore look at the renormalization of vertices in a more systematic manner.

The correct normalization of the on-shell pion-nucleon vertices is known from the low-energy theorem[5] for πN-scattering. The πN-Lagrangian which gives this theorem in the renormalized tree approximation can be written as

$$L_{\pi N} = -\ 2f_\pi g_A M_N :\bar{\psi}_{in} i\gamma_5 \vec{\tau}\cdot\vec{\pi}_{in}\psi_{in}: + 2f_\pi^2 g_A^2 M_N :\bar{\psi}_{in}\psi_{in}\vec{\pi}^2_{in}: +$$

$$+\ f_\pi^2 (g_A^2 - 1) :\bar{\psi}_{in}\gamma_\mu \vec{\tau}\psi_{in}(\vec{\pi}_{in} \times \partial^\mu \vec{\pi}_{in}): \quad . \qquad (43)$$

Comparison with eq.(37) shows that the renormalized four-vertex has a momentum-dependent term. This derivative coupling has the effect that diagram (a) needs more subtractions and an undetermined parameter appears. If we neglect the derivative coupling term of eq.(43), the expressions (40) for ξ and η are simply multiplied by g_A^4. The numbers obtained in this way agree reasonably well with experimental data. However, no argument for neglecting the derivative term is known.

Direct $\pi\pi$-interaction.

As already mentioned the pion loop diagram gives contributions $\alpha_{1\pi}$ and $\alpha_{2\pi}$ which cannot be determined by standard renormalization theory. A heuristic argument can be given for the relation $\alpha_{1\pi} = \alpha_{2\pi}$. This follows if we choose the subtraction parameters in such a way that the scattering amplitude is, in a given order of perturbation theory, of minimal growth for large c.m. energies, keeping t (or u) fixed. Let us apply this idea to the isospin I = 1 amplitude. The contribution from the pion loop is

$$T^{(1)}_{\pi\pi} = \frac{f_\pi^4}{6\pi^2}(u-t)[s\cdot\log(-s)+t\log(-t)+u\log(-u)-3s\log\frac{\alpha_{1\pi}}{\alpha_{2\pi}}]. \tag{44}$$

The real part has the following behaviour for $s \to \infty$:

$$\text{Re } T^{(1)}_{\pi\pi} \to \text{const. } s^2 \quad \text{if } \alpha_{1\pi} \neq \alpha_{2\pi}$$
$$\text{Re } T^{(1)}_{\pi\pi} \to \text{const. } s.\log s \text{ if } \alpha_{1\pi} = \alpha_{2\pi} \quad .$$

The nucleon contributions do not destroy the argument since they increase at most like $s\cdot\log s$. If we accept this choice, the $\pi\pi$-contribution to the parameter ξ is $\xi_\pi = -\frac{1}{18}$. As a consequence, the contribution of the $\pi\pi$-interaction to the P-wave is very small and is negligible compared to the πN-interaction.

An analogous argument for the parameter η_π cannot be given. Values for η_π (and for ξ_π) have been obtained by the superpropagator method. This is also based on heuristic arguments and yields values for renormalization parameters by giving a special definition to the contribution from a sum of an infinite set of diagrams. Without going into details, let me mention that recently Ecker and Honerkamp[12] have proposed a covariant definition of super-propagators and obtained reasonable values for ξ_π and η_π.

In view of the various uncertainties which I have discussed it is, in my opinion, not yet possible to draw definite conclusions from these one-loop calculations. However, the results obtained may encourage further applications.

REFERENCES

1. S. L. Adler and R. Dashen, Current Algebras, W. A. Benjamin, Inc., New York 1968.
F. Gürsey in "Particles, Currents, Symmetries". Springer-Verlag, Wien, New York 1968.
S. Weinberg in "Lectures on Elementary Particles and Quantum Field Theory", 1970 Brandeis Summer Institute, Vol. I. MIT Press. Cambridge, Mass. 1970.
M. Weinstein, Springer Tracts in Modern Physics, Vol. 60. Springer-Verlag Berlin-Heidelberg-New York 1971.
B. W. Lee, Chiral Dynamics, Gordon and Breach, Science Publishers Inc., New York 1972.

2. F. Gürsey, Nuovo Cimento 16, 230 (1960).

3. M. Gell-Mann and M. Levy, Nuovo Cimento 16, 703 (1960).

4. S. Weinberg, Phys. Rev. 166, 1568 (1968).

5. R. Dashen and M. Weinstein, Phys. Rev. 183, 1261 (1969).

6. S. Weinberg, Phys. Rev. Letters 16, 879 (1966).

7. S. Weinberg, Phys. Rev. Letters 17, 616 (1966).

8. H. J. Schnitzer, Phys. Rev. D2, 1621 (1970), and further references given there.

9. H. Lehmann, Phys. Letters 41B, 529 (1972).

10. L. S. Brown and R. L. Goble, Phys. Rev. D4, 723 (1971).

11. Proceedings of the XVI International Conference on High Energy Physics, Vol. I, published by National Accelerator Laboratory, Batavia, Illinois, USA.

12. L. S. Brown and R. L. Goble, Phys. Rev. Letters 20, 346 (1968).

13. H. Lehmann, DESY-preprint 72/54.

14. G. Ecker and J. Honerkamp, CERN preprint TH 1573, 1972.

15. G. Ecker and J. Honerkamp, private communication.

16. J. L. Basdevant and B. W. Lee, Phys. Rev. D2, 1680 (1970).

FIGURE CAPTIONS

Fig. 1. δ_0^2 and δ_0^0 for $\sqrt{s} = m_\rho$.

Fig. 2. $\delta_0^2(\sqrt{s})$ with $\delta_0^0(m_\rho) = 80^0$.

Fig. 3. Pion loop diagram.

Fig. 4. Nucleon loop diagrams.

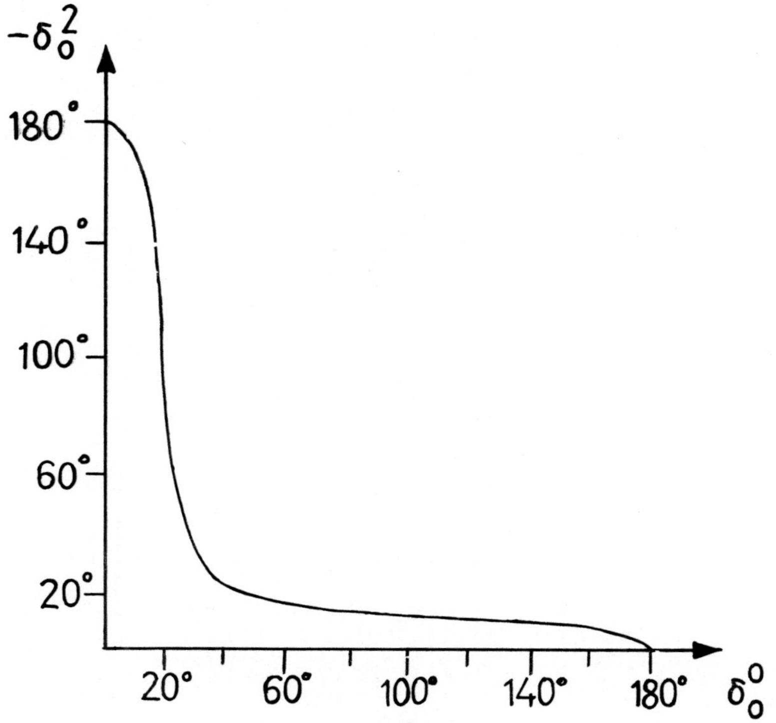

$-\delta_0^2$
180°
140°
100°
60°
20°
20°
60°
100°
140°
180°
δ_0^0

Fig.1

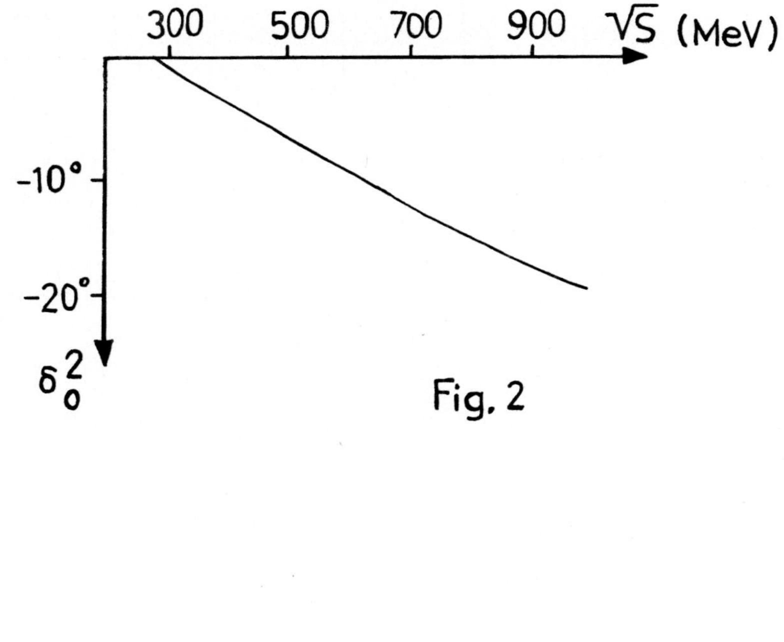

300
500
700
900
$\sqrt{s}$ (MeV)
-10°
-20°
δ_0^2

Fig. 2

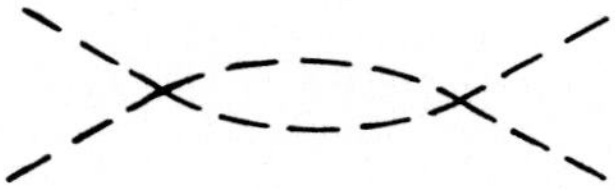

Fig. 3

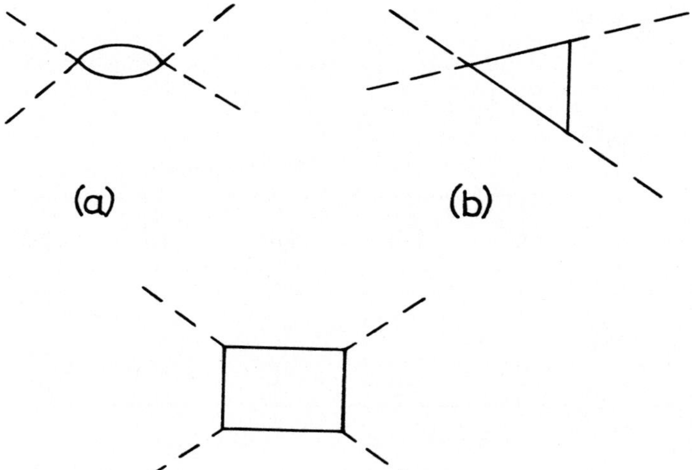

(a) (b) (c)

Fig. 4

Acta Physica Austriaca, Suppl.XI, 167–198 (1973)

THE INFRARED PROBLEM IN ELECTRON SCATTERING*

BY

O. STEINMANN
Schweizerisches Institut für Nuklearforschung
Villigen, Switzerland

1. INTRODUCTION

The subject of these lectures is the problem of the field theoretical description of electron scattering. We know that scattering and production processes are usually described with the help of an S-matrix, but we know also that this description is fraught with difficulties if applied to electrons, or more generally any charged particles, as long as their electromagnetic interactions cannot be forgotten. If we try, for example, to define an S-matrix in QED (= quantum electrodynamics) in the canonical way as

$$S = \lim_{\substack{t\to\infty \\ t'\to-\infty}} e^{iH_o t}\, e^{-iH(t-t')}\, e^{-iH_o t'}, \qquad (1)$$

$H = H_o + H_{int}$ the Hamiltonian of the theory, we find that this limit does not exist, even if the ultraviolet divergences are duly removed by some renormalization or

*Lecture given at XII. Internationale Universitätswochen für Kernphysik, Schladming, February 5 - 17, 1973.

regularization procedure. There still remain what is known as "infrared divergences". The axiomatic way of defining S also does not work. In the known proofs of asymptotic conditions it is assumed that the particles under consideration belong to isolated one-particle hyperboloids in the energy-momentum spectrum of the relevant superselection sector. This is not the case for electrons. A state with one electron and any number of photons with arbitrarily small momenta can have the same quantum numbers as a single electron: the continuum starts right at the one-particle hyperboloid. Hence the asymptotic conditions cannot be proved and the axiomatic definition of S breaks down. From perturbation theory we learn that this is not merely a problem of unsuitable mathematical techniques, the asymptotic conditions are indeed not satisfied.

However, the experimental physicist notes nothing of these difficulties. For him the electron is a completely well behaved particle, which can be treated with the same methods and notions as any other particle. If anything, he prefers charged particles to the uncharged ones, because they are easier to observe. Therefore it must be possible to set up a scattering formalism which works for electrons.

Many authors have worked on this problem, trying to find suitable generalizations of the methods which are successful in the normal case, i.e. the case without infrared troubles. I mention only the recent works by Kulish and Faddeev [1], who start from the canonical formalism, and by Kibble [2], who generalizes the LSZ approach. The earlier literature can be traced from the references in these two papers. None of the existing solutions is completely satisfying.

I shall try to approach the problem from a new direction. I take my clue from the fact, mentioned above, that no particular problems arise in connection with charged particles in the experiment. This suggests that we may possibly learn something useful by following experimental procedures rather more closely than is usually done by theoreticians. We do this by considering scattering events as events taking place in space-time, studied with the help of localized detectors, called "counters" for short.

Before starting to explain in detail how this is done, I wish to make a few remarks limiting the scope of my enterprise.

1) I will not consider processes in which photons participate as observed particles. The problems connected with the photon are of a different nature from those connected with the electron, and the latter are more than sufficient to fill three lecture hours.

2) I shall forget about all spin complications, i. e. I shall essentially proceed as though both electron and photon had spin zero. This is no serious mutilation since the infrared problems are known to be spin independent [3]. I will therefore not waste time with writing down innumerous spinor indices and γ-matrices. I shall also drop factors 2π and the like, except in places where they are of qualitative importance.

3) The level of mathematical rigour will be low. Approximations will often be based more on intuition (or just plain luck) than on epsilontics, and limits

will be interchanged freely. Our results constitute therefore by no means the final establishment of a scattering formalism in QED. They are of heuristic value only, they may give an indication on the direction in which we ought to proceed.

A more detailed account of the formalism will hopefully be published in the not-too-distant future [4].

2. THE FORMALISM

We work in the framework of QED, considered as a Wightman field theory. The fundamental fields are the familiar electron-positron fields $\psi, \overline{\psi}$, in the Heisenberg representation, and the electromagnetic field. In order to really have a Wightman theory we ought to work directly with the field strength $F_{\mu\nu}$ and dispense with the potentials A_μ, thereby avoiding the necessity of working with an indefinite metric or other similarly unpleasant notions. Such a formulation of QED has not yet been developed, though I think that it can be done. Fortunately this problem is unimportant for our purpose, because we have no explicit interest in photons.

We must first discuss the states we want to consider. How do we describe the initial state of a scattering process? Since no simple asymptotic condition holds we cannot use asymptotic states in the LSZ sense. Instead we must work with states prepared at a finite time, in accordance with the experimental facts of life.

We shall explain the formalism with the example

of elastic electron-electron scattering. Let me therefore not try to define the most general initial state, but only a special case, that of a state with two electrons of momenta p_1, p_2, prepared in neighbourhoods of the origin and of the point a, $a^0 = 0$, respectively. We make the ansatz

$$\Phi = \int d^4p_1\, d^4p_2\, \tilde{f}(-p_1)\tilde{g}(-p_2)e^{iap_2}\tilde{\psi}(p_1)\tilde{\psi}(p_2)\Omega \ . \tag{2}$$

Ω is the vacuum state. $\tilde{\psi}(p)$ is the Fourier transform of the interacting electron field $\psi(x)$. $\tilde{f}$, $\tilde{g}$ are "wave functions" with the following properties.

i) Let p_1, p_2 lie on the forward mass shell: $p_1^2 = p_2^2 = m^2$, $p_{1,0} > 0$, $p_{2,0} > 0$, with $m > 0$ the electron mass. We assume that $\tilde{f}$ and $\tilde{g}$ have compact supports with a diameter $d_2 \ll m$, containing p_1 and p_2 respectively. d_2 is chosen smaller than the experimental momentum resolution (what this is will become clear later on), so that p_1 and p_2 are indeed, within experimental accuracy, the 4-momenta of the two electrons. The minus signs in the arguments of $\tilde{f}$, $\tilde{g}$ are due to $\tilde{\psi}(p)$ creating, according to the usual conventions, a state with momentum $-p$. We assume that $(p_1+p_2)^2$ lies roughly in the middle of the elastic interval $(4m^2, 9m^2)$, so that we can forget about threshold complications.

ii) $\tilde{f}$, $\tilde{g}$ are assumed to be <u>smooth</u>. This means not only that they are infinitely differentiable, but that they show no violent oscillations or sharp kinks. More exactly we demand that $\left|\frac{\partial \tilde{f}}{\partial p}\right|$ is nowhere larger by orders of magnitude than $\frac{\mathrm{Max}|\tilde{f}|}{d_2}$, and the same for $\tilde{g}$. Under this condition we find that

$$f(x) = \int d^4p \, \tilde{f}(p) \, e^{-ipx} \tag{3}$$

is large around the origin and decreases rapidly for x becoming large. There exists a bounded neighbourhood N of the origin, with diameter d_1, outside of which f is negligibly small with respect to a given experimental situation. Again, this point will be formulated a bit more precisely later on. We say that f is "essentially localized" in N, or that N is the "essential support" of f. In the same way we find that

$$g_a(x) = \int d^4p \, \tilde{g}(p) \, e^{-ip(x-a)} \tag{4}$$

is essentially localized in a neighbourhood of diameter d_1 of the point a. The diameters d_1 and d_2 are connected by a kind of uncertainty relation, they cannot be made arbitrarily small simultaneously.

That the ansatz (2) indeed describes a two-particle state with the desired characteristics will become apparent in the course of our investigation.

We want to analyze the space-time behaviour of the state Φ, or of similarly constructed n-particle states, with the help of counters. For want of time I can only describe briefly the mathematical formalism used to represent counters, without discussing the physical motivation of this description. For a somewhat more detailed discussion I refer to [5]. You may roughly think of a counter as a massive test body which can interact with electrons, and whose having done so can easily be ascertained. Due to its large mass a counter is well localizable in space. We assume that it is equally well localized in time, i. e. that it is switched on only during a time interval of a size comparable to its spatial extension. This is a slight improvement on reality, but certainly a permitted idealization.

A counter is described by a quasilocal bounded field $C(x)$ with the following properties.

a) $C(x)$ transforms under the translation group with the same unitary representation as the electron field:

$$C(x+a) = U(a)C(x)U^*(a), \tag{5}$$

U unitary.

b) $$||C(x)|| \leq 1 . \tag{6}$$

c) $$C(x) = C^*(x)\Omega = 0 . \tag{7}$$

d) The Fourier transform $\tilde{C}(p)$ of $C(x)$ has a compact support of diameter $d_2 \ll m$ centred at the origin.

e) $C(x)$ is essentially localized in a neighbourhood N_x of x, with diameter d_1. This means that $C(x)$ can be approximated within experimental accuracy by an element of the local field algebra of N_x, i.e. by a function of field operators with argument in N_x. It implies that $[C(x), C(y)]$ is negligibly small if x and y are situated such that all points in N_x are space-like with respect to all points in N_y. What "negligibly small" means will be explained presently.

We cannot demand that $C(x)$ be itself a local observable, because this would contradict conditions c) and d).

The operator $C(x)$ describes a counter localized

in N_x. Its action on a state Φ can be understood as follows. We use the statistical interpretation of quantum mechanics, so that Φ stands for an ensemble of identical systems which have been prepared in a certain prescribed way. We expose these systems to our counter. Some will trigger the counter, some won't. We consider only those systems further which have triggered the counter. They form a new ensemble Φ', whose mathematical representative is the vector

$$\Phi' = C(x)\Phi. \tag{8}$$

The probability that a system from Φ triggers $C(x)$ is

$$w(x) = ||C(x)||^2 , \tag{9}$$

where we have assumed that the initial state vector Φ was normalized to 1. The boundedness condition b) ensures that this probability is never larger than one.

The argument p of $\tilde{C}(p)$ is the momentum transfer from the electron to the counter in the counting process. Condition d) means that it is small, i.e. that the electron is not appreciably deviated in the counter. The momentum of the electron after counting is uncertain within an interval of length d_2, so that d_2 is the experimental momentum resolution mentioned in connection with the wave functions $\tilde{f}$, $\tilde{g}$.

$w(x)$ can only be measured within a finite accuracy η determined by the experimental situation (background problems, counter reliability, etc.). Hence we are only interested in the value of w within this accuracy: in our calculations we can neglect contributions which are much smaller than η. This is what we meant when we said that f(x) is negligibly small outside N and that C(x) can be approximated bx a local observable: replacing f by zero outside N will change w(x) by a negligible amount, and similarly for C(x).

These considerations can immediately be generalized to an arrangement of several counters. We place counters at the points $x_1,..,x_n$ (i.e. in neighbourhoods of these points) with distances $||x_i-x_j||$* which are large compared to d_1, and hence to m^{-1}, and subject the systems in Φ to these counters. The systems that have triggered all the counters form the new state

$$\Phi' = T[C(x_1)...C(x_n)]\Phi \quad . \tag{10}$$

T is the time-ordering operator. It must be applied because it does not make sense to subject a state first to a counter in a certain space-time region and afterwards to a counter localized in an earlier region. The probability that all counters are triggered is

$$w_\Phi(x_1,..,x_n) = ||T[C(x_1)..C(x_n)]\Phi||^2 \tag{11}$$

Obviously a state Φ can only be analyzed after it has been prepared. If we apply (11) to the 2-electron state (2), which is prepared at time t = 0, we must demand $x_i^{\,o} > d_1$ for the counter positions x_i. Neglecting the tails of f, g outside their essential supports we can then include the electron fields in the time ordering:

$$w(x_1,..,x_n,a) = \int dp_1 dp_2 dp_1' dp_2' \tilde{f}^*(-p_1)\tilde{f}(-p_1')\tilde{g}^*(-p_2)\tilde{g}(-p_2')e^{ia(p_2'-p_2)}.$$

---

* $||x||$ is the Euclidian length of the 4-vector x.

$$\cdot \prod_{1}^{n}\{d\ell_j d\ell_j' e^{ix_j(\ell_j-\ell_j')}\} \quad W(p_1,..,p_2', \ell_1,..,\ell_n'), \qquad (12)$$

with

$$W(p_1,..,\ell_n')=(\Omega,\tilde{T}^*(\ell_1,..,\ell_n,p_1,p_2)\tilde{T}(\ell_1',...,\ell_n',p_1',p_2')\Omega).\,(13)$$

Here $\tilde{T}(\ell_1,..,p_2)$ is the Fourier transform of the time ordered product $T[C(x_1)...C(x_n)\psi(y_1)\psi(y_2)]$.

We do not know W, hence we cannot calculate the exact form of w. However, since the counters are placed at large distances from each other and from the region of preparation of Φ, we need only know the asymptotic behaviour of w for large $||x_i||$, $||x_i-a||$, $||x_i-x_j||$. This behaviour is determined by the local singularities of W, and these may be accessible without the explicit knowledge of W. Indeed, in the normal case with isolated one-particle states we can determine the relevant singularities from the general axioms, without knowing anything about the dynamics, if only we assume that W is maximally well behaved, i.e. that it is smooth outside the singularities that are physically compulsory. Under our assumption that p_1+p_2 is far away from thresholds the relevant singularities are the one-particle singularities, which can be determined (in the normal case) by a generalization of Zimmermann's discussion of Green's functions [6]. Starting from this knowledge one can then discuss the asymptotics of w and finds results which are equivalent to the familiar LSZ scattering formalism.

In QED things are more complicated. True, the relevant singularities are still the one-particle singularities, their position is the same as in the normal case. Their character, however, can no longer be derived from the axioms, it becomes a dynamical property dependent on the special form of the interaction. Since we have no exact solution of QED we must rely at this point on perturbation theory, but in an improved version. Naive finite-order perturbation theory is known to completely misrepresent the infrared behaviour [7], and realistic expressions can only be obtained by "summing over soft photons".

In considering vacuum expectation values of products of fields, like the Green's functions or our function W, we find that the ultraviolet and the infrared difficulties are nicely separated. We shall therefore ignore ultraviolet divergences, with the tacit understanding that they are removed by renormalization. The only exception to this rule is the fact that the electron field renormalization, as defined traditionally, is infrared divergent. This can be remedied by suitably changing the renormalization condition for the electron field, the Källén renormalization condition. For lack of time I will not go into this.

In any finite order of perturbation theory the function $W(p_1,..,p_2',\ell_1,..,\ell_n')$ is a sum over graphs, the "probability graphs" (p-graphs for short) or "unitarity graphs", of a similar structure to Feynman graphs (see Fig. 1). They are distinguished by consisting of two halfs, joined along a middle line (dash-dotted in Fig.1). Each half contains two external electron lines (straight lines) and n counter lines (wiggly lines), carrying unprimed variables in the left half, primed variables in the right half. The interior of the black box of Fig.1 consist of an ordinary Feynman graph. Each counter line is attached to an electron line in a three-line vertex. Counter lines do not occur as internal lines.

An integrand is associated with this graph according to the following rules.

i) To the right of the middle line we have the ordinary Feynman rules, with $+i\varepsilon$ propagators. The electron-counter vertices carry a vertex function of the form (p, q the two electron momenta, ℓ the counter momentum)

$$\delta^4(q-p+\ell)\ V(q,p)$$

where V is a smooth function with essential support in $||q-p|| \leq d_2$ at least for q, p in the vicinity of the mass shell $p^2 = q^2 = m^2$.

ii) To the left of the middle line we have the complex-conjugate of the Feynman rules, i.e. $-i\varepsilon$ propagators and V replaced by V^*.

ii) Lines crossing the middle line carry mass shell δ-functions as propagators, i.e. $\delta^m_\pm(k) = \Theta(\pm k_o)\delta(k^2-m^2)$ for electron lines, $\delta_\pm(k) = \Theta(\pm k_o)\delta(k^2)$ for photon lines. (As mentioned in the introduction we disregard spin and therefore drop the numerators of the propagators.) The sign alternative depends on the direction of the line. No vertices lie on the middle line.

In order to take care of the changed renormalization condition we ought to consider our expressions as the limit $\Lambda \to 0$ of the corresponding expressions in a theory with a finite photon mass Λ and add a factor $[Z(\Lambda)]^{-2}$ for each electron line crossing the middle, with $\lim_{\Lambda\to 0} Z(\Lambda) = 0$. For saving time I disregard this subtlely, anticipating the fact that in the final result the unpleasant factors Z^{-2} drop out neatly.

As mentioned before these finite-order expressions give the wrong sort of singularities. In order to obtain more realistic expressions we must sum over soft photons. We do this by a method used by Kibble [7] for the Green's functions, i.e. for ordinary Feynman graphs.

The soft photons are photons with small 4-momentum, more exactly with 4-momentum in a small neighbourhood of the origin which we choose as

$$K = \{k: |k_0| \leq 2K, |\underline{k}| \leq K\} , \qquad (14)$$

with $0 < K \ll d_2 \ll m$. The final result will be independent of the choice of K if we neglect terms of order K/d_2 or $(K/m)^\alpha$, $\alpha > 0$. Hard photons are photons with momenta outside K.

We take note of this distinction by introducing two types of photon lines in the p-graphs: the soft lines whose momenta are integrated over K and the hard lines with momenta integrated over CK. A graph without soft lines is called a "core graph". The problem is to sum over all the graphs which are obtained from a given core graph by the insertion of any number of soft photon lines. We shall carry this out for a special example in the following section.

3. ELECTRON-ELECTRON SCATTERING

We consider the function $W(p_1,\ldots,\ell'_n)$ defined in (13). To fix the ideas we choose n = 4. We are interested in the one-particle singularities of W, since they determine the asymptotics of $w(x_1,\ldots,x_4,a)$. There are various core graphs which give rise to singularities leading to an asymptotically non-negligible contribution to w. Among them we single out for treatment a typical and instructive example, the core graphs of the form shown in Fig. 2.

The black boxes represent any Feynman graphs without soft lines. The variables $p_1,..,p_7,s_8,\ell_1,...,\ell_4$ are the momenta of the lines shown explicitly. They satisfy momentum conservation, so that

$$p_3 = p_1+\ell_1,\ p_4 = p_2+\ell_2,\ p_6 = s_8-\ell_4,$$

$$p_3+p_4 = p_5+p_6, \quad p_7 = p_5+\ell_3, \tag{15}$$

and the same for the primed variables. s_8 is the integration variable associated with the closed loop shown explicitly in the symmetrically completed Fig.2. The remaining variables shown in Fig. 2 become relevant only after insertion of soft lines and will be explained later on.

We shall show that this graph and the associated soft-line graphs represent events where the initial electrons trigger a counter each, then scatter from each other, and the two emerging electrons again hit a counter each. Other relevant graphs belong to events where two or more counters are triggered by the same electron, or to events where hard bremsstrahlungs photons emerge besides the scattered electrons.

Let us see what happens if soft photon lines are added to a core graph of the form of Fig. 2. A soft line, at least one of whose ends is attached to a line inside one of the black boxes, leads to a contribution to W which is smaller than that of the naked core graph by a factor K/m. These soft photons can therefore be neglected. Not ignored can be, however, the soft lines both of whose ends lie on one of the electron lines shown explicitly in Fig. 2. We are interested in the singularities of W at $p_i^2 = m^2$ or $s_i^2 = m^2$. Consider a soft photon with momentum k connecting the p_i with the p_j line (i=j is permitted). The k-integral over K contains the factor

$$\frac{1}{k^2}\,\frac{1}{p_i^2-m^2-2p_ik+k^2}\,\frac{1}{p_j^2-m^2+2p_jk+k^2}$$

and diverges at the origin for $p_j^2 = p_i^2 = m^2$. Hence this soft photon line leads to a mass shell singularity and cannot be neglected. We call these lines "critical".

Consider a core graph of the form of Fig. 2 and insert any number of critical soft photon lines. Each of the explicit electron lines (the p_i and s_8 lines) is divided by the end points of the soft lines into a number of segments. The variables r_i, q_i of Fig. 2 are the momenta of the extremal segments, s_7 and s_8 the momenta of the segments crossing the middle line. To each segment belongs a propagator of the general form $[r_i^2 - m^2 + 2(r_i,\sum k_j) + (\sum k_j)^2]^{-1}$ where $\sum k_j$ runs over a certain number of soft photons. Instead of r_i we may have p_1 or p_2. We note that the term $(\sum k_j)^2$ is of order K and can therefore be dropped as negligible. The first-order term $2(r_i,\sum k_j)$ cannot be neglected because we are especially interested in the region where the zero-order term $r_i^2-m^2$ vanishes or is small. After this approximation it is possible to sum over all graphs obtained from our core graph by adding soft photon lines in all possible ways. The result is a quite complicated expression. Before writing it down I wish to introduce a convenient notation.

The variables p_i, r_i, q_i, ℓ_i and other variables to be used later occur in pairs: primed and unprimed. We define the symbol $\mathring{x}$ to mean "x or x'". The equation $\mathring{x}=\mathring{y}$ means x=y and x'=y', $f=f(\mathring{x})$ means that f depends on x and x', and $d\mathring{x} = dx\, dx'$.

We obtain (note that the variables $\mathring{p}_i$, $\mathring{\ell}_i$, $\mathring{s}_i$ are not independent, but related by (15)):

$$w(x_i,a) = \int \prod_1^7 \{d\mathring{p}_j e^{-i(\xi_j,p_j-p_j')}\} \tilde{f}^*(-p_1)\tilde{f}(-p_1')\tilde{g}^*(-p_2)\tilde{g}(-p_2') \cdot$$

$$\cdot\ \delta^4(p_3+p_4-p_5-p_6)\delta^4(p_3'+p_4'-p_5'-p_6')X(\mathring{p}_1,\ldots,\mathring{p}_7), \tag{16}$$

$$\xi_1=x_1,\ \xi_2=x_2-a,\ \xi_3=-x_1,\ \xi_4=-x_2,\ \xi_5=x_3,\ \xi_6=x_4,\ \xi_7=-x_3, \tag{17}$$

$$X(\mathring{p}_i)=\int d^4s_7d^4s_8\delta_-^m(s_7)\delta_-^m(s_8)(s_7^2-m^2)^2(s_8^2-m^2)^2\int \prod_1^6 d^4\mathring{q}_j \prod_3^8 d^4\mathring{r}_j\ \cdot$$

$$\cdot\ \delta^4(q_3+q_4-r_5-r_6)\delta^4(q_3'+q_4'-r_5'-r_6')M^*(q_3,q_4;-r_5,-r_6)\ \cdot$$

$$\cdot\ M(q_3',q_4';-r_5',-r_6')\prod_V\{\delta^4(q_V-r_V+\ell_V)\delta^4(q_V'-r_V'+\ell_V')N^*(q_V-r_V)$$

$$\cdot\ N(q_V'-r_V')\}\Delta(p_1,q_1;p_2,q_2;r_3,q_3;..;r_6,q_6;r_7,s_7;r_8,s_8|$$

$$|\ p_1',q_1';..;r_8',s_8), \tag{18}$$

$$\Delta(\ldots) = \int \prod_{1}^{8} \{d^4 \overset{\circ}{y}_j d\overset{\circ}{\sigma}_j \Theta(\sigma_j)\Theta(\sigma_j')\} \cdot$$

$$\cdot \exp i[\sum_{1}^{2}(q_j-p_j,y_j)+\sum_{3}^{6}(q_j-r_j,y_j)+\sum_{7}^{8}(s_j-r_j,y_j) -$$

$$- \text{the same primed}]$$

$$\cdot \exp - i[\sum_{1}^{2}\sigma_j(p_j^2-m^2)+\sum_{3}^{8}\sigma_j(r_j^2-m^2) - \text{the same primed}]$$

$$\cdot \exp -\alpha i \ [- \int \frac{d^4k}{k^2-i} LL^* + \int \frac{d^4k}{k^2+i} L'L'^* + 4\pi i \int d^4k \delta_+(k) LL'^*], \tag{19}$$

$$L = \sum_{1}^{2} e^{-iky_j} \frac{e^{-2i\sigma_j p_j k}-1}{p_j k} + \sum_{3}^{6} e^{-iky_j} \frac{e^{-2i\sigma_j r_j k}-1}{r_j k} +$$

$$+ \sum_{7}^{8} e^{-iky_j} \frac{1-e^{2i\sigma_j s_j k}}{s_j k} , \tag{20}$$

L' the same written in the primed variables. M and N are the vertex functions belonging to the black boxes in Fig. 2, M to the 4-electron vertex, N to the counter vertices. M and N are smooth, slowly varying functions. N(q,-r) moreover has an essential support of diameter d_2 in the variable q-r, at least for q and r close to the mass shell. This follows from the support condition for $\tilde{C}$. The product $\prod_V$ in (17) runs over the four counter vertices, q_V, r_V, ℓ_V stand for the variables adjacent to vertex V. α is the fine structure constant. The k-integrals are understood to be taken over K, also in all expressions to come. Factors 2π and similar trifles have been omitted, with the exception of the factor 4π in the last line of (19). This factor is essential because a cancellation between the three integrals of this line will play an important role.

That the complicated expressions (16)-(20) indeed sum the soft photon insertions into the core graph under consideration, within the approximation mentioned above, can be seen by expanding $\exp -i\alpha[\ldots]$ in (19) into a power series in α. The terms of order α^n give the graphs with n critical soft photon lines. The $(k^2-i\varepsilon)^{-1}$ integral comes from the soft photons in the left half of the graph, the $(k^2+i\varepsilon)^{-1}$ term from the right half and the δ_+ part from the lines crossing the middle.

The α^o-term Δ_o in the singularity function Δ, i.e. the contribution of the core graph itself, is a product of factors each belonging to one of the explicit electron lines of the core graph. The factors are of the form

$$\frac{\delta^4(q_i-r_i)}{q_i^2-m^2-i\varepsilon}$$

etc. Δ generalizes these one-particle poles known from the normal case. Δ does not factorize like Δ_o. Also it does not contain δ-factors. But it still has a small essential support around the origin, with a diameter of order K, in the difference variables $p_1-q_1,\ldots,r_i-q_i,\ldots$

Because of this small support we can replace in (17) the variables $\mathring{q}_i$ and $\mathring{r}_j$, $j=1,\ldots,6$, by $\mathring{p}_j$, $\mathring{r}_7$ and $\mathring{r}_8$ by s_7 and s_8, in the arguments of the slowly varying functions M and N. The factors M, M^*, N, N^* can then be drawn in front of the $\mathring{q}_j$, $\mathring{r}_j$ integral. Their product will be called $G(\overset{o}{p}_1,\ldots\overset{o}{p}_6,s_7,s_8)$. G is a smooth function with the support properties of its factors N, N^*.

Moreover, the small support of Δ is entirely due to the factor $\exp i[\sum_1^2(q_j-p_j,y_j)+\ldots]$ in (18), the remaining factors give only a weak dependence on the various arguments. We can therefore also replace $\overset{o}{r}_j$ by $\overset{o}{p}_j$ or s_j in L, and $\overset{o}{r}{}_j^2$ by $p_j^2+2(\overset{o}{p}_j,\overset{o}{r}_j-\overset{o}{p}_j)$, or the same with s_j instead of p_j, in $\exp -i[\ldots+\sum\sigma_j(r_j^2-m^2)\ldots]$. This last approximation corresponds exactly to the neglection of $O(K^2)$ terms in the denominators of the electron propagators which made summation over soft photons possible. For the same reason as there we cannot neglect the $O(K)$ terms $2(\overset{o}{p}_j,\overset{o}{r}_j-\overset{o}{p}_j)$.

After these approximations we insert (19) into (18), (18) into (16), and (16) into (12), to obtain $w(x_1,\ldots,x_4,a)$, or rather the contribution of our special graph to w. We find that the integrations over $\overset{o}{q}_j$, $\overset{o}{r}_j$, $\overset{o}{p}_7$, and then over $\overset{o}{y}_j$, can be carried out explicitly, with the result

$$w(x_1,\ldots,a)=\int\prod_1^6\{d\overset{o}{p}_j e^{-i\xi_j(p_j-p_j')}\}\tilde{f}(-p_1)\tilde{f}(-p_1')\tilde{g}(-p_2)g(-p_2')\cdot$$

$$\cdot\,\delta^4(p_3+p_4-p_5-p_6)\delta^4(p_3'+p_4'-p_5'-p_6')\bar{X}(\overset{o}{p}_1,\ldots,\overset{o}{p}_6), \tag{21}$$

with

$$\bar{X}(\overset{o}{p}_i)=\int ds_7 ds_8 \delta_-^m(s_7)\delta_-^m(s_8)(s_7^2-m^2)^2(s_8^2-m^2)^2 G(\overset{o}{p}_i,s_j)\cdot$$

$$\cdot\int\prod_1^8\{d\overset{o}{\sigma}_j\Theta(\sigma_j)\Theta(\sigma_j')\}\;\cdot$$

$$\cdot \exp -i\left[\sum_1^6 \sigma_i (p_i^2-m^2) + \sum_7^8 \sigma_j (s_j^2-m^2) \text{ - the same primed}\right] \cdot$$

$$\cdot F(\overset{\circ}{p}_i, s_j, \overset{\circ}{\sigma}_j) \ , \qquad (22)$$

$$F(\overset{\circ}{p}_i, s_j, \overset{\circ}{\sigma}_j) = \exp -i\alpha\left[- \int \frac{dk}{k^2-i\varepsilon} \bar{L}\bar{L}^* + \int \frac{dk}{k^2+i\varepsilon} \bar{L}'\bar{L}'^* + 4\pi i \int dk\, \delta_+(k)\, \bar{L}\bar{L}'^* \right], \qquad (23)$$

$$\bar{L} = e^{-2ik(\sigma_3 p_3 + \sigma_5 p_5)} \frac{e^{-2i\sigma_1 p_1 k}-1}{p_1 k} + e^{-2ik(\sigma_4 p_4 + \sigma_5 p_5)} \frac{e^{-2i\sigma_2 p_2 k}-1}{p_2 k}$$

$$+ e^{-2i\sigma_5 p_5 k} \frac{e^{-2i\sigma_3 p_3 k}-1}{p_3 k} + e^{-2i\sigma_5 p_5 k} \frac{e^{-2i\sigma_4 p_4 k}-1}{p_4 k} + \frac{e^{-2i\sigma_5 p_5 k}-1}{p_5 k}$$

$$+ e^{-2i\sigma_5 p_5 k} \frac{1-e^{2i\sigma_6 p_6 k}}{p_6 k} + \frac{1-e^{2i\sigma_7 s_7 k}}{s_7 k} + e^{-2ik(\sigma_5 p_5 - \sigma_6 p_6)} \cdot$$

$$\frac{1-e^{2i\sigma_8 s_8 k}}{s_8 k} \ , \qquad (24)$$

$\bar{L}'$ the same in the primed variables.

We see from (21) that the asymptotic behaviour of w depends on the singularities of $\bar{X}$, more exactly on the singularities in the variables $(p_j - p_j')$, after integration over $(p_j + p_j')$. Such singularities can only come from the $\overset{\circ}{\sigma}_j$-integration in (22), since the integrand of (22) is a smooth function of $\overset{\circ}{p}_i$. We must therefore know the asymptotic behaviour of F for $\overset{\circ}{\sigma}_j \to \infty$. This behaviour can be found with the help of some diligence. For lack of time I cannot go into this. Let me just mention the general type of result one finds. The square bracket in the exponent of (23) increases, in general, logarithmically for $\overset{\circ}{\sigma}_j \to \infty$, the coefficient multiplying the logarithm being negative imaginary. For some special values of the variables p_j there occurs also a linearly rising term, which is, however, cancelled by mass renormalization. The remainder is bounded and smooth, in particular non-oscillatory. F itself is also non-oscillatory. As a result $\bar{X}$ shows $p_j - p_j'$ singularities only at $p_j - p_j' = 0$. The explicit $\overset{\circ}{p}_j$ dependence of F being weak we can replace p_j' by p_j in the argument of F without changing anything essential. We make this replacement in the definition (22) of $\bar{X}$, and replace moreover the squares $\overset{\circ}{p}{}_j^2$ in the exponents $\overset{\circ}{\sigma}_j(p_j^2 - m^2)$ by $\overset{\circ}{\tau}_j$. The result is a function of $\overset{\circ}{p}_j$ and $\overset{\circ}{\tau}_j$ which we call $Y(\overset{\circ}{\tau}_1, .. \overset{\circ}{\tau}_6, \overset{\circ}{p}_1, .. \overset{\circ}{p}_6)$. For fixed $\overset{\circ}{\tau}_i$ it is a smooth function of $\overset{\circ}{p}_j$. We obtain, using the δ-factors of (21) to integrate over $\overset{\circ}{p}_6$:

$$w(x_i, a) = \int \prod_1^6 d\overset{\circ}{\tau}_j \int \prod_1^5 d\overset{\circ}{p}_j \prod_1^6 \{\delta(\tau_j - p_j^2)\,\delta(\tau_j' - p_j'^2)\}\; e^{-i\sum_1^5 (\bar{\xi}_j, p_j - p_j')} \cdot$$

$$\cdot\ \tilde{f}^*(-p_1) \ldots \tilde{g}(-p_2')\, Y(\overset{\circ}{\tau}_j, \overset{\circ}{p}_j), \tag{25}$$

with

$$\overset{\circ}{p}_6 = \overset{\circ}{p}_3 + \overset{\circ}{p}_4 - \overset{\circ}{p}_5,$$

$$\bar{\xi}_1 = x_1, \quad \bar{\xi}_2 = x_2 - a, \quad \bar{\xi}_3 = x_4 - x_1, \quad \bar{\xi}_4 = x_4 - x_2, \quad \bar{\xi}_5 = x_3 - x_4 . \tag{26}$$

We consider the $\overset{\circ}{p}_j$-integral for fixed $\overset{\circ}{\tau}_i$. This integral is exactly of the type we should meet at this point in the normal case of isolated one-particle states, if we applied our procedure to that case. The $\overset{\circ}{\tau}_j$ integral would not occur, the parameters $\overset{\circ}{\tau}_j$ would take the value m^2. The asymptotic behaviour of this integral can be obtained from a stationary phase argument which I will not explain in any sort of detail. For simplicity we do not let the x_i and a tend to infinity independently. We multiply all these variables with a common scaling parameter λ , and look for the asymptotically leading term in $w(\lambda x_i, \lambda a)$ for $\lambda \to \infty$, with x_i, a held fixed. For macroscopic x_i, a, and under our smoothness assumptions, this term is a good approximation to w already for $\lambda = 1$.

The manifold $\overset{\circ}{p}{}_j^2 = \overset{\circ}{\tau}_j$, $j = 1,..,6$, is suitably parametrized. In terms of these parameters $\{u_\nu\}$ the $\overset{\circ}{p}_j$-integral in (25) takes the form

$$\bar{w}(\lambda x_i, \lambda a, \overset{\circ}{\tau}_j) = \int \Pi du_\nu \phi(u_\nu, \overset{\circ}{\tau}_j)\, e^{i\lambda E(u_\nu, x_i, a)} \tag{27}$$

with ϕ and E smooth functions. The asymptotically relevant terms come from the points where the exponent E is stationary and can be determined with a finite amount of work. The result is as follows: $\bar{w}$ decreases fast and becomes unmeasurably small at $\lambda = 1$ for almost all values of x_i, a. It remains measurable at $\lambda = 1$ only if the geometrical arrangement of x_i,a, is such that it corresponds to the following classical scattering event: a particle passing through O with 4-momentum p_1 hits approximately the counter at x_1, a particle through a with momentum p_2 hits x_2, then the two particles continue in essentially unchanged directions and meet approximately in some point, where they scatter, observing 4-momentum conservation, such that the scattered particles hit the remaining counters at x_3 and x_4. Under these conditions we find

$$\bar{w} \sim \lambda^{-14} e^{i\lambda\sum_1^6(\alpha_j\tau_j-\alpha_j'\tau_j')} \tilde{f}^*(-\bar{p}_1)\tilde{f}(-\bar{p}_1)\tilde{g}^*(-\bar{p}_2)\tilde{g}(-\bar{p}_2') \cdot$$

$$Y(\overset{\circ}{\tau}_j,\overset{\circ}{\bar{p}}_j), \qquad (28)$$

where the 4-momenta $\overset{\circ}{\bar{p}}_j$ and the $\overset{\circ}{\alpha}_j$ solve the equations

$$\bar{\xi}_1 = -\overset{\circ}{\alpha}_1\overset{\circ}{\bar{p}}_1 , \quad \bar{\xi}_2 = -\overset{\circ}{\alpha}_2\overset{\circ}{\bar{p}}_2$$

$$\bar{\xi}_3 = -\overset{\circ}{\alpha}_3\overset{\circ}{\bar{p}}_3 - \overset{\circ}{\alpha}_6\overset{\circ}{\bar{p}}_6$$

$$\bar{\xi}_4 = -\overset{\circ}{\alpha}_4\overset{\circ}{\bar{p}}_4 - \overset{\circ}{\alpha}_6\overset{\circ}{\bar{p}}_6$$

$$\bar{\xi}_5 = -\overset{\circ}{\alpha}_5\overset{\circ}{\bar{p}}_5 + \overset{\circ}{\alpha}_6\overset{\circ}{\bar{p}}_6$$

$$\overset{\circ}{\bar{p}}{}_j^2 = \overset{\circ}{\tau}_j , \quad \overset{\circ}{\bar{p}}_3 + \overset{\circ}{\bar{p}}_4 = \overset{\circ}{\bar{p}}_5 + \overset{\circ}{\bar{p}}_6 . \qquad (29)$$

These equations have always a solution, and this solution is unique if the signs of all $\overset{\circ}{\alpha}_i$ are prescribed. If all $\overset{\circ}{\alpha}_i$ are positive, then $\overset{\circ}{\bar{p}}_i$ lie in the support of $\tilde{f}^*\ldots Y$ precisely in the geometrical situation explained above.

The factor λ^{-14} also corresponds to this geometrical picture, it is given by the decrease of available solid angles with increasing λ in the various shooting exercises of which the event consists.

We must yet insert $\bar{w}$ into the expression (25) for w and carry out the $\overset{\circ}{\tau}_j$ integrations. The factor λ^{-14} evidently comes in front of the integral. The full λ-

dependence of the integrand is then contained in the exponential factor $\exp\{i\lambda\Sigma(\alpha_j\tau_j-\alpha_j'\tau_j')\}$. The integrand has a small essential support around $\tau_j = \tau_j' = m^2$, due to the properties of $\tilde{f}$, $\tilde{g}$, G. We can therefore replace the slowly varying functions $\overset{\circ}{\alpha}_j$ and $\overset{\circ}{\bar{p}}_j$ by their values at the mass shell $\overset{\circ}{\tau}_j = m^2$. There we have $\alpha_j' = \pm\alpha_j$, $\bar{p}_j' = \pm p_j$. We combine the above exponential with the $\overset{\circ}{\tau}_j$-dependent factor in Y, to obtain

$$\exp -i[\ \sum_1^6(\sigma_j-\lambda\alpha_j)(\tau_j-m^2)-\sum_1^6(\sigma_j'-\lambda\alpha_j')(\tau_j'-m^2)\ ]\ .$$

By introducing $\overset{\circ}{\omega}_j = \overset{\circ}{\sigma}_j - \lambda\overset{\circ}{\alpha}_j$ as new integration variables we shift the λ-dependence into the factor $\Pi\Theta(\overset{\circ}{\omega}_j+\lambda\overset{\circ}{\alpha}_j)$ · · $F(\bar{p}_j, s_j, \overset{\circ}{\omega}_j+\lambda\overset{\circ}{\alpha}_j, \overset{\circ}{\sigma}_{7,8})$, F given by (23):

$$F = \exp -i\alpha[-\int\frac{dk}{k^2-i\varepsilon}\bar{L}\bar{L}^* + \ldots]\ .$$

For $\overset{\circ}{\alpha}_j < 0$ we note that $\Theta(\overset{\circ}{\omega}_j+\lambda\overset{\circ}{\alpha}_j) \to 0$ if $\lambda\to\infty$. Because of this, w decreases fast with increasing λ if at least one $\overset{\circ}{\alpha}_j$ is negative. Of interest to us is therefore only the case of $\overset{\circ}{\alpha}_j > 0$ for all j. In this case we have $\Theta(\ldots) \to 1$, and we can insert this limiting value already at the start without changing the result. We are then left to deal with F.

In the product $\bar{L}\bar{L}^*$ we find three types of terms, those only containing $\bar{p}_i$ variables, those only containing s_j, and the mixed cross terms. We remember also the factors $\delta_-^m(s_j)$, $(s_j^2-m^2)^2$ of the integrand. Because of the δ_-^m we are only interested in the value of the integrand on the mass shell of s_j, and because of the $(s_j^2-m^2)^2$ we are, more exactly, interested in the singularity of the integrand on the s_j mass shell. This is determined by the asymptotic behaviour of F for $\overset{\circ}{\sigma}_{7,8} \to \infty$, which can be found from the known explicit form of F. It is at this point that a cancellation between the three terms in the exponent of F becomes essential, as promised earlier. I have not the time to show explicitly what happens, I can only state the results, which are as follows. We examine the square bracket in the exponent of F. Our considerations indicate that we must determine the asymptotically leading term for $\overset{\circ}{\sigma}_{7,8} \to \infty$ before letting λ tend to ∞. If we do this we find that the mixed $\bar{p}$-s term goes to zero, so that F factorizes into two factors depending on $\{\bar{p}_j\}$ and $\{s_j\}$ respectively. The second factor neatly cancels the renormalization factor Z^{-4} that we have mentioned at one time but never wrote down explicitly.

The $\bar{p}$-$\bar{p}$-terms in the first integral of [...] are a sum of terms of the form

$$\frac{1}{(\bar{p}_i k)(\bar{p}_j k)} e^{-2i[\sum_{I_1} (\omega_j+\lambda\alpha_j)\bar{p}_j k - \sum_{I_2} (\omega_j+\lambda\alpha_j)\bar{p}_j k]}$$

where $I_{1,2}$ are certain subsets of $\{1,..,6\}$. Similar terms occur in the two other integrals, with ω_j' replacing ω_j everywhere in the second integral, in the I_2 sum in the third integral. Taking the three integrals together and studying the limit $\lambda \to \infty$ we find that only the λ-independent terms survive i.e. those with $I_1 = I_2$. Let $-i\eta(\overset{\circ}{\omega}_j)$ be the sum of these terms, hence the limit of $[\ldots]_{\bar{p}\bar{p}}$ for $\lambda \to \infty$. The $\overset{\circ}{\omega}_j$ integral over the $\overset{\circ}{\omega}_j$-dependent factors is

$$\chi(\overset{\circ}{\tau}_j) = \int \prod_1^6 d\overset{\circ}{\omega}_j \; e^{-i(\tau_j - m^2)\omega_j} e^{i(\tau_j' - m^2)\omega_j'} \; e^{-\alpha\eta(\overset{\circ}{\omega}_j)} \; . \qquad (30)$$

The ω-derivatives of η can be estimated. The Nth derivatives are small of order K^N, hence the Nth momenta of χ are small of the same order. This means that χ has an essential support with a diameter of order K around the mass shell $\overset{\circ}{\tau}_j = m^2$. Due to the smoothness of the remaining factors χ acts essentially as a product of δ-functions:

$$\chi(\overset{\circ}{\tau}_j) \sim \prod_1^6 \{\delta(\tau_j - m^2)\delta(\tau_j' - m^2)\} \; . \qquad (31)$$

Inserting this into our $\overset{\circ}{\tau}_j$-integral we obtain the final result

$$w(\lambda x_1, \lambda a) \sim \lambda^{-14} \tilde{f}^*(-\bar{p}_1)\tilde{f}(-\bar{p}_1)\tilde{g}^*(-\bar{p}_2)\tilde{g}(-\bar{p}_2)$$

$$\cdot \int ds_7 ds_8 \delta_-^m(s_7)\delta_-^m(s_8) G(\bar{p}_j, s_j) \; . \qquad (32)$$

The vectors $\bar{p}_j$ where defined in (29). In particular, $-\bar{p}_1$ is the mass shell vector parallel to x_1, $-\bar{p}_2$ the mass shell vector parallel to x_2-a. Only if these vectors are in the supports of $\tilde{f}$ and $\tilde{g}$ respectively is w non-negligible for macroscopic x_1.

G was defined as product of counter vertices N, N^* and the two scatter vertex functions M, M^*. This factorization means that the various counting processes and the scattering are independent from one another, which is, of course, what we expect. The factors NN^* describe the counters: their efficiency and geometry, etc. Of main interest to us is the factor

$$M^*(\bar{p}_3,\bar{p}_4;-\bar{p}_5,-\bar{p}_6)\; M(\bar{p}_3,\bar{p}_4;-\bar{p}_5,-\bar{p}_6)\,. \qquad (33)$$

(We do not write this as $|M|^2$ because M and N are matrices if we take spin into account.) This factor describes the dependence of the scattering probability on the momenta of the particles involved. Applying our precedure to the normal LSZ case we should find in this place the product S^*S, S the relevant S-matrix element. Hence M can be considered as a generalized S-matrix element.

Remember that what we have calculated is the asymptotic contribution of a core graph of the form shown in Fig. 2 to the triggering probability w. M is the four-electron vertex function in this core graph. We can now assume that the form (32) remains valid after summation over all core graphs of the form considered, i.e. that the exact w contains an asymptotically relevant contribution of the form (32). This is so if the exact four-particle vertex function M, the sum over all possible core graphs with four external electron lines, has the smoothness properties that we used in our derivation.

A definition of this exact M which makes sense irrespective of whether the perturbation series converges or not can be given as follows. The method we used for finding the singularities of W can also be applied to the electron Green's functions, in particular to

$$\tilde{\tau}(p_1,p_2;p_3,p_4) = (\Omega,\tilde{T}(p_1,p_2;p_3,p_4)\Omega) \quad , \tag{34}$$

where

$$T(x_1,x_2;x_3,x_4) = T[\psi(x_1)\psi(x_2)\bar{\psi}(x_3)\bar{\psi}(x_4)].$$

In fact, these Green's functions are the objects to which Kibble applied the method in the first place [8]. For any core diagram we obtain a contribution to $\tilde{\tau}$ of the form

$$\tilde{\tau}_c(p_1,p_2;p_3,p_4) = \delta^4(p_1+\ldots+p_4)M(p_1,p_2;p_3,p_4)s(p_1,\ldots,p_4), \tag{35}$$

where M is the vertex function occurring in (33) and s is the singularity function

$$s(p_1,\ldots,p_4) = \int_0^\infty \prod_{j=1}^{4}\{d\sigma_j e^{i\sigma_j(p_j^2-m^2)}\}\exp[-i\alpha\int\frac{d^4k}{k^2+i\varepsilon}\cdot \left|\sum_1^4 \frac{e^{-2i\sigma_j p_j k}-1}{p_j k}\right|^2] \tag{36}$$

which generalizes the pole product $\prod_1^4(p_j^2-m^2+i\varepsilon)^{-1}$ of the normal case. If we assume that (33) holds exactly, outside of perturbation theory, with M a smooth function, and if we make similar assumptions also for the counter vertex functions, then (32) also holds exactly: the term

(32) is one of a finite number of contributions to w which are non-negligible for macroscopic counter separations. Note that only the values of M on the mass shell enter the expression (32). The definition (35) of M is an obvious generalization of the reduction formula for the S-matrix element for elastic scattering known from the normal case.

About the properties of M I can hardly say anything. I can only say that M is not Lorentz invariant. This is so because the probability w includes processes in which soft photons are emitted, but excludes emission of hard photons, and this distinction soft-hard is not invariant. I have no idea about what becomes of unitarity. In order to investigate this problem one would have to consider also processes involving photons as observed particles, and this I have not done.

CONCLUSIONS

I hope that I have convinced you that looking at scattering processes as events in space-time is a natural way for establishing a scattering formalism in QED. The results we obtain in this way (the values of scattering probabilities) are, of course, the same as in the traditional procedure, but their derivation is physically more transparent and less in the nature of a cook book recipe. That the mathematics of the approach compares unfavorably, by being entirely non-rigorous, with what one is used to in the axiomatic approach of the non-infrared case, lies in the nature of things. In the normal case we can go far using only very general assumptions, but in the infrared case things become essentially model dependent. We cannot avoid tackling dynamical problems, and nobody has yet found out how to do this in a rigorous way for QED or any other realistic quantum field theory.

REFERENCES

1. P. P. Kulish and L. D. Faddeev, Theor. and Math. Phys. (USSR) 4, 745 (1970).
2. T. W. B. Kibble, J. Math. Phys. 9, 315 (1968); Phys. Rev. 174, 1882 (1968); Phys. Rev. 175, 1624 (1968).
3. S. Weinberg, Phys. Rev. 140, B516 (1965).
4. O. Steinmann, "Scattering of Infraparticles", SIN preprint (1973).
5. O. Steinmann, Comm. Math. Phys. 7, 112 (1968).
6. W. Zimmermann, Nuovo Cim. 13, 503 (1959); Nuovo Cim. 16, 690 (1960).
7. F. Bloch and A. Nordsieck, Phys. Rev. 52, 54 (1937).
8. T. W. B. Kibble, Phys. Rev. 173, 1527 (1968).

FIGURE CAPTIONS

Fig. 1 General form of a p-graph.

Fig. 2 A relevant core graph for electron-electron scattering. The graph must be completed symmetrically on the right-hand side, with

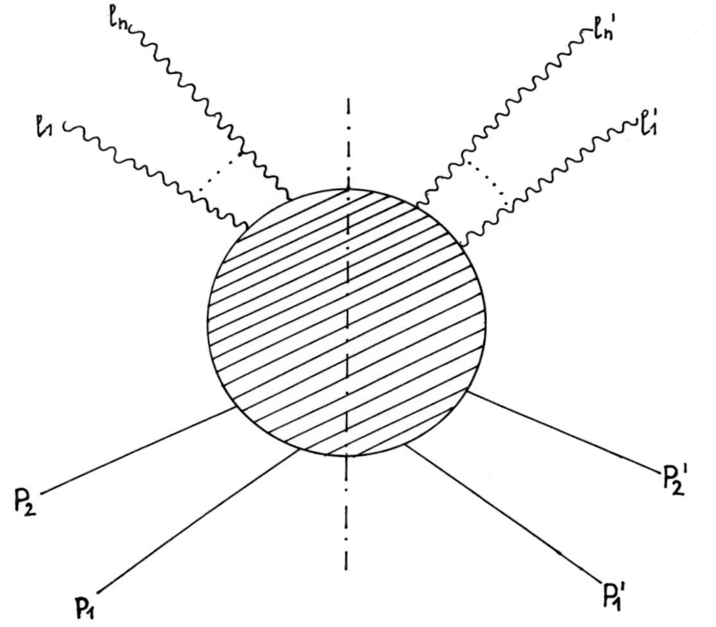

Fig.1

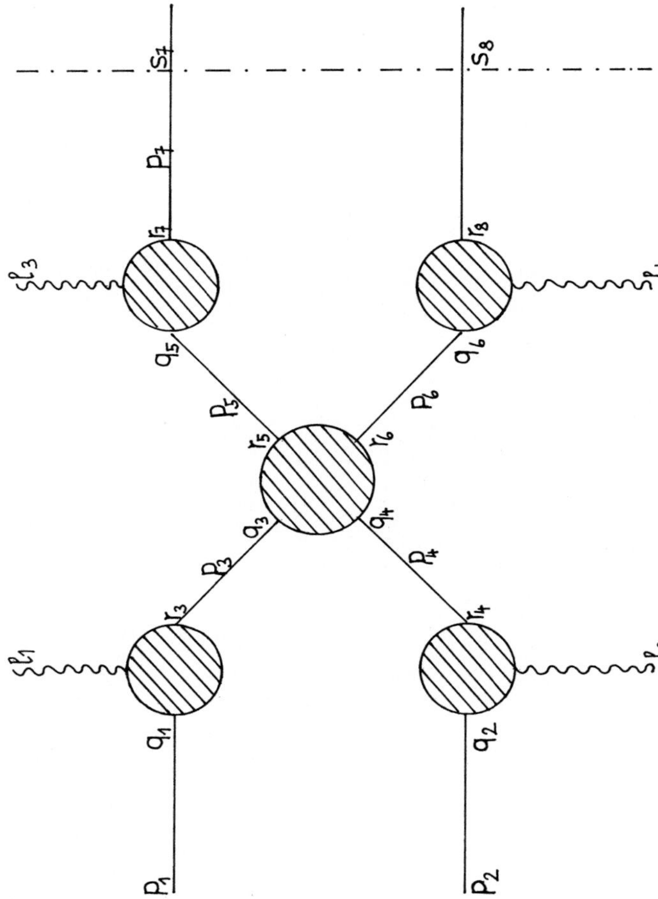

p1
q1
r3
p3
q3
r5
p5
q5
r7
p7
s7
ℓ1
ℓ3
p2
q2
r4
p4
q4
r6
p6
q6
r8
s8
ℓ2
ℓ4

Fig. 2

Acta Physica Austriaca, Suppl.XI, 199–240 (1973)

INFRARED SINGULARITIES IN THEORIES WITH SCALAR MASSLESS PARTICLES*

BY

K. SYMANZIK
DESY, Hamburg

INTRODUCTION

The best known theory with massless particles is quantum electrodynamics (QED). The masslessness of the photons leads to infrared (UR) singularities the Green's functions have near the electron mass shell. These singularities[1] are gotten correctly only by summing over all orders of perturbation theory as far as the soft-photon effects are concerned. I will describe UR singularities in theories with scalar neutral self-coupled massless particles. Also here the singularities are gotten correctly only by summing over all orders of perturbation theory, and again it suffices to do so with respect to the soft-massless-particle effects. Apart from this, the singularities have no relation nor resemblence to those in QED. They play a role in the quantum field theoretical treatment of some problems[2] in statistical mechanics, but not in

*Lecture given at XII. Internationale Universitätswochen für Kernphysik, Schladming, February 5 - 17, 1973.

particle physics since we know no such massless particles. The reason for studying such singularities goes back to large-momenta-behaviour problems as I will briefly explain at the end.

First I will discuss the theory with massless particles alone, which is a particular case of Φ^4 theory, and will then show that the UR singularities derived in that theory are the same in theories that have in addition massive particles. Some discussion and concluding remarks are at the end.

I. CONSTRUCTION OF ZERO-MASS Φ^4 THEORY

I.1 Renormalization conditions.

We consider the Green's functions

$$q^{-L}\langle(\Phi(x_1)\ldots\Phi(x_{2n})N_2(\Phi^2(y_1))\ldots N_2(\Phi^2(y_L)))_+\rangle$$

of massless Φ^4 theory, whereby $N_2(\Phi^2)$ is Zimmermann's[3] finite normal product, normalized, however, such as to exist in the massless theory which in general does not tolerate subtractions to be done at zero momenta as will be discussed in detail. The connected one-particle-irreducible parts of the Green's functions, amputated on all Φ-arguments, are Zimmermann's[3] proper functions, which we call vertex functions (VFs). Their Fourier transforms are introduced by

$$(2\pi)^4\ \delta(\sum p+\sum q)\,\Gamma_o(p_1..p_{2n},\ q_1...q_L;\ U^2,V)\ =$$

$$= \int \Pi dx\, e^{ipx}\, \Pi dy\, e^{iqy}\ .$$

$$\cdot\ q^{-L} \langle (\Phi(x_1) \ldots \Phi(x_{2n}) N_2(\Phi^2(y_1)) \ldots N_2(\Phi^2(y_L)))_+ \rangle^{prop}$$

and

$$\Gamma_o(p(-p),;\ U^2,V) = -\ G(p)^{-1}$$

the negative inverse propagator. U^2, V indicate the renormalization conditions used:

$$\Gamma_o\ (OO,;\ U^2,V) = 0 \qquad \text{(I.1a)}$$

$$\Gamma_o(p(-p),;\ U^2,V)\ \Big|_{p^2=-U^2} = -\ iU^2 \qquad \text{(I.1b)}$$

$$\Gamma_o(p_1 \ldots p_4,;\ U^2V)\ \Big|_{\text{s.pt.}^*\ \text{to}-U^2} = -\ iV \qquad \text{(I.1c)}$$

---

* $p_1 \ldots p_4 \Big|_{\text{s.pt.to}-\alpha^2}$ means $p_i p_j = -\frac{1}{3}\alpha^2(4\delta_{ij} - 1)$

$$\Gamma_o(\tfrac{1}{2}q\ \tfrac{1}{2}q,\ (-q);\ U^2,V)\Big|_{q^2=-U^2} = 1 \qquad \text{(I.1d)}$$

$$\Gamma_o(\ ,q(-q);\ U^2,V)\ \Big|_{q^2\ =\ -\ U^2} = 0\ . \qquad \text{(I.1e)}$$

In the perturbation theoretical construction, one has

to avoid imposing normalizations at points where the functions considered are UR singular, which is the reason for the momenta choices in (I.1b-e). The location (and strength) of the UR singularities of these functions will be discussed at length later. Thus, if one has an unrenormalized Feynman integral on which a final subtraction is to be performed, nonzero subtraction momenta must be employed in general. Nevertheless the loss of manifest covariance can be kept to a minimum, or even avoided, by tricks of the variety Appelquist[4] used to perform subtractions in the finite-mass theory on the mass shell. Results of Blanchard and Sénéor[5] imply that the Γ_o functions specified by (I.1) are to all orders in perturbation theory tempered distributions in Minkowski space.

Simple calculations give

$$\Gamma_o(p(-p),;U^2,V) = \tag{I.2a}$$

$$= ip^2-2c_oV^2\ ip^2 \ln\,[U^{-2}(-p^2-i\varepsilon)] + O(V^3)$$

$$\Gamma_o(p_1\dots p_4,;\ U^2,V) = -\ i\ V\ - \tag{I.2b}$$

$$-\ i\ \tfrac{1}{3}b_o\ V^2\ \{\ln[\tfrac{3}{4}U^{-2}(-(p_1+p_2)^2\ -i\varepsilon)]\ +$$

$$+\ \ln[\tfrac{3}{4}U^{-2}(-(p_1+p_3)^2\ -\ i\varepsilon)]\ +$$

$$+\ \ln\ [\tfrac{3}{4}\ U^{-2}(-(p_1+p_4)^2\ -\ i\varepsilon)]\}\ +\ \ O(V^3)$$

$$\Gamma_o(p_1p_2,\ (-p_1-p_2);\ U^2,V) = \tag{I.2c}$$

$$= 1 + \tfrac{1}{3}b_o\ V\ \ln[U^{-2}(-(p_1+p_2)^2\ -\ i\varepsilon)] + O(V^2)$$

$$\Gamma_o(\ ,q(-q);\ U^2,V) = \tag{I.2d}$$

$$= \tfrac{1}{3}b_o \ln [U^{-2}(-q^2 - i\varepsilon)] + O(V)$$

and all other VFs

$$\Gamma_o(p_1 \ldots p_{2n}, q_1 \ldots q_L;\ U^2, V) = O(V^n)$$

are obtained by skeleton expansions. The constants in (I.2) are

$$b_o = 3(32\pi^2)^{-1}\ , \qquad c_o = (2^{11}3\pi^4)^{-1} \tag{I.3}$$

and are gotten simplest from the absorptive parts of those functions, using that the n-massless-particle phase space is

$$(n!)^{-1}\int \prod_{i=1}^{n} [dk_i (2\pi)^{-3}\ \Theta(k_{io})\,\delta(k_i^2)\ \ (k_i^2)]\ \cdot$$

$$\cdot\ (2\pi)^4\ \delta(p - \sum_{i=1}^{n} k_i) = \tag{I.4}$$

$$= 2^{5-4n}\ \pi^{3-2n}[(n-2)!\ (n-1)!\ n!]^{-1}\Theta(p^2)\Theta(p_o)\,(p^2)^{n-2}\ .$$

In (I.2) one sees the UR divergences one would have encountered in an attempt to impose renormalization conditions at zero momenta.

I.2 Renormalization group equations

The momentum square $-U^2$ used in (I.1) is not intrinsic, such that, according to the theory of finite renormalizations[6], its change can be compensated by V-

and normalization change, and in the case of (I.2d) an additive term since that function involves a final subtractive renormalization besides the multiplicative ones of the N_2 operators. This means

$$\Gamma_o(p_1\cdots p_{2n},\ q_1\cdots q_L;\ U^2,V) =$$

$$= Z_1(U^{-1}U_o,\ V)^n\ Z_2(U^{-1}U_o,V)^L \cdot \tag{I.5}$$

$$\cdot\,\Gamma_o(p_1\cdots p_{2n},\ q_1\cdots q_L;\ U_o^2,\ V(U^{-1}U_o,V)) +$$

$$+\ i\ \delta_{no}\ \delta_{L2}\ f(U^{-1}\ U_o,V)$$

with, from (I.2),

$$Z_1(U^{-1}U_o,V) = 1-4\ c_oV^2\ \ln(U^{-1}U_o) + O(V^3) \tag{I.6a}$$

$$Z_2(U^{-1}U_o,V) = 1 + \tfrac{2}{3}b_oV\ \ln(U^{-1}U_o) + O(V^2) \tag{I.6b}$$

$$V(U^{-1}U_o,V) = V + 2b_oV^2\ \ln(U^{-1}U_o) + O(V^3) \tag{I.6c}$$

$$f(U^{-1}U_o,V) = \tfrac{2}{3}\ b_o\ \ln(U^{-1}U_o) + O(V)\ . \tag{I.6d}$$

Differentiating (I.5) with respect to U_o^2 at $U_o^2 = U^2$ gives

$$\{U^2[\partial/\partial U^2] + \hat{\beta}\ (V)[\partial/\partial V] - 2n\hat{\gamma}(V) +$$

$$+ L(2\hat{\gamma}(V) + \hat{\eta}(V))\}\ \Gamma_o(p_1\cdot\cdot p_{2n},\ q_1\cdots q_L;U^2,V) \equiv \tag{I.7}$$

$$\equiv \hat{O}p_{2n,L}\ \Gamma_o(\cdots\cdots) = -\ i\ \ \delta_{no}\ \delta_{L2}\ \hat{\kappa}(V)$$

with*

---

* The notation chosen for these functions is the one of Ref. 7.

$$\hat{\beta}(V) = b_o V^2 + b_1 V^3 + \ldots \qquad (I.8a)$$

$$\hat{\gamma}(V) = c_o V^2 + c_1 V^3 + \ldots \qquad (I.8b)$$

$$\hat{\eta}(V) = h_o V + h_1 V^2 + \ldots, \; h_o = \frac{1}{3} \mathbf{b}_o \qquad (I.8c)$$

$$\hat{\kappa}(V) = k_o + k_1 V + \ldots\ldots, \; k_o = \frac{1}{3} p_o \, . \qquad (I.8d)$$

For later reference, we give explicitly, as follows from (I.7) and (I.1c),

$$\hat{\beta}(V) = 4 V\hat{\gamma}(V) + \qquad (I.9)$$

$$+\frac{1}{2}i \sum_{j=1}^{3} p_j[\partial/\partial p_j]\Gamma_o(p_1p_2p_3(-p_1-p_2-p_3),; U^2,V)\Big|_{s.pt.to-U^2}$$

where it has been used that this VF is homogeneous in momenta of order zero.

To integrate (I.7) we introduce*

---

* In (I.10c-d) a convenient choice of the integration constants has been made.

$$\hat{\rho}(V) = \int^{V} dV' \; \hat{\beta}(V')^{-1} =$$

$$= - b_o^{-1} V^{-1} - b_o^{-2} b_1 \ln V + \text{const.} + ..V + \ldots \qquad (I.10a)$$

$$\hat{a}(V) = \exp\left[2 \int_{o}^{V} dV' \; \hat{\beta}(V')^{-1} \hat{\gamma}(V')\right] =$$

$$= 1 + 2 b_o^{-1} c_o V + \ldots \qquad (I.10b)$$

$$\hat{h}(V) = \exp\left[\int^{V} dV' \; \hat{\beta}(V')^{-1} \hat{\eta}(V')\right] =$$

$$= V^{\frac{1}{3}} (1 + .. V + \ldots) \qquad (I.10c)$$

$$\hat{k}(V) = \int^{V} dV' \; \hat{\beta}(V')^{-1} \hat{a}(V')^2 \hat{h}(V')^2 \hat{\kappa}(V') =$$

$$= - V^{-\frac{1}{3}} (1 + \ldots . V + \ldots ..) \qquad (I.10d)$$

and

$$V(\lambda) = \hat{\rho}^{-1}(\ln\lambda^2 + \hat{\rho}(V)) \qquad (I.11a)$$

such that for any C^∞ function $F(V)$

$$F(V(\lambda)) = \sum_{r=0}^{\infty} (r!)^{-1} (\ln\lambda^2)^r [\hat{\beta}(V) \partial/\partial V]^r F(V) \qquad (I.11b)$$

as the series expansion in increasing powers of $\ln\lambda^2$.

In (I.10) we have taken V to obey $0 < V < V_\infty$ where V_∞ is the first positive zero of $\hat{\beta}(V)$, if there is such a zero of that function which starts out positive. The reason for $V > 0$ is that the existence of the theory with $V < 0$ is somewhat doubtful, as the Hamiltonian of the Φ^4 theory should have a (formally, at least) positive bare coupling constant, and it appears difficult to imagine that this may lead to a negative (renormalized) V. The reason for $V < V_\infty$ is that only then can we really compute the strength of UR singularities, as we shall see in sect. II.3.

With (I.10-11), (I.7) integrates to

$$\begin{aligned} &\Gamma_o(p_1 \ldots p_{2n}, q_1 \ldots q_L; U^2, V) = \\ &= \hat{a}(V)^{n-L}\, \hat{a}(V(\lambda))^{-n+L} \hat{h}(V)^{-L}\, \hat{h}(V(\lambda))^L \cdot \\ &\cdot \Gamma_o(p_1 \ldots p_{2n}, q_1 \ldots q_L; U^2 \lambda^2, V(\lambda)) - \\ &- i\delta_{no}\, \delta_{L2}\, \hat{a}(V)^{-2}\, \hat{h}(V)^{-2}\, [\hat{k}(V) - \hat{k}(V(\lambda))] \, , \end{aligned} \tag{I.12}$$

and comparison with (I.5) allows to read off the expressions for V, Z_1, Z_2, and f, with $U^{-1} U_o = \lambda$. (I.12) shows that, due to (I.11b), normalization momentum change, which is in view of

$$\begin{aligned} &\Gamma_o(p_1 \ldots p_{2n}, q_1 \ldots q_L; U^2 \lambda^2, V) = \\ &= \lambda^{4-2n-2L}\, \Gamma_o(\lambda^{-1} p_1 \ldots \lambda^{-1} p_{2n}, \lambda^{-1} q_1 \ldots \lambda^{-1} q_L; U^2, V) \end{aligned} \tag{I.13}$$

equivalent to overall scaling of momenta, produces to all finite orders of perturbation theory polynomials in $\ln\lambda^2$ only.

Assume now we would parametrize the Γ_o differently from (I.1), e.g. by choosing the normalization momenta differently, given a normalization momentum squared $-\bar{U}^2$, but doing this so that the new functions $\bar{\Gamma}_o$ as functions of the new $\bar{V}$ keep in (I.2a-d) the forms $ip^2 + O(\bar{V}^2)$, $-i\bar{V} + O(\bar{V}^2)$, $1 + O(\bar{V})$, and $O(1)$, respectively. (Think, e.g., of using in (I.1c) a nonsymmetric normalization point, with momenta proportional to $\bar{U}$ and Euclidean, or Minkowskian with spacelike partial sums, to keep $\bar{V}$ real, or, for general Minkowskian normalization momenta, take the imaginary part of $\bar{\Gamma}_o(..)$ as $-i\bar{V}$) Then there will be renormalization group equations of the form (I.7), with functions $\bar{\beta}(\bar{V}) = \bar{b}_o \bar{V}^2 + \ldots$ etc. On the other hand, again by renormalization theory[6], the $\bar{\Gamma}_o$ must be related to the Γ_o by equations similar in form to (I.5), with (I.6a) replaced by $Z_1(U^{-1}\bar{U},V) = 1 + O(V^2)$ etc. Operating on that relation with $Op_{2n,L}$ and using the partial differential equations (PDEs) for the $\bar{\Gamma}_o$, one finds that the resulting differential relations for the Z_1 etc. can be integrated to $Z_1 = 1 + O(V^2)$ etc. only if

$$\bar{b}_o = b_o,\ \bar{b}_1 = b_1,\ \bar{h}_o = h_o,\ \bar{k}_o = k_o$$

hold. From the first equality, we will draw an amusing consequence in sect. IV.

II. INFRARED SINGULARITIES

II.1. Location of UR singularities.

For generic external momenta, the VFs are finite functions. Unless we indicate explicitly otherwise, we will in the following discuss Euclidean momenta only. For such momenta, starting from Wightman axioms and using some mild technical assumptions that are certainly satisfied in perturbation theory, Ruelle[8] has shown that Green's functions, and consequently also VFs, are (imaginary-valued) real-analytic functions in our model, provided the mass spectrum of the theory (apart from the vacuum) is positive. If there are zero-mass particles, then VFs cease to be regular at Euclidean points if a partial sum of momenta vanishes, which identifies these singularities as UR ones. They are threshold singularities from intermediate states whose momentum spectrum reaches to zero.

Not all UR singularities are equally bad, however. It may happen that the VF considered is finite and continuous at that point, and this also to all orders of perturbation theory. Such momenta we call, together with all generic momenta, nonexceptional. E.g. if a Φ-momentum is zero, all other momenta generic, due to the smallness of three-particle phase space (see(I.4) and (I.2a)), we have a nonexceptional momenta set. If, however, an "even" partial sum of momenta vanishes, which means an even number of Φ-plus any number of $N_2(\Phi^2)$ momenta, then there are two-particle intermediate states, leading in view of (I.4) to a logarithmic singularity, as in (I.2b-d). The condition that also in perturbation theory, we should at nonexceptional momenta have finiteness and continuity, is introduced because it may happen, and we will prove this in sect.II.3 for the function $\Gamma_o(p-r)(-p-r)r\ r,;\ U^2,V)$ at $r \to 0$, that the exact VF is continuous at some point (in the example, it vanishes there) while all separate orders of perturbation theory are logarithmically singular there. This requires precautions since we want all our manipulations and deductions to be verifiable in all orders of perturbation theory, except those concerning the <u>true</u> behaviour of the function considered.

For Minkowskian momenta, Ruelle[8] has shown that Green's functions, and consequently also VFs, are regular unless a partial sum of momenta lies on the momentum spectrum of the theory (the vacuum excluded). In the massless case, this means that for regularity all partial sums of momenta must be spacelike. However, nonregularity at all other points does not follow. At all finite orders of perturbation theory, mass thresholds remain at zero mass, and one has in perturbation theory nonregularity, and, therefore possibly exceptionality, only at lightlike partial sums of momenta, and at momenta sets where a nontrivial Coleman-Norton diagram[9] with lightlike internal momenta can be drawn. I will not discuss Minkowskian momenta further, since they require a more careful investigation which I have not done yet. Note, however, that the aforementioned results of Blanchard and Sênéor[5] do constrain the possible UR singularities in Minkowski space.

II.2. Structure of UR singularities

We want to analyze the behaviour of the VFs near

Euclidean exceptional momenta, which, according to sect.II.1., are related to two-particle intermediate states. The difficulty is to sum over arbitrarily many such intermediate states lying behind each other. This can be accomplished by manipulating the two-particle Bethe-Salpeter (BS) equations implicit in the definitions of all VFs until a separation into singular and nonsingular (more precisely, continuous) parts is achieved.

For the BS equation we use the matrix notation

$$I = B + B\,G\,I = B + I\,G\,B \quad . \tag{II.1}$$

Here I is the four-point VF (Φ arguments only), B the BS kernel (which in the present model involves a logarithmically divergent constant such that for finiteness one should first consider a regularized theory and go to the limit only in the subtracted equations, which are all equations except (II.1) itself) and G stands for the pair of propagators. Momenta are indicated as

$$\Gamma((-\tfrac{1}{2}q-p)(-\tfrac{1}{2}q+p)(\tfrac{1}{2}q+p')(\tfrac{1}{2}q-p')) \leftrightarrow p\overset{q}{I}p'$$

$$(2\pi)^4[\delta(p-p')+\delta(p+p')]G(-\tfrac{1}{2}q-p)G(-\tfrac{1}{2}q+p) \leftrightarrow \overset{q}{G}$$

$$(2\pi)^4[\delta(p-p')+\delta(p+p')] \leftrightarrow 1$$

and are suppressed if internal (and then integrated over with a factor $\frac{1}{2}(2\pi)^{-4}$ supplied), or if external (left relative, right relative, and throughgoing or total) and kept general. In (II.1) it is understood that internal relative momenta fit unless a momentum associated with the factor to the left or right is indicated to be fixed (usually at zero) such that that factor, if the relative momentum is fixed, is constant in the internal-momentum integration, indicated for clarity additionally by a bracket <or> . For a G-link, this amounts to let arguments coalesce in coordinate space.

The formulae that follow will not all be UR finite, e.g. $\mathring{I}$ and $\mathring{I}^{o}$ are not, and $\mathring{G}$ is not if an integration is involved. Thus, one should more precisely read the throughgoing momentum in those cases first as small but nonzero and let it go to zero only in the final equations for which we shall discuss the UR behaviour in detail.

From (II.1) we have

$$GI = \mathring{G}\mathring{I} + (G-\mathring{G})I + \mathring{G}(I-\mathring{I}) \tag{II.2}$$

and

$$\begin{aligned} I-\mathring{I} &= B-\mathring{B} + (I-\mathring{I})GB + \mathring{I}(G-\mathring{G})B + \mathring{I}\mathring{G}(B-\mathring{B}) = \\ &= (1 + \mathring{I}\mathring{G})(B-\mathring{B})(1+GI) + \mathring{I}(G-\mathring{G})I \end{aligned} \tag{II.3}$$

where again (II.1) is used to obtain the last line. Inserting (II.3) into (II.2)

$$\begin{aligned} 1 + GI &= 1 + \mathring{G}\mathring{I} + (G-\mathring{G})I + \\ &+ (1 + \mathring{G}\mathring{I})\mathring{G}(B-\mathring{B})(1+GI) + \mathring{G}\mathring{I}(G-\mathring{G})I = \\ &= (1 + \mathring{G}\mathring{I})[1 + \mathring{G}(B-\mathring{B})(1+GI) + (G-\mathring{G})I] \; . \end{aligned} \tag{II.4}$$

Furthermore from (II.1)

$$\mathring{I} - \mathring{I}^{o} = \mathring{B}-\mathring{B}^{o} + \mathring{I}\mathring{G}(\mathring{B}-\mathring{B}^{o})$$

such that

$$\mathring{I} = (\mathring{I}^{o} + \mathring{B}-\mathring{B}^{o})[1 - \mathring{G}(\mathring{B}-\mathring{B}^{o})]^{-1} \qquad (II.5)$$

and therefrom

$$1 + \mathring{G}\mathring{I} = (1 + \mathring{G}\mathring{I}^{o})[1 - \mathring{G}(\mathring{B}-\mathring{B}^{o}]^{-1} \qquad (II.6)$$

inverses everywhere defined by expansion. Thus from (II.4)

$$1 + GI = (1 + \mathring{G}\mathring{I}^{o})\ (1 + W) \qquad (II.7)$$

where

$$W = -1 + [1-\mathring{G}(\mathring{B}-\mathring{B}^{o})]^{-1}[1+\mathring{G}(B-\mathring{B})(1+GI)+(G-\mathring{G})I]\ . \qquad (II.8)$$

Multiplying (II.7) from the left by the bare vertex γ yields

$$\Gamma = \mathring{\Gamma}^{o}\!<(1 + W) \qquad (II.9)$$

where

$$\Gamma_{o}((\tfrac{1}{2}q + p)(\tfrac{1}{2}q-p),\ (-q)) \leftrightarrow \overset{q}{\Gamma}{}^{p}\ .$$

Here Γ is normalized by (I.1d), such that for the RHS of (II.9), in view of (I.2c), an interpretation as described before is essential: since the RHS is independent of the throughgoing momentum denoted by zero and implicit in W, the two factors on the RHS must develop a factorizing UR singularity cancelling between them. In sect.II.3 we will determine that singularity precisely.

The UR divergence of $<W$ stems from the $<\mathring{G}$ in its definition; all other factors $\mathring{G}$ standing elsewhere in the expression obtained by developing the inverse have to their left a factor vanishing at zero relative momentum of first order such that no UR divergence develops. (II.7) can be rewritten

$$1 + GI = \overset{o}{G}[\overset{oo}{I}(\overset{oo}{\Gamma})^{-1}]\Gamma + 1 + W \qquad \text{(II.10)}$$

which is the simplest Wilson expansion[10] formula, to be used for left relative momentum large, the others finite. In coordinate space, (II.10) is, in Zimmermann's[3] notation

$$N_o \; (A(x+\xi) \; A(x-\xi)) =$$

$$= \underline{E}(\xi) \; \tfrac{1}{2}\underline{N}_2(A^2(x)) + N_2(A(x+\xi) \; A(x-\xi)) \; , \qquad \text{(II.11)}$$

where the underlinings are ours and indicate normalization change, relative to Zimmermann's[3] convention, as discussed before.

Since GI, W, Γ, and $\overset{o}{G}$ in (II.10) are finite in momentum space, the square bracket must be finite too:

$$\overset{oo}{I} \; (\overset{oo}{\Gamma})^{-1} \; \equiv \; \underline{\overset{oo}{I}} \qquad \text{(II.12)}$$

is a UR finite function, for which we will give below an explicit expression. However, the coordinate space form (II.11) of (II.10) is not yet UR finite, due to the necessary integration involving the leftmost $\overset{o}{G}$ in W. A variant* of (II.10) finite in coordinate space is

*The author thanks H.-J. Thun for a helpful discussion in this connection.

$$1 + GI = [(1 + \overset{\circ\circ}{GI})^{\circ}(\overset{\circ}{\Gamma}^{\circ})^{-1}]\,\Gamma + 1 + W - 1^{\circ}<1 - 1^{\circ} < W \tag{II.13}$$

which corresponds to Zimmermann's[3]

$$N_o(A(x+\xi)A(x-\xi)) = \underline{H}(\xi)\tfrac{1}{2}\underline{N}_2(A^2(x)) + M_2(A(x+\xi)A(x-\xi))$$

where

$$M_2(A(x+\xi)A(x-\xi)) = N_2(A(x+\xi)A(x-\xi)) - N_2(A^2(x)) \ .$$

The aforementioned UR divergence is eliminated in the difference (since in (II.8) only the leftmost $\overset{\circ}{G}$ is dangerous), and the underlining has the same meaning as before. We will stay in momentum space, however, where the simpler equ. (II.10) is UR finite.

We return to (II.3), which we write as

$$I = \overset{\circ}{I}[1 + \overset{\circ}{G}(B-\overset{\circ}{B})(1 + GI) + (G-\overset{\circ}{G})I] + (B-\overset{\circ}{B})(1 + GI)$$

and insert (II.5), obtaining

$$I = \underline{\overset{\circ}{I}}^{\circ}\Gamma + (\overset{\circ}{B}-\overset{\circ}{B}^{\circ})(1+W) + (B-\overset{\circ}{B})(1 + GI) \tag{II.14}$$

which is UR finite (also in coordinate space). It yields an expression for $\underline{\overset{\circ}{I}}^{\circ}$ upon use of (I.1d):

$$\underline{\overset{\circ}{I}}^{\circ} = \overset{u}{I}^{\circ} - (\overset{\circ}{B}-\overset{\circ}{B}^{\circ})\,\overset{u}{W}^{\circ} - \overset{u}{(B-\overset{\circ}{B})}\overset{uu}{(1 + GI)} \tag{II.15}$$

where the U on top means that the throughgoing momentum has square-U^2. As the derivation shows, the RHS of (II.15) is independent of the direction of that momentum. The alternative definition, alluded to implicitly in the formula (II.12), is recovered from (II.14) due to

$$\overset{p_o}{I} = \underline{\overset{o}{I}}{}^{o}\ \overset{p_o}{\Gamma} + O(p) \tag{II.16}$$

for $p \to 0$, $\overset{p_o}{\Gamma}$ vanishing as $(\ln p^2)^{-\frac{1}{3}}$ thereby as we shall prove in sect. II.3. (II.16) also shows the UR singularity structure of the function on the LHS.

Interchanging in (II.4) the "zero" momentum and the through momentum and multiplying from the left by the bare vertex, we obtain

$$\overset{o}{\Gamma} = \Gamma[1 - G(B-\overset{o}{B})(1 + \overset{o}{G}\overset{o}{I}) - (G-\overset{o}{G})\overset{o}{I}] \ . \tag{II.17}$$

Inserting here the transpose of (II.14) taken at through momentum "zero" yields

$$\overset{o}{\Gamma} = \Gamma(1 - R) - (\Pi - \overset{o}{\Pi})\ {}^{o}\underline{\overset{o}{I}} \tag{II.18}$$

with

$$R = G(B-\overset{o}{B}) + [G-\overset{o}{G} + G(B-\overset{o}{B})\overset{o}{G}]\ (1+\overset{o}{W}{}^{T})(\overset{o}{B} - {}^{o}\overset{o}{B})$$

and

$$\Pi - \overset{o}{\Pi} = \Gamma[G-\overset{o}{G} + G(B-\overset{o}{B})\overset{o}{G}]\ \overset{o}{\Gamma}{}^{T} =$$

$$= \Gamma G\ \overset{o}{\gamma}{}^{T} - \gamma\overset{o}{G}\ \overset{o}{\Gamma}{}^{T} = \Gamma G\ \gamma^{T} - \overset{o}{\gamma}\overset{o}{G}\ \overset{o}{\Gamma}{}^{T}$$

where (II.9) is used. (II.18) shows that

$$\overset{o}{\Gamma} - \overset{o}{\Pi}\ {}^{o}\underline{\overset{o}{I}} \equiv \underline{\overset{o}{\Gamma}} \tag{II.19}$$

is finite, where the LHS is to be understood similarly as in (II.12). In view of (I.2e) we have from (II.18) the precise definition

$$\underline{\overset{o}{\Gamma}} = \overset{u}{\Gamma}\,(1 - \overset{u}{\mathbf{R}}) \qquad \text{(II.20)}$$

and for small momentum p(II.18) reads

$$\overset{p}{\Gamma} = \underline{\overset{o}{\Gamma}} + \overset{p}{\Pi}\;{}^{o}\underline{\overset{o}{I}} + O\,(p) \qquad \text{(II.21)}$$

displaying the UR structure of the LHS, and giving meaning to (II.19).

Finally, we subtract from (II.14) the same equation with zero through momentum and obtain, using (II.18),

$$\begin{aligned} I - \overset{o}{I} = {} & \underline{\overset{o}{I}}{}^{o}\,\Gamma\,R + \underline{\overset{o}{I}}{}^{o}\,(\Pi - \overset{o}{\Pi})\;{}^{o}\underline{\overset{o}{I}} + \\ & + (\overset{o}{B} - \overset{o}{B}{}^{o})(W - \overset{o}{W}) + (B - \overset{o}{B})\,(1 + GI)\quad . \end{aligned} \qquad \text{(II.22)}$$

This suggests the definition, in the sense explained before,

$$\overset{o}{I} - \underline{\overset{o}{I}}{}^{o}\;\overset{o}{\Pi}\;{}^{o}\underline{\overset{o}{I}} \equiv \underline{\overset{o}{I}} \qquad \text{(II.23)}$$

with the UR finite form, from (II.22) and (I.2e)

$$\begin{aligned} \underline{\overset{o}{I}} = {} & \overset{u}{I} - \underline{\overset{o}{I}}{}^{o}\,\overset{u}{\Gamma}\,\overset{u}{R} - (\overset{o}{B} - \overset{o}{B}{}^{o})\,(\overset{u}{W} - \overset{o}{W}) - \\ & - (\overset{u}{B} - \overset{o}{B})\,(1 + \overset{uu}{GI})\,. \end{aligned} \qquad \text{(II.24)}$$

For small through momentum p, (II.22) takes the form

$$\overset{p}{I} = \overset{\circ}{\underline{I}} + \overset{\circ\circ}{\underline{I}}{}^{\circ}\, \overset{p}{\Pi}\; {}^{\circ}\overset{\circ}{\underline{I}} + O(p) \qquad \text{(II.25)}$$

displaying the UR structure of the LHS, and giving meaning to (II.23).

II.3. Strength of UR singularities.

Formulae (II.16, 21, 25) show that we need to know the small-p behaviour of $\overset{p}{\Gamma}{}^{o}$ and $\overset{p}{\Pi}$ to learn the UR singularity of all VFs considered so far, as the momenta sets we have been considering are the only exceptional ones for those functions. From (I.12) we find

$$\Gamma_o(\tfrac{1}{2}p\ \tfrac{1}{2}p,\ -p;\ U^2,\ V) = \qquad \text{(II.26)}$$
$$= \hat{h}(V)^{-1}\hat{h}(V(\lambda))\ \Gamma_o(\tfrac{1}{2}p\ \tfrac{1}{2}p,\ -p;\ U^2\ \lambda^2,\ V(\lambda))\ .$$

We now choose $\lambda^2 = U^{-2}(-p^2)$. In view of (I.13) the last Γ_o then depends on p only through the coupling constant. Now for $\lambda \to 0$ from (I.10a) and (I.11a) for $0 < V < V_\infty$ as explained there,

$$V(\lambda) = b_o^{-1}\ (\ln\lambda^{-2})^{-1} - b_o^{-3}\ p_1(\ln\lambda^{-2})^{-2}\ln\ln\lambda^{-2} +$$
$$+ O((\ln\lambda)^{-2}) \qquad \text{(II.27)}$$

by straightforward computation. Now for the last Γ_o in (II.26) we may use (I.1d), and for $\hat{h}(V(\lambda))$ the expansion indicated in (I.10c). The result is

$$\overset{p_o}{\Gamma} = \hat{h}(V)^{-1} \, b_o^{-\frac{1}{3}} \, [-\ln(-p^2 - i\varepsilon)]^{-\frac{1}{3}} +$$

$$+ O((\ln p^2)^{-\frac{4}{3}} \ln \ln p^2) \qquad \text{(II.28)}$$

which at first follows for $p^2 \to 0$ from negative values, but the analytic structure of the function allows to extend this to all small p^2.

Setting in (I.12) $n = 0$, $L = 2$ we obtain by analogous reasoning, whereby the inhomogeneous term is the essential part and (I.10d) is used,

$$\overset{p}{\Pi} = -i \, \hat{a}(V)^{-2} \, \hat{h}(V)^{-2} \, b_o^{\frac{1}{3}} \, [-\ln(-p^2 - i\varepsilon)]^{\frac{1}{3}} +$$

$$+ O((\ln p^2)^{-\frac{2}{3}} \ln\ln p^2) \; .$$

It is instructive to consider the absorptive parts of these two functions:

$$2 \, \mathrm{Im} \, \overset{p_o}{\Gamma} \approx -i \, \tfrac{2}{3}\pi \, \hat{h}(V)^{-1} \, b_o^{-\frac{1}{3}} \, \Theta(p^2) \, [-\ln p^2]^{-\frac{4}{3}} =$$

$$= \hat{a}(V)^{-2} \, [\hat{h}(V)^{-1} \, b_o^{-\frac{1}{3}} \, (-\ln p^2)^{-\frac{1}{3}}] \; .$$

$$\cdot [-i \, \hat{a}(V)^2 \, b_o^{-1} \, (-\ln p^2)^{-1}] \, \{\tfrac{1}{2} \, q^{-3} \, \pi^{-1} \, \Theta(p^2)\} \; .$$

The last bracket is the two-particle phase space from (I.4). The first square bracket is (II.28) again, and the second is the non-exceptional small-momenta form of the four-point vertex following from (I.2) and (II.27). The factor $\hat{a}(V)^{-2}$ in front is due to the following form of the propagator for small momenta, obtained from (I.12) for n = 1, L = 0 and (II.27)

$$- \Gamma_o(p(-p),; U^2, V)^{-1} = G(p) =$$

$$= \hat{a}(V)^{-1} i (p^2 + i\varepsilon)^{-1} \cdot \qquad \text{(II.29)}$$

$$\cdot \{1 + 2 b_o^{-2} c_o[-\ln(-p^2-i\varepsilon)]^{-1} + O((\ln p^2)^{-2} \ln\ln p^2)\} .$$

Also for this function we can identify the structure of the absorptive part:

$$2 \text{ Re } G(p) \approx 2\pi\hat{a}(V)^{-1}[\delta(p^2)+2b_o^{-2}c_o(p^2)^{-1}\Theta(p^2)(\ln p^2)^{-2}] =$$

$$= 2\pi \hat{a}(V)^{-1} \delta(p^2) +$$

$$+ \hat{a}(V)^{-3}|i(p^2)^{-1}|^2|-ib_o^{-1} \hat{a}(V)^2(-\ln p^2)^{-1}|^2 \cdot \qquad \text{(II.30)}$$

$$\cdot \{\tfrac{1}{6}2^{-8} \pi^{-3} \Theta(p^2)p^2\}$$

the last bracket being the three-particle phase factor, the other factors being obvious. Finally,

$$2 \text{ Re } \overset{p}{\Pi} \approx \tfrac{2}{3}\hat{a}(V)^{-2}\hat{h}(V)^{-2} b_o^{\frac{1}{3}} \Theta(p^2) [-\ln p^2]^{-\frac{2}{3}} \approx$$

$$\qquad \text{(II.31)}$$

$$\approx \hat{a}(V)^{-2}|\hat{h}(V)^{-1}b_o^{-\frac{1}{3}}[-\ln(-p^2-i\varepsilon)]^{-\frac{1}{3}}|^2 \cdot \{\tfrac{1}{2}2^{-3}\pi^{-1}\Theta(p^2)\}$$

where the absolute square of the RHS of (II.28) appears.

Actually, also the small-momenta behaviour of the functions $\underline{\overset{\circ}{I}}{}^{\circ}$, $\underline{\overset{\circ}{\Gamma}}$, and $\underline{\overset{\circ}{I}}$ can be gotten precisely. To this end one derives

$$[\hat{O}p_{4,0} - \hat{\eta}(V)]\ \underline{\overset{\circ}{I}}{}^{\circ} = 0$$

$$\hat{O}p_{2,1}\ \underline{\overset{\circ}{\Gamma}} = i\ \hat{\kappa}(V)\ \ {}^{\circ}\underline{\overset{\circ}{I}}$$

and

$$\hat{O}p_{4,0}\ \underline{\overset{\circ}{I}} = i\ \hat{\kappa}(V)\ \underline{\overset{\circ}{I}}{}^{\circ}\ \ {}^{\circ}\underline{\overset{\circ}{I}}$$

which follow from (II.14), (II.18) with (II.19), and (II.22) with (II.23), respectively. Integrating these PDEs, as (I.7) was integrated to (I.12), and using the same procedure as before, one finds

$$p\underline{\overset{\circ}{I}}{}^{\circ} \approx -i\ \hat{a}(V)^{2}\ \hat{h}(V)\ b_{o}^{-\frac{2}{3}}\ [-\ln(-p^{2}-i\varepsilon)]^{-\frac{2}{3}}$$

$$\underline{\overset{\circ}{\Gamma}}p \approx q\ \hat{h}(V)^{-1}\ b_{o}^{-\frac{1}{3}}\ [-\ln(-p^{2}-i\varepsilon)]^{-\frac{1}{3}}$$

$$\lambda p\underline{\overset{\circ}{I}}\lambda q \approx -\ 2i\ \hat{a}(V)^{-2}\ b_{o}^{-1}\ (\ln\lambda^{-2})^{-1}$$

and

$$\lambda p\underline{\overset{\circ}{I}}\lambda q \approx -\ 2i\ \hat{a}(V)^{-2}\ b_{o}^{-1}\ (\ln\lambda^{-2})^{-1}$$

i.e., these functions vanish for small momenta with various inverse-logarithmic strength.

As we showed in examples, the true UR behaviour

found is consistent with dispersion relations and unitarity, and this is so also if more accurate expressions are used, i.e. higher inverse powers of logarithms are retained. While from the point of view of perturbation theory, infinite summations of binomial series outside their convergence radius would be involved to obtain the formulae we used, the behaviour we deduced from the renormalization group equations must be the true one if the properties under finite renormalizations used in the derivation, and the asymptoticity of perturbation expansions for small coupling constants (here $V(\lambda) \to 0$ for $\lambda \to 0$), hold in the exact theory. While the existence of the theory itself has not been proven yet, certainly the results obtained in constructive field theory so far (see B. Simon's lecture) do not suggest the mentioned assumptions to be violated if the theory considered does exist.

III. THEORIES WITH MASSLESS AND MASSIVE PARTICLES

III.1. Parquet approximation.

To investigate the asymptotic behaviour of the meson-meson scattering amplitude in pseudoscalar meson theory, Diatlov, Sudakov, and Ter-Martirosian[11] introduced in 1956 the parquet approximation (PA) for the four-pion amplitude. That approximation is closely related to the large-momenta approximations Landau and coworkers[12] used to discredit the consistency of QED. It is known[13] that these approximations are not adequate ones for large momenta problems, but they are so for UR behaviour problems*, as seems first to have been observed by V. G. Vaks.[14] Larkin and Khmel'nitskii[2] used renormalization group methods to prove this, and we will do so too, extending somewhat the work of these authors in the next section.

The PA four-point VF Γ_p is defined by the crossing symmetric BS equation Fig. 1. The notation is self-explanatory, the function marked i being two-particle irreducible in all three channels. The propagator is the free massless one $i(k^2 + i\varepsilon)^{-1}$. In Fig. 1c, the PA in the narrow sense uses only the first term on the RHS, a constant in momentum space. We shall see that inclusion of higher skeleton terms does not affect the UR-singularities.

The BS equation Fig. 1a is to be soved upon subtraction to eliminate the undetermined constant in the BS kernel, similarly as remarked in sect. II.2. With the subtracted BS kernel given, Γ_p is fully determined by prescribing its value at some point in momentum space, whereby the eliminated constant in B is implicitly determined. We choose

$$\Gamma_p(p_1 \ldots p_4,;\ U^2,V)\Big|_{\text{s.pt.to } -U^2} = -iV\ . \qquad \text{(III.1)}$$

The letters below the functions in Fig.1 indicate the orders of approximation in a possible solution procedure with Fig. 2 the starting approximation. Hereby

*We do not have in mind here an application of renormalization group methods to the ordinary UR singularities in QED, as in §43.3 of Ref. 13. We shall comment on ordinary QED in section IV.

the approximate I_p will not be crossing symmetric, for which reason we used dotted lines to fix the orientation. The simplest approximation procedure is, of course, iteration to generate power series in V with for I_p in each order crossing symmetric coefficient. For our purposes, the technique of solving the equs. Fig. 1, given the condition (III.1), is arbitrary.

The normalization momentum square $-U^2$ in (III.1) is not intrinsic, and, since it is the only mass parameter in the PA approximation, its change is described by

$$\Gamma_p(p_1 \ldots p_4,; U^2, V) = \tag{III.2}$$

$$= \Gamma_p(p_1 \ldots p_4; U_o^2, i\Gamma_p(p_1' \ldots p_4'; U^2, V)\Big|_{s.pt.to\ -U_o^2}),$$

which means that the analog of Z_1 in (I.5) is unity. The reason is that in contrast to full massless Φ^4 theory where (I.1b-c) need be imposed, condition (III.1) alone determines the PA function I_p completely. Differentiating (III.2) with respect to U_o^2 at $U_o^2 = U^2$ yields

$$\{U^2[\partial/\partial U^2] + \beta_p(V)[\partial/\partial V]\}\ \Gamma_p(p_1 \ldots p_4,; U^2, V) \equiv$$

$$\equiv Op_p\ \Gamma_p(\ldots) = 0 \tag{III.3}$$

where

$$\beta_p\ (V) =$$

$$= \frac{1}{2} i \sum_{j=1}^{3} p_j [\partial/\partial p_j] \cdot \Gamma_p (p_1 p_2 p_3 (-p_1-p_2-p_3),;U^2,V) \Big|_{s.pt.to-U^2} =$$

$$= b_o V^2 + b_{1p} V^3 + \ldots \quad . \qquad \text{(III.4)}$$

Comparison with (I.9) easily shows that b_o is the one of (I.3) but

$$b_1 = b_{1p} + 4 c_o \ . \qquad \text{(III.5)}$$

(III.3) allows to write (III.2) in the form

$$\Gamma_p(p_1 \ldots p_4,;U^2,V) = \Gamma_p(p_1 \ldots p_4,;U^2\lambda^2, V_p(\lambda)) \qquad \text{(III.6)}$$

with $V_p(\lambda)$ defined in analogy to (I.11a), such that in place of (II.27)

$$V_p(\lambda) = b_o^{-1}(\ln\lambda^{-2})^{-1} - b_o^{-3} b_{1p} (\ln\lambda^{-2})^{-2} \ln\ln\lambda^{-2} +$$

$$+ O((\ln\lambda)^{-2}) \qquad \text{(III.7)}$$

wherefrom the behaviour of Γ_p under scaling down of the momenta follows:

$$\Gamma_p(\lambda p_1 \ldots \lambda p_4,;U^2,V) \approx -i b_o^{-1}(\ln\lambda^{-2})^{-1} \qquad \text{(III.8)}$$

differing from Φ^4 theory by $\hat{a}(V)$ being replaced by one due to its absence in the propagator, cp. (II.29).

As in Φ^4 theory, by the formulae of sect. II.2 but B read as B_p, G as the free-propagator pair, and I as I_p we can define all PA VFs, upon fixing normalizations

$$\Gamma_p(p_1p_2, q_1; U^2, V) \equiv \Gamma_p , \qquad (\Gamma_p^{u_i} = 1) , \qquad \text{(III.9a)}$$

$$\Gamma_p(,q(-q); U^2, V) \equiv \Pi_p , \qquad (\Pi_p^{u} = 0) , \qquad \text{(III.9b)}$$

the higher functions being given by skeleton expansions. The arbitrariness inherent in the choice (III.9) leads to extra terms in the PDEs involving any vertex type arguments:

$$[Op_p + L\eta_p(V)]\Gamma_p(p_1 \ldots p_{2n}, q_1 \ldots q_L; U^2, V) =$$
$$= -i\, \delta_{no}\, \delta_{L2}\, \kappa_p\,(V) \qquad \text{(III.10)}$$

where

$$\eta_p\,(V) = \tfrac{1}{3}b_o\, V + \ldots V^2 + \ldots \qquad \text{(III.11a)}$$

$$\kappa_p\,(V) = \tfrac{1}{3}b_o + \ldots V + \ldots \qquad \text{(III.11b)}$$

Herefrom it follows that the strength of UR singularities of PA functions is as found in sect. II.3 for Φ^4 functions, with $\hat{a}(V)$ replaced by one, $\hat{h}(V)$ by $\hat{h}_p(V)$ and $\hat{k}(V)$ by

$\hat{k}_p(V)$ defined in analogy to (I.10c-d). It is now clear that the higher terms on the RHS of Fig.1c, leading in $\beta_p(V)$ to terms $O(V^4)$ would have no other effect than changing the normalization terms $\hat{h}_p(V)$ and $\hat{k}_p(V)$.

III. 2. Application of parquet approximation.

We now prove that in a renormalizable theory that has, besides the scalar neutral massless particles of the Φ^4 type, also massive particles, the UR singularities of the VFs are the same* ones as in pure massless Φ^4 theory or, equivalently, as of PA functions, up to normalization factors such as $\hat{a}(V)$ etc.

Let M^2 and G stand for the "large" masses and for the coupling constants related to renormalization functions with fields to these "heavy" particles. With respect to two-massless-particle intermediate states, still equs. Fig. 1 hold, the propagator being the fully corrected one, however, and the i-kernel being subject only to the condition of two-massless-particle-irreducibility in each channel, thus in particular being two-particle-reducible in general with respect to all other possible particle pairs. This means that in general the new BS equations are not UV convergent upon subtraction; this can be remedied, if necessary, by displaying in Fig. 1 only the two-mass-particle links to not too large internal momenta, the "UV-cutoff" parameter being then apparently another large one of the M type but ultimately disappearing again since only artificially introduced. For brevity, we will ignore it henceforth.

*That this should be so was suggested to me by K.G. Wilson, who also pointed out the relevance of the work of Larkin and Khmel'nitskiĭ.[14]

It is now desirable to be as close as possible to the PA scheme, which is achieved by separating the corrections to the massless propagator from the pure pole term and adding their contribution to the irreducible kernel. (We will later see how effective that is in a simple calculation.) Also the pure pole term should, for ease of comparison, be normalized to be $i(k^2 + i\varepsilon)^{-1}$ as in the PA, which means to throw a factor, like the $\hat{a}(V)^{-1}$ of (II.29), onto the four point function, the BS- and the i kernel. Again we skip the insignificant details hereto. What remains is that the four-point VF $\Gamma(p_1 \ldots p_4,; U^2, V, M^2, G)$ depends, besides the parameters M^2 and G entering through the i kernel, on U^2 and V as in the PA, the strength of self-coupling of the massless-particle field being, by renormalization theory, a free parameter, and the convention determining it is arbitrary. So (III.3) becomes now

$$\{U^2[\partial/\partial U^2]+\beta(V,M^{-2}U^2, G)[\partial/\partial V]\} \cdot \Gamma(p_1 \ldots p_4,; U^2, V, M^2, G) = 0 \qquad \text{(III.12)}$$

with, due to (III.1) which we keep,

$$\beta(V, M^{-2}U^2, G) = \frac{1}{2} i \sum_{j=1}^{3} p_j[\partial/\partial p_j]\Gamma(p_1 p_2 p_3(-p_1-p_2-p_3),; U^2, V, M^2, G)\Big|_{\text{s.pt.to-}U^2} \qquad \text{(III.13)}$$

where we have used that Γ is a homogeneous function of order zero of p_i, M, and U, and that

$$M^2[\partial/\partial M^2]\Gamma(p_1 \ldots p_4,;U^2,V,M^2,G)\Big|_{s.pt.to-U^2} = 0$$

due to (III.1).

We now consider letting $U \to 0$. Then in (III.12) the momenta dependence of Γ near the arbitrarily small normalization momentum enters, the large masses and related coupling constants fixed. This means that the i-kernel, the only input remaining from the full theory, will depend on the features of the theory, as far as the massive particles are concerned, only via the small parameter $M^{-1}U$, which is most implausible to induce a momentum variation of that kernel. This could certainly not happen in perturbation theory with respect to G, the massive-particle couplings. This accepted, however, the i-kernel may be approximated by a constant, which is just the first term on the RHS of equ. Fig. 1c, the value of that constant being formally determined by V rather than G and $M^{-1}U$. Then, as a consequence,

$$\lim_{U\to 0}\beta(V,\ M^{-2}\ U^2,\ G) = \beta_p(V) \qquad \text{(III.14)}$$

or, what actually suffices for our purpose, also

$$\lim_{U\to 0}\lim_{V\to 0} V^{-2}\beta(V,\ M^{-2}\ U^2,\ G) = b_o \quad . \qquad \text{(III.15)}$$

This already gives the UR singularity structure of the PA approximation, up to normalization factors as mentioned before, in view of the prominent occurrence of b_o in all relevant equations of sect. II.3. A calculation hereto is given in the next section.

The argument that leads to (III.14) is similar to the familiar one used in the older approach to renormalization group consequences: to proceed from the exact renormalization group equations (see Ref. 13) to usable relations one had to assume that for large normalization momenta, the momenta dependence near the normalization point was insensitive towards the true, relatively arbitrarily small, mass. It is now known that this holds to all orders of perturbation theory at generic (i.e., nonexceptional) momenta, but a satisfactory demonstration (see, e.g., Ref. 15) is not completely trivial. It is likely that also of (III.14) a formal proof in perturbation theory can be given but may not be trivial either. Why in (II.13) $\beta_p(V)$ appears where $\hat{\beta}(V)$ might have been expected will be explained in sect. III.3.

(III.14) holds also in the σ-model in the Goldstone mode (see, e.g., Ref. 16 for its treatment in perturbation theory). However, here the masslessness of some particles is of dynamical origin, namely related to the conserved axial vector current, rather than imposed through an ad hoc renormalization condition. This has the consequence that the relevant solution of (III.12) in the small-momenta limit is $I \equiv 0$, corresponding to $V = 0$ in $\beta_p(V)$ of (III.14). In fact, the four-pion amplitude vanishes in the σ-model of order[17] $f_\pi^{-2}\ p^2$, p a characteristic momentum, rather than of order $[\ln p^2]^{-1}$, as $p \to 0$. In the terms used in this section, the amplitude is itself only a correction to zero of order $M^{-2}\ U^2$. One could say that the PA has two solutions, the self-coupled one with small-momenta behaviour (III.8), and the free one with self-coupling suppressed completely, and that axial-vector current conservation enforces the second solution for Goldstone particles.

III. 3. Correction to the parquet approximation.

The argument used in the last section will now be illustrated by a simple calculation. We show that a nonleading UR singularity in Φ^4 theory that differs from the one in the PA can be gotten correctly by use of the PDEs we have been discussing.

(I.12) and (II.27) give

$$\Gamma_o(\lambda p_1 \ldots \lambda p_4; U^2, V) =$$

$$= \hat{a}(V)^2[-ib_o^{-1}(\ln\lambda^{-2})^{-1} + ib_o^{-3}b_1(\ln\lambda^{-2})^{-2}\ln\ln\lambda^{-2}] +$$

$$+ O((\ln\lambda)^{-2}) \qquad \text{(III.16)}$$

as $\lambda \to 0$, for nonexceptional momenta. The factor $\hat{a}(V)^2$ stems from (II.29), so a deviation from the PA result occurs first in the second term in the square bracket, due to $b_1 \neq b_{1p}$ from (III.5).

Comparing equs. Fig. 1 with the Φ^4 ones, where the i-kernel has to contain all non-PA effects, we observe that the propagator correction shows up first in the contribution depicted in Fig. 3. It gives a term $O(V^4)$, apparently merely competing with the higher neglected terms on RHS of Fig. 1c, but the propagator correction is the only one differing in effect from the main $O(V^2)$ term by one inverse logarithm only, which is related to the fact that it involves the maximum number, two, of UR-divergent integrations in order $O(V^4)$ while the neglected term on the RHS of Fig.1c has in that order only one, the final one.

We showed already in (II.30) how the propagator correction could be estimated correctly in terms of PA concepts. It remains to evaluate its contribution to (III.13). This contribution is

$$\begin{aligned} &\Delta\beta(V,\ U_o^{-2}\ U^2) = \\ &= 3i\ \frac{1}{2} \sum_{j=1}^{3} [p_j\ \partial/\partial p_j]\ (-iV)^2 \cdot \\ &\cdot\ (2\pi)^{-4} \int dk\ i\ [(k+p_1+p_2)^2 + i\varepsilon]^{-1} \cdot \\ &\cdot\ 2i(k^2+i\varepsilon)^{-1}\ b_o^{-2}\ c_o\ [\ln(U_o^2(k^2+i\varepsilon)^{-1})]^{-1}\Big|_{s.pt.to\ -\ U^2} \end{aligned} \qquad \text{(III.17)}$$

the factor three stemming from the sum over three channels. The U_o^2 occurring here is the parameter needed to parametrize the Φ^4 theory itself, not to be confounded with U^2. Rotating to Euclidean momenta and carrying out the differentiation yields

$$\begin{aligned} &\Delta\beta(\ldots) = 3\ V^2\ b_o^{-2}\ c_o\ (2\pi)^{-4}\ 2\ \cdot \\ &\cdot\ \int dk[(p_1+p_2)(k+p_1+p_2)][(k+p_1+p_2)^2]^{-2} \cdot \end{aligned}$$

$$\cdot \left[k^2 \ln(U_o^2 (k^2)^{-1}]^{-1}\right|_{\text{s.pt. to } U^2} =$$

$$= 6\, V^2\, b_o^{-2}\, c_o (2\pi)^{-4} \cdot \int dk\, [p(k+p)][(k+p)^2]^{-2} \cdot$$

$$\cdot (k^2)^{-1} \left[\ln(U_o^2 \tfrac{3}{4} U^{-2}) - \ln k^2]^{-1}\right|_{p^2 = 1}$$

$$\approx 6\, V^2\, b_o^{-2}\, c_o\, [\ln(U_o^2 \tfrac{3}{4} U^{-2})]^{-1} (2\pi)^{-4} .$$

$$\cdot \int dk\, [p(k+p)][(k+p)^2]^{-2} (k^2)^{-1} =$$

$$= 6 \cdot 2^{-4}\, \pi^{-2}\, V^2\, p_o^{-2}\, c_o\, [\ln(U_o^2 \tfrac{3}{4} U^{-2})]^{-1} =$$

$$= 4\, V^2\, b_o^{-1}\, c_o\, [\ln(U_o^2 \tfrac{3}{4} U^{-2})]^{-1}$$

using (I.3). Inserting this correction to the PA into (III.12), we need to integrate

$$U^{-2}\, dU^2 = \tag{III.18}$$

$$= \{b_o V^2 + b_{1p} V^3 + 4V^2\, b_o^{-1}\, c_o\, [\ln(U_o^2 \tfrac{3}{4} U^{-2})]^{-1} + \ldots\}^{-1}\, dV$$

for small U and V, the dots indicating smaller corrections than of interest. Now the last term in the curly bracket can be treated as a small perturbation, which is equivalent to insert there the integral of the lowest-order approximation $\ln U^2 = - b_o^{-1} V^{-1}$ + const. This then shows that, to the accuracy considered, (III.18) is equivalent to replacing in the curly bracket b_{1p} by b_1, in view of (III.5), and omitting the last term, recovering the Φ^4 result to this accuracy as was to be shown.

It is now also clear why in (III.14), which is verified for $\beta_p(V)$ plus the computed $\Delta\beta(V, U_o^{-2} U^2)$, $\beta_p(V)$ rather than $\hat{\beta}(V)$ appears: we were separating PA computations from all other computations, which include the feedback from deductions from the PA approximation. That no extra coupling constant V_o, in analogy to U_o, had to be introduced was due to the V_o-independence of the propagator correction actually used.

The manner in which the calculation in this section was conducted shows that only the soft-massless-particle effects needed to be considered to obtain, apart from mere normalization factors, the correct UR singularities, and this even in finer detail. This can be paraphrased by saying that all hard (and massive) particle graphs are dressed by soft-massless-particle lines, and thereby get soft themselves, the latter because the only remnant of the non-soft feature of the theory in e.g. Φ^4 theory is in (III.16) in the factor $\hat{a}(V)^2$, which is completely accounted for by the factor $\hat{a}(V)^{-1}$ in (II.29), that is, renormalizing to the free massless propagator removes it.

IV. CONCLUDING REMARKS

We have analysed UR singularities in massless Φ^4 theory, and in theories that have in addition to such massless scalar neutral particles also massive particles, and found in the latter case the alternative of either having the same UR singularities as in interacting Φ^4 theory, or else non-selfcoupling as in the Goldstone case.

The tool we used was merely renormalization group technique, and that the decisive parameter b_o in (I.8a) is positive. At the end of sect. I.2 we showed that b_o is independent of the choice of normalization momenta, e.g., of the four-point vertex. Now if we had $V < 0$, the statements of sect. II.3 on true UR behaviour would have been replaced by statements on "true UV behaviour" for $|V|$ sufficiently small. This shows that if $V > 0$ for the choice (I.1c) of normalization momenta, the four-point VF should stay negative imaginary for all Euclidean momenta, and its imaginary part should stay negative for all Minkowskian momenta since also these may be taken as some renormalization momenta. For some further discussion, see Ref. 18.

Since the renormalization group technique applies to any renormalizable theory with only massless particles, our results have straightforward extensions. Consider e.g. QED. The electrons being massive, the considerations of sect. III.2 apply, and the associated solution with photons only is, of course, the theory of free photons, the true photon-photon scattering amplitude involving the electron mass similarly as in the Goldstone case f_π is involved, only that the photon-photon amplitude vanishes at zero momenta to fourth rather than second order. Again it is a conserved current which here suppresses the photon self-coupling.

If also the electrons are made massless, the renormalization group technique leads to nontrivial results.[15] Some technical points concerning renormalization in this case have been discussed elsewhere.[19] We take the photon propagator in the socalled generalized Landau gauge, and find

$$D_{F\mu\nu}(k) = [-g_{\mu\nu} + (1+c)(k^2)^{-1} k_\mu k_\nu] \cdot$$
$$\cdot\ i(k^2+i\varepsilon)^{-1}\{[e^2 \tfrac{1}{12} \pi^2 \ln(m^2(-k^2-i\varepsilon)^{-1})]^{-1} + \dots\} \tag{IV.1}$$

and

$$S_F(p) = i \not{p} (p^2+i\varepsilon)^{-1}\{[\ln(m^2(-p^2-i\varepsilon)^{-1})]^{\frac{3}{4}c} + \dots\} \tag{IV.2}$$

dots signifying terms logarithmically smaller at small momenta. e in (IV.1) is the charge parameter in the Ward identity; the physical charge of the electron (defined from the electron-photon vertex at zero photon momentum) is zero.[15] The form of the photon propagator is easiest understood as resulting from the massless-electron-positron pair contribution to the self energy: due to that masslessness, strict second-order-vanishing as required for a discrete photon cannot be achieved. m is a mass parameter entering, just as in sect. I U entered, due to need of normalizing away from momenta zero. The electron propagator is gauge dependent and has a pole in Landau gauge proper (c = O) while it has in ordinary QED a pole in Yennie gauge.[1] The forms (IV.1) and (IV.2) should remain, up to factors, if we have also some massive charged particles in the theory, by the argument of sect. III.2.

Since we know no massless scalar particles, our results on UR singularities have no physical application. I was led to this topic by considerations on large-momenta behaviour. For generic (technically, non-exceptional) momenta, the renormalization group gives all information obtainable on the formal level, that is, without really solving e.g. the problem of existence of a Gell-Mann-Low eigenvalue (see K.G. Wilson's lectures). However, the physical interest is not in that "deep Euclidean"[20] situation, but where only some invariants (squares of partial sums of momenta) are large, others not. Scaling momenta down, we have some invariants fixed, others arbitrarily small together with the masses. If the masses could be set zero hereby, we would have a true UR-behaviour situation. In the cases of real interest, like scattering amplitudes with momentum transfer of the order of the masses but total energy large, the masses themselves cannot be set zero. Thus the present study is propaedeutic in character rather than real physics.

REFERENCES

1. R. Jackiw, L. Soloviev, Phys. Rev. 173, 1485 (1968).
   T. W. B. Kibble, Phys. Rev. 173, 1527 (1968).
   J. K. Storrow, Nuovo Cimento 54A, 15 (1968).

2. A. I. Larkin, D. E. Khmel'nitskiǐ, Zh. E. T. F. 56, 2087 (1969), transl. JETP 29, 1123 (1969).

3. W. Zimmermann, in "Lectures on Elemementary Particles and Quantum Field Theory", 1970 Brandeis Summer Institute, Eds. S. Deser, M. Orisaru, H. Pendleton, Cambridge Mass.: MIT Press 1971. Techn. Repts. 9/72

and 11/72, New York University, April 1972.

4. T. Appelquist, Thesis, Cornell University, June 1968.

5. P. Blanchard, R. Sênêor, Th. 1420, CERN, Nov. 1971.

6. K. Hepp, Theorie de la rênormalisation, Berlin: Springer 1969.

7. K. Symanzik, DESY 73/6, Hamburg, March 1973.

8. D. Ruelle, Nuovo Cimento 19, 356 (1961).

9. S. Coleman, R. F. Norton, Nuovo Cimento 38, 438 (1965)

10. K. G. Wilson, Phys. Rev. 179, 1499 (1969).

11. I. T. Diatlov, V. V. Sudakov, K. A. Ter-Martirosian, Zh. E. T. F. 32, 767 (1957), transl. JETP 5, 631 (1957).

12. L. D. Landau, A. Abrikosov, L. Halatnikov, Suppl. al Nuovo Cimento 3, 80 (1956).

13. N. N. Bogoliubov, D. V. Shirkov, Introduction to the Theory of Quantized Fields, New York: Intersc. Publ., 1959.

14. V. G. Vaks, Zh. E. T. F. 40, 792 (1961), transl. JETP 13, 556 (1961).

15. K. Symanzik, in Springer Tracts in Modern Physics, 57, 222 (1971).

16. K. Symanzik, Lett. al. Nuovo Cimento 2, 10 (1969).

17. E.g. S.L. Adler, R. F. Dashen, "Current Algebras", New York: Benjamin Inc., 1968.

18. K. Symanzik, Lett. al Nuovo Cimento 6, 77 (1973).

19. K. Symanzik, Lectures on Lagrangian Quantum Field Theory, University of Islamabad, January 1968.

(DESY T-71/1, February 1971).

20. S. Coleman, Lectures at Erice Summer School on Subnuclear Physics, July 1971.

FIGURE CAPTIONS

Fig. 1 Crossing symmetric Bethe-Salpeter equation

Fig. 2 Starting approximation for calculation Fig.1.

Fig. 3 Propagator correction to the irreducible kernel.

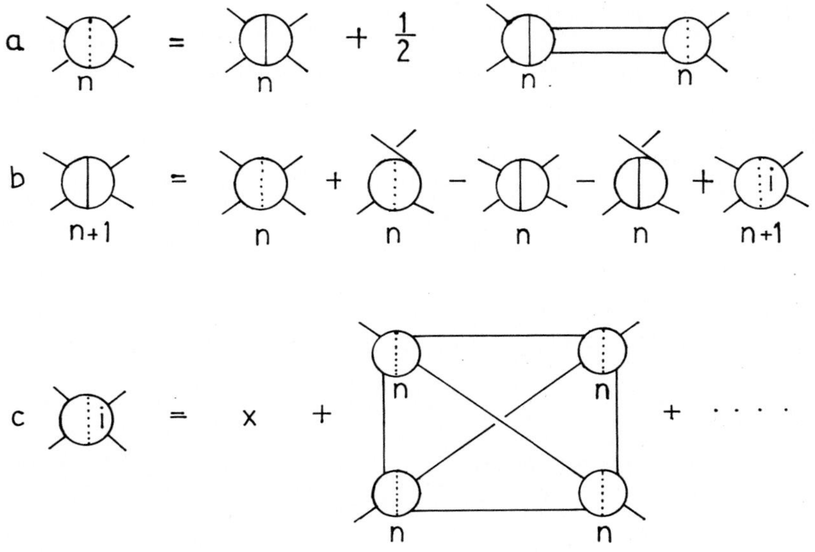

Fig.1

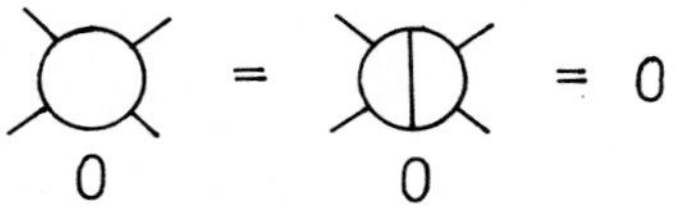

Fig.2

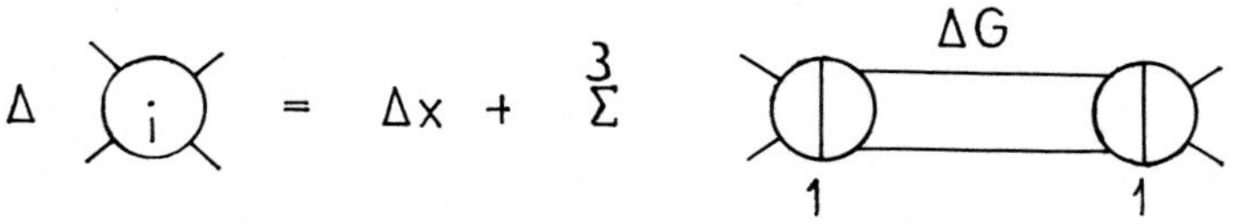

Fig. 3

Acta Physica Austriaca, Suppl.XI, 241–315 (1973)

CONFORMAL INVARIANT EUCLIDEAN QUANTUM FIELD THEORY*

BY

I. T. TODOROV
Laboratory of Theoretical Physics
Joint Institute for Nuclear Research

ABSTRACT

The paper presents a review of conformal covariant quantum field theory with anomalous dimensions. An emphasis is made on the Euclidean formulation of conformal invariance and skeleton perturbation theory.

CONTENTS

---
*Lecture Notes given at XII. Internationale Universitätswochen für Kernphysik, Schladming, February 5 - 17, 1973.

I. INTRODUCTION

Let me start with a rough scheme which indicates the place of conformal invariant quantum field theory (QFT) among various related theoretical and experimental developments in recent years

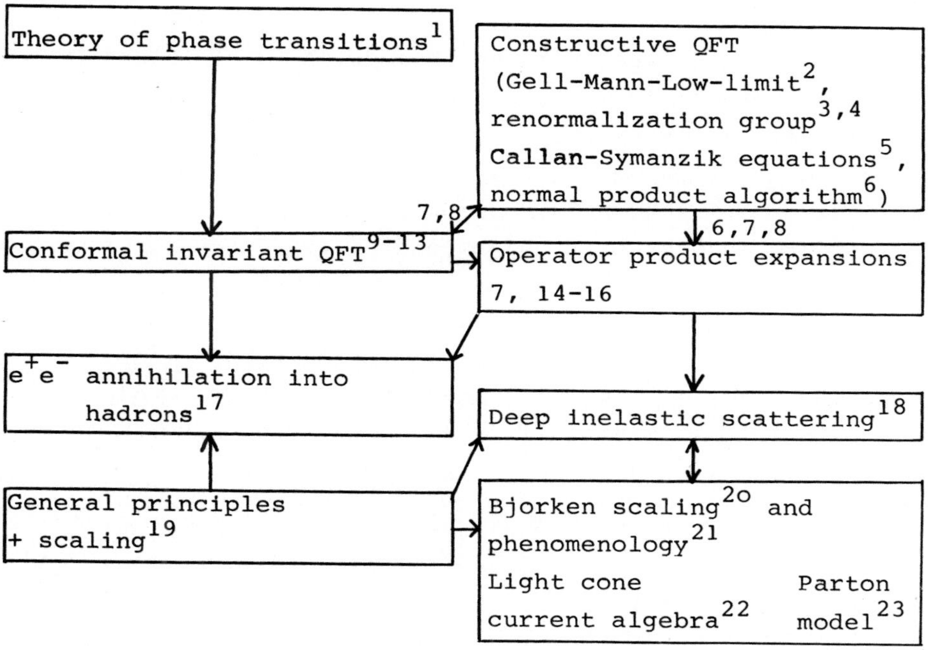

Fig. 1

Scheme of interrelations among various theoretical approaches to the few observed scaling phenomena. (Figures refer to the bibliography at the end of the paper.)

It is noteworthy that some of the ideas of scale

and conformal invariance came from the theory of phase transitions[1]. (There have been several instances in the past of methods of QFT successfully applied in statistical mechanics; we are happy now to have a return.)

Since the conformal group includes dilations

$$x \to \rho x, \qquad p \to \frac{1}{\rho} p, \qquad \rho > 0 \qquad (I.1)$$

it could only appear in a theory with no dimensional parameters (such as masses or dimensional coupling constants). Moreover, nontrivial conformal invariant Green functions exist only if at least some of the fundamental fields have anomalous dimensions[7] (which bar the existence of corresponding free particle states). Thus, any exactly scale invariant theory can give us at best the short distance behaviour of off shell Green functions, not any direct information about on mass shell S-matrix elements. That is why the rectangle of conformal invariant QFT in our scheme is only indirectly connected with the deep inelastic scattering phenomenology.

It is known that global conformal invariance is inconsistent with the causal order of space-time events[24] and violates locality in any nontrivial QFT[25]. There are two equivalent ways out of this difficulty. First, one may restrict oneself to the consideration of infinitesimal special conformal transformations[26, 11, 25] (see Appendix A). Alternatively, one can impose the requirement of global Euclidean conformal invariance on the (analytically continued to pure imaginary times) Schwinger functions[27, 28]. We adopt here this second point of view.

The material is organized as follows. Generalities about Euclidean QFT are summarized in Sec. II.1. Appendix B provides a pedagogical complement to this brief review by considering free fields' operator formalism in Euclidean 4-space. Sec. II.2 defines the conformal transformation properties for Euclidean fields. The infinitesimal field transformation laws in Minkowski space are given in Appendix A.

In Chapter III we derive O(5,1) and γ_5-invariant expressions for the Euclidean Schwinger 2- and 3-point functions. The conformal inversion R is used extensively both in the derivation and in the final presentation of the results. Sec. III.3 deals with the case of S-functions involving conserved current. Ward identity for the 3-point function of the stress energy tensor (and a scalar field) is written down in Appendix A.2. It contains an anomalous Schwinger term (as given in ref. 12).

Chapter IV is devoted to skeleton diagram expansions and bootstrap equations. The proof of absence of ultraviolet and momentum independent infrared divergences in a certain range of anomalous dimension is only sketched since it is presented in all details in the original paper of G. Mack and the author[11]. An emphasis is made on the symmetry $d \leftrightarrow 4-d$ in the Feynman rules. Sec. IV.1 and Appendix C review Symanzik's work on conformal invariant calculations.

The last Chapter V contains a brief survey of several problems related to conformal invariant QFT. A general discussion of results and prospects is given in Sec. V.4.

II. EUCLIDEAN QUANTUM FIELD THEORY AND CONFORMAL INVARIANCE

II.1. Synopsis of Euclidean QFT

In the words of Symanzik[27] "Euclidean QFT is ordinary (relativistic local) QFT looked at from a particular angle, namely, 90° in the complex time plane".

Consider the time-ordered Green functions

$$\tau(x_1,\ldots, x_n) = \langle T\phi_1(x_1)\ldots\phi_n(x_n)\rangle_0 \qquad \text{(II.1)}$$

of n local fields $\phi_1,\ldots\phi_n$ (not necessarily different). The functions

$$\tau_z(\underline{x}_1,t_1,\ldots,\underline{x}_n,t_n) = \tau(\underline{x}_1,zt_1,\ldots,\underline{x}_n,zt_n),\ n = 2,3\ldots, \qquad \text{(II.2)}$$

defined for real z and reducing to (II.1) for z = 1, admit analytic continuation for noncoinciding arguments ($x_a \neq x_b$ for $a \neq b$) in the lower half plane z (as a consequence of the spectrum condition). The Schwinger s-functions are defined by

$$s(\underline{x}_1,-t_1,\ldots,x_n,-t_n) = \tau_{-i}(\underline{x}_1,t_1,\ldots,\underline{x}_n,t_n). \qquad \text{(II.3)}$$

We shall denote the Euclidean 4-vector ($\underline{x}$, -t) again by x; since the component $x_4 = -t$ of this vector corresponds to the pure imaginary time-component $x_0 = -it$ of the Minkowski vector x, we can write $x_4 = -ix_0$.

The s-functions are invariant under the Euclidean group $E(4) = T_4.SO(4)$, T_4 being the (normal) subgroup of translations in four space. They are analytic functions of $x_{a\mu}$ for noncoinciding arguments. (This analyticity should be constrasted with the corresponding properties of Wightman functions or τ-functions in Minkowski space which are only boundary values of analytic functions but not themselves analytic for real arguments.)

The s-functions could also be considered as analytic continuation of Wightman functions. We have,

$$s(\underline{x}_1,\tau_1,\ldots,\underline{x}_n,\tau_n) = w(\underline{x}_1,-i\tau_1,\ldots,\underline{x}_n,-i\tau_n),$$
$$\lim_{\varepsilon_a-\varepsilon_{a+1}\,0} s(x_1,-\varepsilon_2+it_2)\ldots,\underline{x}_n,-\varepsilon_n+it_n) = w(\underline{x}_1,t_1,\ldots\underline{x}_n,t_n). \tag{II.4}$$

As an example consider the 2-point s-function for a free (real or complex) scalar field ϕ of mass m. We have

$$\langle T\,\phi(x)\phi^*(y)\rangle_0 \rightarrow s(x,y) = s(x-y,0)$$

where

$$s(x,0)=\Delta_+(x)\equiv\int\frac{e^{ipx}}{m^2+p^2}(dp)=\frac{1}{(4\pi)^2}\int_0^\infty\frac{d\alpha}{\alpha^2}\exp(-\alpha m^2-\frac{x^2}{4\alpha})=\frac{m}{4\pi^2\sqrt{x^2}}\cdot$$

$$\cdot K_1(m\sqrt{x^2}), \tag{II.5}$$

$$\Delta_+(x)\approx\frac{1}{4\pi^2x^2} \quad \text{for} \quad m\sqrt{x^2}\rightarrow 0;$$

here

$$px = \underline{px} + p_4x_4, \qquad (x^2=\underline{x}^2+x_4^2), \qquad (dp)=\frac{d^4p}{(2\pi)^4}. \tag{II.6}$$

We note that both the function s(x,0) and its Fourier transform are positive. This property also holds in a theory of interacting spinless fields and allows a generalization to the n-point functions (see Symanzik[27]. and references therein).

In the case of a free Dirac field we have

$$<T\psi(x)\tilde{\psi}(y)>_0 \to s_{\psi\tilde{\psi}}(x,y) = \int \frac{m+i\not{p}}{m^2+p^2} e^{ip(x-y)}(dp) = -s_{\tilde{\psi}\psi}(y,x) \equiv S_+(x-y),$$

$\not{p} = \gamma p$.

(II.7)

Concluding this section we shall write down the transformation properties of Schwinger functions under the discrete transformations P(or I_s)$\cdot$C and T. We shall derive the rules for the simple case of a complex scalar field, and shall only present the results for a Dirac field.

Consider the general τ-function of a spinless complex field $\phi(x)$,

$$\tau(x_1,\ldots,x_k;y_1,\ldots,y_\ell) = <T\phi(x_1)\ldots\phi(x_k)\phi^*(y_1)\ldots\phi^*(y_\ell)>_0 \ .$$

(We will not use charge conservation in the subsequent discussion, so that we can allow for $k \neq \ell$.) Weak locality and spectral conditions imply TCP-invariance which says that

$$\tau^+(-y_1,\ldots,-y_\ell;-x_1,\ldots,-x_k) = \tau(x_1,\ldots,x_k;y_1,\ldots y_\ell) \ ,$$

where τ^+ is by definition the expectation value of the antichronological product $T^+\phi(-y_1)\ldots\phi^*(-x_k)$. Assuming P-invariance and introducing the variable z (II.2) we obtain

$$\tau_z(x_1,\ldots,x_k;y_1\ldots,y_\ell) = \overline{\tau^+_{-\bar{z}}(y_1,\ldots,y_\ell;x_1\ldots,x_k)} \ .$$

Continuing analytically to z = - i we find

$$s(x_1,\ldots,x_k;y_1,\ldots,y_\ell) = \overline{s(y_1,\ldots,y_\ell;x_1,\ldots,x_k)} \ .$$

A straightforward application of charge conjugation invariance gives

$$s(x_1,\ldots,x_k;y_1,\ldots,y_\ell) = s(y_1,\ldots,y_\ell;x_1,\ldots,x_k).$$

Finally, if T (or CP) is valid, we obtain the reality condition

$$s(x_1,\ldots,x_k;y_1,\ldots,y_\ell) = \overline{s(x_1,\ldots,x_k;y_1,\ldots,y_\ell)}.$$

The corresponding properties for the 2n-point function

$$s(x_1,\ldots,x_n;y_1,\ldots,y_n) \leftrightarrow \langle T\psi(x_1)\ldots\psi(x_n)\tilde{\psi}(y_1)\ldots\tilde{\psi}(y_n)\rangle_o$$

for a pair of Dirac fields ψ and $\tilde{\psi}$ read:

$$P: s(x_1,\ldots,x_n;y_1,\ldots,y_n) = \underbrace{\gamma_5\otimes\ldots\otimes\gamma_5}_{n}\, s^*(y_1,\ldots,y_n;$$

$$x_1,\ldots,x_n)\ \underbrace{\gamma_5\otimes\ldots\otimes\gamma_5}_{n},$$

$$(\gamma_5^*=\gamma_5),$$

$$C: s(x_1,\ldots,x_n;y_1,\ldots,y_n) = C^{-1}\otimes\ldots\otimes C^{-1}\ {}^t s(y_1,\ldots y_n;x_1\ldots,x_n)\cdot$$

$$\cdot\, C\otimes\ldots\otimes C,$$

where the superscript ^{t} in ^ts stands for transposition.

II.2. CONFORMAL COVARIANT EUCLIDEAN FIELDS

In order to describe the transformation properties of the Schwinger functions it is convenient to write them down as expectation values of products of auxiliary Euclidean (E) fields and to postulate the transformation law for each field. Such fields do in fact exist and can be defined through the Gel'fand-Segal-Wightman-construction (using the Osterwalder-Schrader positivity property of the s-functions). We have reviewed a simple example of free E-fields in Appendix B. For the time being, however, we need not go into this problem since we shall only use here the E-fields as a short-hand for writing down s-function transformation properties.

We shall postulate the maximal possible symmetry for a scale invariant (massless) EQFT. Apart from internal symmetries (e.g. various charge conservations) it consists of O(5,1) invariance (including reflections) and of γ_5-invariance. We shall assume throughout invariance of the Euclidean vacuum $|0>_E$ under all these transformations. Combined with the E-fields transformation law this will ensure the covariance of the s-functions.

Before setting the field transformation rules we note that the special conformal transformation*

$$x'_\mu = \frac{x_\mu + c_\mu x^2}{1+2cx+c^2x^2} \qquad \text{(II.8)}$$

can be written down as a superposition of two conformal inversions R,

$$(Rx)_\mu = - \frac{x_\mu}{x^2} \qquad \text{(II.9)}$$

and a translation $(T_{-c}x)_\mu = x_\mu - c_\mu$:

$$x' = RT_{-c}\, Rx \ . \qquad \text{(II.10)}$$

---

*Eq.(II.8) is obtained from the corresponding Minkowski space transformation law (A.1) by noticing that the signs of the scalar products are reversed in the transition to Euclidean coordinates.

(The inversion R defined by (II.9) differs in sign from the corresponding transformation, say $R^{(11)}$, of ref.11. If we use the (proper) Euclidean transformation $I_4x=-x$, then $R = I_4R^{(11)}$.) Thus, it is sufficient to verify E(4), dilatation and R-covariance, in order to ensure the full O(5,1) covariance. (We note that space reflection is connected to R-inversion by a continuous SO(5,1) transformation, and hence, need not be considered separately.)

Now, we are ready to give the transformation rules for the E-fields.

The E(4) transformation law looks identical to the Poincaré covariance law (A.3). The only difference is that the finite dimensional representation $V(\Lambda)$ in the right-hand side (RHS) of (A.3) is unitary in the Euclidean case (just as well as $\Lambda \in O(4)$). The representation of scale transformations is identical to the one in Minkowski space given by (A.6). It remains to write down the rules for the conformal inversion in Euclidean metrics. We postulate the following R-covariance laws for a (pseudo) scalar, a spinor and a vector field of dimension d:

a) If $\phi(x)$ is a spinless E-field transforming under space reflection I_s either as a scalar or as a pseudoscalar

$$U(I_s)\phi(x)U^{-1}(I_s) = \varepsilon_s\, \phi(-\underline{x},x_4) = \varepsilon_s\, \phi(I_s x),$$
$$\varepsilon_s = \pm 1 \qquad \text{(II.11)}$$

then we set

$$U(R)\phi(x)U^{-1}(R) = \frac{\varepsilon_s}{(x^2)^d}\, \phi(Rx). \qquad \text{(II.12)}$$

b) For a Dirac E-field we set

$$U(I_s)\psi(x)U^{-1}(I_s) = i\eta_s\gamma_4\psi(I_s x),$$

$$U(I_s)\tilde{\psi}(x)U^{-1}(I_s) = -i\bar{\eta}_s\tilde{\psi}(I_s x)\gamma_4, \quad \eta_s^4 = 1; \tag{II.13}$$

$$U(R)\psi(x)U^{-1}(R) = \eta_s \frac{\not{x}}{(x^2)^{d+1/2}}\psi(Rx),$$

$$U(R)\tilde{\psi}(x)U^{-1}(R) = \bar{\eta}_s \tilde{\psi}(Rx)\frac{\not{x}}{(x^2)^{d+1/2}} . \tag{II.14}$$

The transformation law (II.14) is fixed by assuming (II.13) up to a sign ambiguity (but no free i factors!). It can be deduced using the O(5,1) version of the manifestly covariant formalism of refs. 9,11, and observing that R corresponds to a reflection of the first five coordinates on the cone

$$\sum_{a=1}^{5} \xi_a^2 = \xi_b^2 .$$

c) For a vector or an axial-vector current $J_\mu(x)$ such that

$$U(I_s)J_\mu(x)U^{-1}(I_s) = \varepsilon_s(I_s J)_\mu(I_s x) \tag{II.15}$$

where $\varepsilon_s = 1$ for a vector, $\varepsilon_s = -1$ for an axial vector, we set

$$U(R)J_\mu(x)U(R)^{-1} = \varepsilon_s \frac{V_{\mu\nu}}{(x^2)^d} J_\nu(Rx) , \tag{II.16}$$

where

$$V_{\mu\nu} = V_{\mu\nu}(x) = \delta_{\mu\nu} - 2\frac{x_\mu x_\nu}{x^2}, \quad (V^2)_{\mu\nu} = V_{\mu\sigma}V_{\sigma\nu} = \delta_{\mu\nu}. \tag{II.17}$$

We leave it to the reader to verify that if the "normal products"

$$J_\mu(x) = :\tilde{\psi}(x)\gamma_\mu\psi(x): \quad \text{and} \quad J_{5\mu}(x) =:\tilde{\psi}(x)\gamma_\mu\gamma_5\psi(x):$$

exist in some sense (and $d_\psi = \frac{d}{2}$), then $J_\mu(x)$ and $J_{5\mu}(x)$ satisfy (II.16)(as well as (II.15)) with $\varepsilon_s = 1$ and $\varepsilon_s = -1$, respectively.

Having defined the R-inversion for the basic representations it is easy to extend it to higher rank spinors or tensors. For instance, if $\theta_{\mu\nu}$ is a conformal covariant second rank tensor of dimension d then

$$U(R)\theta_{\mu\nu}(x)U(R)^{-1} = \frac{1}{(x^2)^d} V_{\mu\mu'} V_{\nu\nu'}\theta_{\mu'\nu'} \tag{II.18}$$

where $V_{\mu\nu}$ is given by (II.17).

The discrete γ_5-transformation is defined in the following way for the basic E-fields

$$\psi \to i\gamma_5\psi \quad , \quad \tilde{\psi} \to i\tilde{\psi}\gamma_5 \quad (\gamma_5^2 = 1)$$

$$\phi \to -\phi \quad , \quad J_\mu \to J_\mu \quad . \tag{II.19}$$

We remark that the transformation laws (II.13) (II.14) and (II.18) for $\tilde{\psi}$ are not a consequence of the corresponding transformation laws for ψ, since ψ and $\tilde{\psi}$, considered as E-fields, cannot be obtained from one another through simple Dirac conjugation.

We further notice that the existence of conformal inversion among the symmetry transformations makes it natural to perform a stereographic projection of the Euclidean space R^4 and to regard the E-fields as defined on a 5-dimensional hypersphere[29].

III. CONFORMAL COVARIANT SCHWINGER FUNCTIONS

III.1. O(5,1) covariant 2-point functions

The 2-point function of a single (pseudo-)scalar field is determined from Euclidean and scale invariance, or briefly, DE(4)-invariance, alone. In order to appreciate the role of conformal invariance in this simplest case, we consider a pair of different scalar fields ϕ_1 and ϕ_2 whose dimensions d_1 and d_2 a priori need not coincide. From DE(4)-invariance **it** follows that

$$s_{12}(x_1,x_2) = \langle\phi_1(x_1)\phi_2(x_2)\rangle_E = C\,\frac{1}{x_{12}^{\,d_1+d_2}} \,,$$

$$x_{12} = |x_1-x_2| = \sqrt{(x_1-x_2)^2} \,. \qquad \text{(III.1)}$$

On the other hand, according to (II.12), conformal inversion invariance gives

$$\frac{1}{|x_1|^{2d_1}\,|x_2|^{2d_2}}\, s_{12}\left(\frac{x_1}{x_1^2}, \frac{x_2}{x_2^2}\right) = s_{12}(x_1, x_2) \,.$$

Using that $|Rx_1-Rx_2| = |x_1|^{-1}\,|x_2|^{-1}\,x_{12}$ we find

$$x_{12}^{\,-d_1-d_2} = x_{12}^{\,-d_1-d_2}\,\left|\frac{x_2}{x_1}\right|^{d_1-d_2} \,.$$

This is only possible for

$$d_1 = d_2 \,. \qquad \text{(III.2)}$$

Thus we obtain, as a consequence of conformal invariance, the conservation of scale dimension.

The choice of the normalization constant C is arbitrary except for d = 1 where it is fixed ($C = (2\pi)^{-2}$) by the condition that the residue to the pole of the s-function in momentum space is one. In general, it is convenient (although by no means necessary) to choose the normalization in such a way that

$$s_d(x_1,x_2)=\langle\phi(x_1)\phi(x_2)\rangle_E= \frac{\Gamma(d)}{(4\pi)^2}\left(\frac{4}{x_{12}^2}\right)^d \equiv \Delta_d(x_1-x_2). \quad \text{(III.3)}$$

With this convention the Fourier transform of $\Delta_d(x)$ is given by

$$\tilde{\Delta}_d(p)=\int\Delta_d(x)e^{-ipx}d^4x = \frac{\Gamma(2-d)}{|p|^{2(2-d)}} = \frac{1}{\Gamma(d-1)}\int_0^\infty \frac{\tau^{d-2}d\tau}{\tau + p^2} .$$

(III.4)

The RHS of (III.4) is the Euclidean version of the Källén-Lehmann representation. If for some non-integer d the integral in the RHS diverges, it should be understood via analytic continuation in d. We see that for non-integer d the 2-point function does not correspond to a zero mass particle, but rather to an infraparticle theory in the sense of Schroer[30] with a continuous mass spectrum from zero to infinity. There are no asymptotic free states and hence no on-shell S-matrix in this case.

The most general DE(4)-invariant 2-point function of a pair of Dirac-E-fields ψ and $\tilde{\psi}$ of dimension d' is

$$s_{\psi\tilde{\psi}}(x_1,x_2)=\langle\psi(x_1)\tilde{\psi}_2(x_2)\rangle_E = -\langle\tilde{\psi}(x_1)\gamma_5 \otimes \gamma_5\psi(x_2)\rangle_E =$$

$$= x_{12}^{-2d'}\left(A+B\frac{\not{12}}{x_{12}}\right), \quad \text{(III.5)}$$

where $\not{1}\!\!\not{2} = \not{x}_1 - \not{x}_2$.

The second in the above chain of equalities is a consequence of TCP-invariance of the underlying M-space theory ($\gamma_5^2 = 1$). Using (II.14) we find that the R-invariance condition is

$$A+B\,\frac{\not{1}\!\!\not{2}}{x_{12}} = \frac{\not{x}_1}{|x_1|}\left(A-B\,\frac{x_2^2\not{x}_1-x_1^2\not{x}_2}{|x_1|\,|x_2|\,x_{12}}\right)\frac{\not{x}_2}{|x_2|} = A\frac{\not{x}_1\not{x}_2}{|x_1|\,|x_2|}+B\frac{\not{1}\!\!\not{2}}{x_{12}} .$$

This is only possible for A = 0. Note that the same result follows from γ_5-invariance (II.19). With a suitable choice of the normalization constant B we have

$$s_{\psi\tilde{\psi}}(x,0) = S_{d'}(x) = \frac{\Gamma(d'+1/2)}{(4\pi)^2}\left(\frac{4}{x^2}\right)^{d'}\frac{\not{x}}{|x|} \qquad \text{(III.6)}$$

$$\tilde{S}_{d'}(p) = i\,\Gamma\left(\frac{5}{2}-d'\right)\not{p}\,|p|^{2d'-5} = \frac{i\not{p}}{\Gamma(d-3/2)}\int_0^\infty \frac{\tau^{d'-5/2}}{\tau+p^2}\,d\tau .$$

(III.7)

For canonical dimension, $d' = \frac{3}{2}$, we recover with this choice the s-function for a free 0-mass Dirac field (of (II.7)).

The general DE(4) invariant expression for a real (axial) vector current J_μ of dimension d_J is

$$s_{\mu\nu}(x_1,x_2) = \langle J_\mu(x_1)J_\nu(x_2)\rangle_E = \frac{1}{x_{12}^{2\,d_3}}\left[A\delta_{\mu\nu}+B\,\frac{(12)_\mu(12)_\nu}{x_{12}^2}\right] ,$$

(III.8)

$$(12)_\mu = x_{1\mu} - x_{2\mu} .$$

Imposing R-invariance we find the relation B = -2A. Up to an overall factor which we shall denote by $N_J e^2$ (having in mind application to the electromagnetic current) we obtain the following O(5,1) invariant expression for $s_{\mu\nu}$:

$$s_{\mu\nu}(x,y) = e^2 N_J \Delta_{\mu\nu}(x-y) \qquad \text{where}$$

$$\Delta_{\mu\nu}(x) = \frac{\Gamma(d_J+1)}{(4\pi)^2}\left(\frac{4}{x^2}\right)^{d_J} V_{\mu\nu}(x) \qquad (III.9)$$

and $V_{\mu\nu}(x)$ is given by (II.17);

$$\tilde{\Delta}_{\mu\nu}(p) = [(d_J-1)\delta_{\mu\nu} - 2(d_J-2)\frac{p_\mu p_\nu}{p^2}]\,\frac{\Gamma(2-d_J)}{(p^2)^{2-d_J}}\,. \qquad (III.10)$$

The R-invariance of (III.9) could be verified directly, using the identity

$$V_{\mu\kappa}(x)V_{\kappa\lambda}\left(-\frac{x}{x^2}+\frac{y}{y^2}\right)V_{\lambda\nu}(y) = V_{\mu\nu}(x-y)\,. \qquad (III.11)$$

Similarly, the conformal invariant 2-point function for an antisymmetric tensor field $F_{\mu\nu}(x)$ of dimension d_F is

$$\Delta_{\mu_1\mu_2\nu_1\nu_2}(x) = \langle F_{\mu_1\mu_1}(x)F_{\nu_1\nu_2}(0)\rangle_E = \frac{\Gamma(d_F+1)}{(4\pi)^2}\left(\frac{4}{x^2}\right)^{d_F}\frac{1}{2}\,\cdot$$

$$\cdot\,(V_{\mu_1\nu_1}V_{\mu_2\nu_2} - V_{\mu_1\nu_2}V_{\mu_2\nu_1})\,, \qquad (III.12)$$

$$\tilde{\Delta}_{\mu_1\mu_2\nu_1\nu_2}(p) = \frac{\Gamma(3-d_F)}{(p^2)^{2-d_F}}\,[\frac{1}{p^2}(p_{\mu_1}p_{\nu_1}\delta_{\mu_2\nu_2} + p_{\mu_2}p_{\nu_2}\delta_{\mu_1\nu_1} -$$

$$- p_{\mu_1} p_{\nu_2} \delta_{\mu_2\nu_1} - p_{\mu_2} p_{\nu_1} \delta_{\mu_1\nu_2}) - \frac{1}{2}(\delta_{\mu_1\nu_1} \delta_{\mu_2\nu_2} - \delta_{\mu_1\nu_2} \delta_{\mu_2\nu_1})].$$

(III.13)

We note that the second term in the brackets disappears when we go back to the Wightman function $w_{\mu_1\mu_2\nu_1\nu_2}$ in Minkowski space and let d_F tend to its canonical value ($d_F = 2$):

$$w_{\mu_1\mu_2\nu_1\nu_2}(p) = \frac{2\pi}{\Gamma(d_F-2)} \Theta(p_o)(p_\lambda p^\lambda)_+^{d_F-3} [g_{\mu_2\nu_1} p_{\mu_1} p_{\nu_2} +$$

$$+ g_{\mu_1\nu_2} p_{\mu_2} p_{\nu_1} - g_{\mu_1\nu_1} p_{\mu_2} p_{\nu_2} - g_{\mu_2\nu_2} p_{\mu_1} p_{\nu_1} + \frac{1}{2}(g_{\mu_1\nu_1} g_{\mu_2\nu_2} -$$

$$- g_{\mu_1\nu_2} g_{\mu_2\nu_1}) p^\lambda p_\lambda] \qquad \text{(III.14)}$$

$$\underset{\text{for } d_F \to 2}{\to} 2\pi\Theta(p_o)\delta(p_\lambda p^\lambda)(g_{\mu_2\nu_1} p_{\mu_1} p_{\nu_2} + g_{\mu_1\nu_2} p_{\mu_2} p_{\nu_1} -$$

$$- g_{\mu_1\nu_1} p_{\mu_2} p_{\nu_2} - g_{\mu_2\nu_2} p_{\mu_1} p_{\nu_1}),$$

$(p_\lambda p^\lambda = p_o^2 - \underline{p}^2)$.

Here we have used the notation $\tau_+^\lambda = \Theta(\tau)\tau^\lambda$ and the relation

$$\lim_{\lambda \to 0} \frac{1}{\Gamma(\lambda)} \tau_+^{\lambda-1} = \delta(\tau)$$

(see e.g.[31]). Thus, we recover the well known expression for the 2-point Wightman function of a free electromagnetic field[32]. If we go from the Wightman to the τ-function, then a quasilocal arbitrariness arises from the multiplication of $\Theta(x_o)$ with $w_{\mu_1\mu_2\nu_1\nu_2}(x)$. It is fixed in (III.12) in such a way as to preserve conformal invariance.

We conclude this section with two observations. First, we note that not only was the conformal inversion R the main tool in obtaining the covariant 2-point s-functions, but also the s-space expressions for these functions are given directly in terms of the corresponding finite-dimensional representation of R. This is true for both spinor and tensor functions (cf. (III.6), (III.9) and (III.12)) and was noticed by S.Ferrara, R. Gatto et al. ref. 16.

Our second remark is concerned with the positive definiteness property. Positive definiteness of the metrics in the Hilbert space of physical (M) states implies positivity of the 2-point Wightman distribution $\langle\psi(x_1)\ \psi^*(x_2)\rangle_o$ in momentum space. For a vector (or tensor) field this implies positivity of the corresponding matrix form with respect to the metric $(-g_{\mu\nu})$. For instance, in the case of $w_{\mu\nu}$ corresponding to (III.10) we have to require

$$\xi^\lambda(-g_{\lambda\mu})[(d_J-1)p^2\delta^\mu{}_\nu - 2(d_J-2)p^\mu p_\nu]\xi^\nu =$$

$$= 2(d_J-2)(p\xi)^2 - (d_J-1)p^2\xi^2 \geq 0 \qquad \text{(III.15)}$$

$$(p^2 = p_\lambda p^\lambda = p_o^2 - \underline{p}^2)$$

for every choice of the real M-vector ξ.

On the other hand the distribution

$$\frac{1}{\Gamma(d-1)} \tau_+^{d-2}$$

is only positive for $d \geq 1$ (otherwise the regularization of the corresponding integrals at $\tau = 0$ introduces a subtraction).

A straightforward application of the above rules leads to the following restrictions for the various field dimensions coming from positivity:

$d \geq 1$ (spin 0 field); $d' \geq \frac{3}{2}$ (Dirac field);

$d_J \geq 3$ (vector field); $d_F \geq 2$ (antisymmetric tensor-field).

(III.16)

We note that while for a scalar field the s-function divided by $\Gamma(2-d)$ is positive definite in the wider set of dimensions $d > 0$ n the case of a vector or a tensor field the only dimensions for which both $w_{\mu\nu}$ ... and $s_{\mu\nu}$... are positive are the canonical ones ($d_J = 3$, $d_F = 2$) and then the Γ-factor in the RHS of (III.10) becomes infinite. This explains the necessity for the more intricate positivity condition for Euclidean Green's functions postulated by Osterwalder and Schrader (ref.28).

It is formulated in the following way.

Let $S_+(R^4)$ be the set of tempered distributions with support in the halfspace $x_4 \geq 0$. Then the Osterwalder-Schrader positivity condition for the 2-point Schwinger function says that for any $f \varepsilon S_+(R^4)$, the (finite) matrix

$$\iint \bar{f}(\underline{x},-x_4)\, s_2(x,y)\, f(y)\, d^4x\, d^4y$$

is positive-definite. In order to establish the equivalence of this condition to the positivity of the Fourier transform of the Wightman function we notice that if $f \varepsilon S_+(R^4)$ then

$$\tilde{f}(p) = \int e^{ipx} f(x)\, d^4x$$

is analytic for $\mathrm{Im} p_4 > 0$ (and vanishing for $\mathrm{Im} p_4 \to \infty$). This allows to rotate the p_4 contour of integration (after going to momentum space) thus transforming $\tilde{s}_2$ into $\tilde{w}_2$.

III.2. Covariant 3-point functions[10]

It is in the study of covariant 3-point functions that the condition of γ_5-invariance (II.19) becomes really essential: it reduces the number of covariant structures roughly by half.

First of all we note that there is no γ_5-invariant function of three spinless fields. Thus, the simplest case to consider is the 3-point function of two spinor and one (pseudo) scalar field.

In order to fix the ideas let ψ and $\tilde{\psi}$ be Dirac fields of dimension d' and ϕ be a (neutral) pseudo-scalar field of dimension d. Then from DE(4) invariance we expect the 3-point function

$$s_3(x_1,x_2;x_3) = \langle \psi(x_1)\tilde{\psi}(x_2)\phi(x_3)\rangle_E \qquad \text{(III.17)}$$

to be a superposition of expressions of the type

$$x_{23}^{-2\delta_1}\, x_{13}^{-2\delta_2}\, x_{12}^{-2\delta_3}\, \Big(A + B_1 \frac{\not{x}_{23}}{x_{23}} + B_2 \frac{\not{x}_{13}}{x_{13}} + B_3 \frac{\not{x}_{12}}{x_{12}} +$$

$$+ C_1 \frac{\not{x}_{12}\not{x}_{13}}{x_{12}x_{13}} + C_2 \frac{\not{x}_{12}\not{x}_{23}}{x_{12}x_{23}} + C_3 \frac{\not{x}_{13}\not{x}_{23}}{x_{13}x_{23}} + D \frac{\not{x}_{12}\not{x}_{13}\not{x}_{23}}{x_{12}x_{13}x_{23}}\Big)\gamma_5$$

where $\not{x}_{jk} = \gamma(x_j - x_k)$ (as in (III.5)), $x_{jk} = |x_j - x_k|$, and $\delta_1+\delta_2+\delta_3 = d' + \frac{1}{2}d$. A straightforward application of the requirement of R-invariance (II.12)(II.14) leads to the following relations

$$\delta_1 + \delta_2 = d, \qquad \delta_1 + \delta_3 = \delta_2 + \delta_3 = d' \qquad \text{(III.18)}$$

and

$$A = B_1 = B_2 = C_1 = C_2 = D = 0 \;.$$

On the other hand γ_5-invariance implies that $B_3 = 0$ (for the original amplitude it leads to $B_1 = D = 0$). Thus, we are left with a single invariant structure:

$$s_3(x_1,x_2;x_3) = \frac{g}{\Gamma(4-d'-\frac{1}{2}d)} S_{\frac{1}{2}d}(x_1-x_3)\gamma_5 S_{\frac{1}{2}d}(x_3-x_2)\Delta_{d'-\frac{1}{2}d}(x_1-x_2)$$

(III.19)

where Δ and S are given by (III.3) and (III.6). The RHS of (III.19) can be visualized as a triangular diagram containing two fermion lines with propagators $S_{\frac{1}{2}d}$ and a boson line with propagator $\Delta_{d'-\frac{1}{2}d}$ (see Migdal, ref.10). The equalities (III.18) express then conservation of dimension in each vertex of the diagram.

Similarly, we find the general covariant form of the 3-point function of a spinless charged field $\phi(x)$ (of dimension d) and a vector current $J_\mu(x)$ of dimension d_J, and the same for a Dirac field ψ (of dimension d') and J_μ. The results are

$$s_\mu^\phi(x_1,x_2;x_3) = \langle\phi(x_1)\phi^*(x_2)J_\mu(x_3)\rangle_E =$$

$$= C_\phi \Delta_{d-\delta}(x_1-x_2)\Delta_\delta(x_1-x_3)\Delta_\delta(x_2-x_3)\left[\frac{(13)_\mu}{x_{13}^2} - \frac{(23)_\mu}{x_{23}^2}\right] ;$$

(III.20)

$$s_\mu^{\psi\tilde{\psi}}(x_1,x_2;x_3) = \langle\psi(x_1)\tilde{\psi}(x_2)J_\mu(x_3)\rangle_E =$$

$$= C_1\Delta_{d'-\frac{1}{2}d_J}(x_1-x_2)\ S_{\frac{1}{2}d_J}(x_1-x_3)\gamma_\mu\ S_{\frac{1}{2}d_J}(x_2-x_3) +$$

$$+ C_2\ S_{d'-\delta}(x_1-x_2)\Delta_\delta(x_1-x_3)\Delta_\delta(x_2-x_3)\ \left[\frac{(13)_\mu}{x_{13}^2} - \frac{(23)_\mu}{x_{23}^2}\right] ,$$

(III.21)

where, in both equations, $\delta = \frac{1}{2}(d_J-1)$.

III.3. Conserved currents and Ward identities

The conformal invariant Wightman function for two currents in Minkowski momentum space is (in accord with (III.10))

$$W_{\mu\nu}(p) = N_J e^2 \frac{2\pi}{\Gamma(d_J-1)} \Theta(p_o)(p^2)_+^{d_J-2} [(d_J-1)g_{\mu\nu} - - 2(d_J-2) \frac{p_\mu p_\nu}{p^2}] . \tag{III.22}$$

It satisfies the transversality condition

$$p^\mu W_{\mu\nu}(p) = 0$$

(corresponding to current conservation) if and only if the current dimension is canonical, $d_J = 3$. In this case, however, the Fourier transform (III.10) of $s_{\mu\nu}$ does not have a finite limit. We have the alternative: either first to devide by $\Gamma(2-d_J)$ and then go to the limit $d_J \to 3$ or to give up conformal and even scale invariance and define a regularized expression for $\Delta_{\mu\nu}$ as an "adjoint" homogeneous function" (see Chapt.4, of ref.31)

$$\tilde{\Delta}_{\mu\nu}(p, d_J = 3) = (\delta_{\mu\nu} \; p^2 - p_\mu p_\nu) \ln \frac{p^2}{\tau} . \tag{III.23}$$

$$(\tilde{s}_{\mu\nu}(p) = N_J \; e^2 \; \tilde{\Delta}_{\mu\nu}(p))$$

where τ is an arbitrary constant of dimension of $(\text{mass})^2$. We note that for an appropriate choice of N_J the function $\tilde{s}_{\mu\nu}(p)$ coincides with the large p^2 limit of (B.18) (for $\tau = e^{8/3} m^2 \approx 14.4m^2$). The first possibility leads to $W_{\mu\nu} = 0$ and thus to $j_\mu(x)=0$ (cf.Ref.39).This equality does not seem reasonable for the short distance limit of the current matrixelements since the recent experimental data for e^+e^--annihilation into hadrons[17] indicate that $W_{\mu\nu}$ is fairly well approximated by (III.22) (with $N_J = \frac{1}{12\pi^2}$). Therefore, we shall allow for a violation of scale invariance of the τ-function of two conserved currents.

No such problem arises, however, for the current field (invariant) 3-point functions (III.20) (III.21) provided that the dimensions d(or d') of the fields are not canonical. In this case (for $d_J = 3$) the constants C_ϕ, C_1 and C_2 can be chosen in such a way that the Ward identity,

$$\frac{\partial}{\partial x_{3\mu}} s_\mu(x_1,x_2;x_3) =$$

$$= e[\delta(x_1-x_3)-\delta(x_2-x_3)]\ s(x_1,x_2) \qquad \text{(III.24)}$$

be valid for both (III.20) and (III.21). The proper choice is

$$C_\phi = 2e(4\pi)^2(d-1), \qquad C_1 = e(4\pi)^2(d'-\tfrac{3}{2})(d'-\tfrac{1}{2})\frac{1+C}{2},$$

$$C_2 = e(4\pi)^2\ (d'-\tfrac{1}{2})\ (C+1)$$

where C is another arbitrary constant. With this choice Eq. (III.21) can be rewritten in the form:

$$s_\mu^{\psi\tilde\psi}(x_1,x_2;x_3) = e\Gamma(d'+\tfrac{1}{2})\{\frac{1+C}{2}\ S_+(x_1-x_3)\gamma_\mu\ S_+(x_2-x_3)\ \cdot$$

$$\cdot(\frac{4}{x_{12}^2})^{d'-\frac{3}{2}} + \frac{1-C}{2}\ \frac{\not{x}_{12}}{x_{12}}(\frac{4}{x_{12}^2})^{d'-1}\ [\Delta_+(x_1-x_3)\frac{\overleftrightarrow{\partial}}{\partial x_{3\mu}}\ \Delta_+(x_2-x_3)]\}\ , \qquad \text{(III.25)}$$

where Δ_+ and S_+ are the free fields propagators (given by (II.5) and (II.7) with m = O) and

$$g(x)\overleftrightarrow{\partial}_\mu f(x) = g(x)(\partial_\mu f(x))-(\partial_\mu g(x))f(x).$$

Generalized Ward-Takahashi identities are also satisfied for the conformal invariant 3-point function of two fields and the stress energy tensor $\Theta_{\mu\nu}(x)$ (with $d_\Theta = 4$) (see[12]). They have the advantage of being also applicable for neutral fields. For example, the invariant 3-point s-function of a neutral (pseudo) scalar field ϕ and $\Theta_{\mu\nu}$

$$s_{\mu\nu}(x_1,x_2;x_3)=\langle\phi(x_1)\phi(x_2)\Theta_{\mu\nu}(x_3)\rangle_E =$$

$$= -\frac{4^d\Gamma(d+1)}{8(2\pi)^4}\{x_{12}^{2-2d}\ x_{13}^{-2}\ [\delta_{\mu\nu}(\overleftarrow{\partial}_3\overrightarrow{\partial}_3 - \overleftarrow{\partial}_3^2-\overrightarrow{\partial}_3^2)+\overleftarrow{\partial}_{3\mu}\overleftarrow{\partial}_{3\nu}+$$

$$+ \overrightarrow{\partial}_{3\mu}\overrightarrow{\partial}_{3\nu} - 2(\overleftarrow{\partial}_{3\mu}\overrightarrow{\partial}_{3\nu} + \overleftarrow{\partial}_{3\nu}\overrightarrow{\partial}_{3\mu})]\ x_{23}^{-2} + 3i\pi^2[\delta(x_1-x_3) +$$

$$+ \delta(x_2-x_3)]\ x_{12}^{-2d}\delta_{\mu\nu}\}$$

$$(\partial_3^2 = \sum_{\rho=1}^{4}\partial_{3\rho}\partial_{3\rho}) \qquad \text{(III.26)}$$

satisfied Ward-Takahashi identity with an appropriate Schwinger term (see Eq. (A.19) of Appendix A below).

IV. SKELETON DIAGRAM EXPANSION WITHOUT ULTRAVIOLET DIVERGENCES

IV.1. General form of the conformal covariant 4-point function.

We saw in the preceding chapter that given the fields' dimensions (d,d') conformal (and γ_5) invariance determines the full ("dressed") propagators and 3-point functions up to few multiplicative constants. This is not true for the n-point functions if $n \geq 4$. The reason is that having 4 points, x_1, x_2, x_3, x_4, one can construct two conformal invariant anharmonic ratios

$$h_1 = \frac{x_{13}^2 \; x_{24}^2}{x_{12}^2 \; x_{34}^2} , \qquad h_2 = \frac{x_{14}^2 \; x_{23}^2}{x_{12}^2 \; x_{34}^2} . \tag{IV.1}$$

Any function of these two variables will also be conformal invariant.

Consider as an example the 4-point function of a neutral (pseudo) scalar field $\phi(x)$ of dimension d. On the basis of perturbative calculations Symanzik[33] suggested to write down conformal invariant expressions for s_n as multiple Mellin-Barnes type integrals. For s_4 an expression of this type is

$$s_4(x_1,x_2,x_3,x_4) = \langle\phi(x_1)\phi(x_2)\phi(x_3)\phi(x_4)\rangle_E =$$

$$\Delta_d(x_1-x_2)\Delta_d(x_3-x_4) + \Delta_d(x_1-x_3)\Delta_d(x_2-x_4) + \Delta_d(x_1-x_4) \cdot$$

$$\Delta_d(x_2-x_3) + s^T(x_1,x_2,x_3,x_4), \tag{IV.2}$$

where the truncated (i.e.connected) part s^T of s_4 is given by

$$s^T(x_1,\ldots,x_4) = (x_{12}^2 x_{13}^2 x_{14}^2 x_{23}^2 x_{24}^2 x_{34}^2)^{-\frac{d}{3}} .$$

$$\int_{-i\infty}^{i\infty} d\sigma_1 \int_{-i\infty}^{i\infty} d\sigma_2 K(\sigma_1,\sigma_2)\Gamma(\tfrac{d}{3}+\sigma_1)\Gamma(\tfrac{d}{3}+\sigma_2)\Gamma(\tfrac{d}{3}-\sigma_1-\sigma_2)h_1^{-\sigma_1}h_2^{-\sigma_2}. \tag{IV.3}$$

We can impose consistently a number of desirable properties of s_4 assuming absolute convergence of the integral (IV.6) for positive h_1 and h_2. Locality, i.e. symmetry of s_4 with respect to any permutation of the coordinates, implies that

$$K(\sigma_1,\sigma_2) = K(\sigma_2,\sigma_1) = K(-\sigma_1-\sigma_2,\sigma_2) = K(\sigma_1,-\sigma_1-\sigma_2). \tag{IV.4}$$

The simplest polynomial invariants of the discrete group generated by the substitutions

$$(\sigma_1,\sigma_2) \to (\sigma_2,\sigma_1), \qquad (\sigma_1,\sigma_2) \to (-\sigma_1-\sigma_2,\sigma_2)$$

are $I_1 = \sigma_1^2+\sigma_2^2+\sigma_1\sigma_2$ and $I_2 = \sigma_1\sigma_2(\sigma_1+\sigma_2)$. Eq. (IV.4) implies that K can be considered as a function of these two invariants. As noted in Sec. II.2 T-invariance implies reality of s_4. In terms of K this give

$$\overline{K(\sigma_1,\sigma_2)} = K(\bar\sigma_1,\bar\sigma_2) = K(-\sigma_1,-\sigma_2) . \tag{IV.5}$$

The representation (IV.2) (IV.3) satisfies automatically spectral conditions and cluster decomposition properties[34]. To see the latter we notice that if the Euclidean distances x_{13}, x_{23}, x_{14}, x_{24} are of order R ($R\to\infty$) while x_{12} and x_{34} remain finite (in other words, if we consider two clusters of particles, (1,2) and (3,4)), then

$$s_4(x_1,\dots,x_4) = \Delta_d(x_1-x_2)\Delta_d(x_3-x_4) + O(R^{-\frac{8d}{3}}) . \tag{IV.6}$$

Thus, all Wightman axioms[32,34] except positivity are verified for the above representation, and we are still left with a considerable (functional) freedom. In fact, the particular estimate (IV.6) for the cluster decomposition property shows that (IV.2) (IV.3) does not represent the most general conformal invariant 4-point function, which satisfies the Euclidean version of Wightman axioms (without positivity).

In the following sections we shall attempt to satisfy positivity for the n-point functions by postulating a skeleton diagram expansion which fulfills generalized (off-shell) unitarity relations.

IV.2. Skeleton graph expansion and bootstrap equations

There are two ways which lead to a conformal invariant skeleton diagram expansion.

First, one could start with a renormalizable Lagrangean with a dimensionless coupling, say with the pseudoscalar Yukawa interaction (in M-space)

$$L(x) = \tilde{\psi}(i\partial\!\!\!/-m)\psi + \frac{1}{2}(\partial^{\nu}\phi\partial_{\nu}\phi - \mu^2\phi^2) - g\tilde{\psi}\gamma_5\psi\phi + \text{+ renormalization terms.} \qquad \text{(IV.7)}$$

Assume that the Gell-Mann-Low limit theory[2] (with an infinite field strength and a finite charge renormalization) is realized. Then the short distance behaviour of Green functions is given by dilatation covariant expressions obtained in a skeleton theory with $m = \mu = 0$. It was argued (see Schroer, ref. 6) using the $\lambda\phi^4$ theory as an example that if such scale invariant limit theory does in fact exist, it is automatically also conformal invariant. A consistency condition for such a theory are the renormalized (homogeneous) Schwinger - Dyson equations for the 2- and 3-point functions which will be written below.

The second approach[11,12] which we follow here, starts with the conformal covariant propagators (III.3) (III.6) and 3-point function (III.19) (no reference to any Lagrangean!). We postulate a skeleton diagram expansion with these dressed propagators and vertex function ("skeleton diagram" means a Feynman diagram without self-energy and vertex function corrections). We require that this skeleton expansion satisfies generalized off-shell unitarity condition[35]. It turns out[12] that it is not satisfied identically with respect to the parameters of the theory (g, d, d'), but is equivalent to the renormalized Schwinger-Dyson ("bootstrap") equations mentioned before.

The bootstrap equations for the meson-nucleon and photon (J_μ-) nucleon 3-point functions can be presented graphically in the form[36]

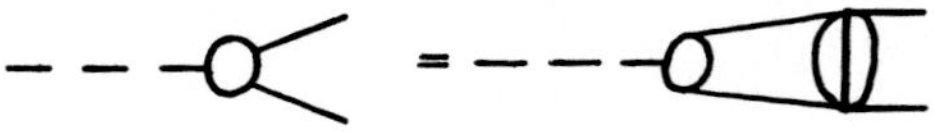

(IV.8a)

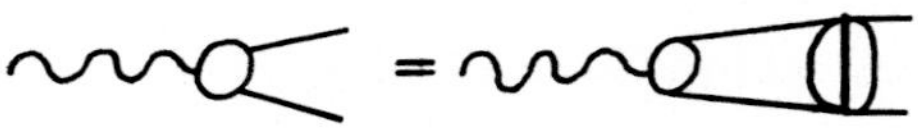

(IV.8b)

Here  stands for the 2-particle irreducible Bethe-Salpeter kernel of the NN-skeleton diagram expansion.

It is important that both sides of eq.(IV.8) are conformal covariant (we shall come back to this point in Sec.IV.4 below). Since the covariant 3-point function is unique (up to a factor) Eqs. (IV.8) reduce to (nonlinear) numerical equations for the parameters g, d and d' of the theory. The second equation, (IV.8b) is to be considered in conjunction with the Ward identity (III.24), instead of the more complicated looking equation for the nucleon propagator. A similar equation can be written down for the 3-point function (III.26) with the stress energy tensor which together with the Ward-Takahashi identity (A 19) is equivalent to the renormalized Dyson equation for the meson propagator[12].

Thus, we have a set of three transcendental equations for the three parameters g, d and d' of the theory. Therefore, we could expect to obtain in principle a theory without any free parameter! The main open theoretical problem of the whole approach is to find an efficient approximation scheme for the actual calculation of all parameters. (At present, the bootstrap equations (IV.8) do not look any easier than the original Gell-Mann-Low[2] or renormalization group[3-5] equations. A perturbative study of such equations for the ϕ^3 theory, in $6 + \varepsilon$ dimensions is given in ref.13).

IV.3. Feynman rules for skeleton diagrams

There are two equivalent ways of writing the Feynman rules for skeleton diagrams. The first possibility is the usual one: we use vertex functions (i.e., amputated 3-point s-functions) and propagators. The second uses full 3-point functions and inverse propagators. It is instructive to compare more closely these two sets of rules, since they exhibit certain symmetry in the space of field dimensions.

First of all, we note that the inverse propagators S^{-1} which satisfy

$$\int S^{-1}(x,z)\,S(z,y)\,d^4z = \delta(x-y) \tag{IV.9}$$

are given by

$$\Delta_d^{-1}(x) = \frac{\Delta_{4-d}(x)}{\Gamma(2-d)\Gamma(d-2)} = \frac{2-d}{\pi}\sin\pi d.\Delta_{4-d}(x),$$

(IV.10)

$$S_{d'}^{-1}(x) = \frac{-S_{4-d'}(x)}{\Gamma(\frac{5}{2}-d')\Gamma(d'-\frac{3}{2})} = \frac{1}{\pi}\sin(d'-\tfrac{1}{2})\pi.S_{4-d'}(x).$$

According to (C.13) (see Appendix C below) the (amputated) vertex function is

$$\Gamma(x_1,x_2;x_3) = \frac{g_\Gamma}{\Gamma(4-\delta_1-\delta_2-\delta_3)} S_{\delta_2}(x_1-x_3)\gamma_5 S_{\delta_1}(x_2-x_3)\Delta_{\delta_3}(x_1-x_2), \tag{IV.11}$$

where $\delta_1 = \delta_2 = 2 - \frac{1}{2}d, \quad \delta_3 = 2 - d' + \frac{1}{2}d$

and g_Γ is proportional to g (see (C.14)). It is related to the complete 3-point function (III.19) by

$$S_3(x_1,x_2;x_3) = \tag{IV.12}$$

$$= \iiint S_{d'}(x_1-y_2)\Gamma(y_1,y_2;y_3)S_{d'}(y_2-x_2)\Delta_d(y_3-x_3)d^4y_1d^4y_2d^4y_3.$$

Now we proceed to the formulation of the two alternative

sets of Feynman rules.

We shall lable by (a) the rules in terms of Γ, Δ_d, $S_{d'}$ and by (b) the rules using S_3, Δ_d^{-1}, $S_{d'}^{-1}$. We shall use identical graphical pictures in both cases. To each dressed vertex function we shall associate a triangular diagram. The diagrams obtained by inserting such "infra-particle triangles" for each vertex of a skeleton graph will be called Migdal diagrams. The two sets of rules are summarized on Fig. 2

	(a)		(b)	
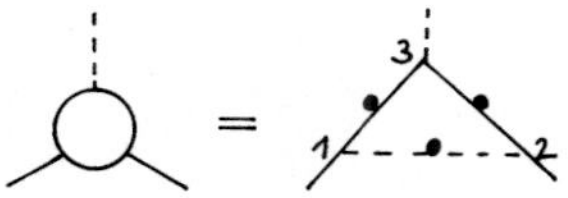	$\Gamma(x_1,x_2;x_3)$	(IV.11)	$S_3(x_1,x_2;x_3)$	(III.19)
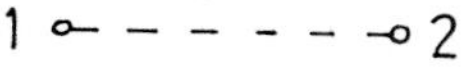	$\Delta_d(x_1-x_2)$	(III.3)	$\Delta_d^{-1}(x_1-x_2)$	(IV.10)
	$S_{d'}(x_1-x_2)$	(III.6)	$S_{d'}^{-1}(x_1-x_2)$	(IV.10)
	$\Delta_{2-d'+\frac{1}{2}d}(x_1-x_2)$		$\Delta_{d'-\frac{1}{2}d}(x_1-x_2)$	
	$S_{2-\frac{d}{2}}(x_1-x_2)$		$S_{\frac{d}{2}}(x_1-x_2)$	
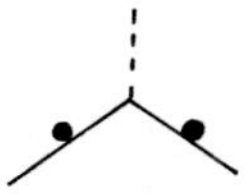	$\dfrac{g_\Gamma \gamma_r}{\Gamma(d' + \frac{d}{2} - 2)}$		$\dfrac{g \gamma_5}{\Gamma(4 - d' - \frac{1}{2}d)}$	

Fig.2. Alternative Feynman rules for Migdal diagrams

The rules (a) reproduce the amputated off-shell amplitudes $\Gamma_n(x_1,\ldots,x_n)$. The rules (b) give the Schwinger (Green) functions $s_n(x_1,\ldots,x_n)$. Γ_n and s_n are related by a 4n-fold integral formula of type (IV.12). Similar rules could be set for the 3-point function with a current or a stress energy tensor (III.26).

It is useful to note that the rules (a) and (b) can be obtained from one another by the substitution

$$d \to 4 - d, \qquad d' \to 4 - d' . \tag{IV.13}$$

This substitution leads to the so-called shadow-representation of the conformal algebra which has the same Casimir operator as the original representation (see Ferrara et al. ref. 16). (The symmetry $d \to 4 - d$ is of the same nature as the well known symmetry $\ell \to -\ell-1$ in the angular momentum plane).

It is also easy to write down the Feynman rules in momentum space. To each continuous (Fermi) line there corresponds a propagator proportional to $\not{p}(p^2)^{-\lambda}$, to each dashed (Bose) line the p-dependent factor is $(p^2)^{-\lambda}$. We have the following rules for the exponents λ:

	$\lambda^{(a)} = \frac{5}{2} - d'$	$\lambda^{(b)} = d' - \frac{3}{2}$
	$\lambda^{(a)} = \frac{d+1}{2}$	$\lambda^{(b)} = \frac{5-d}{2}$
	$\lambda^{(a)} = 2 - d$	$\lambda^{(b)} = d - 2$
	$\lambda^{(a)} = d' - \frac{1}{2}d$	$\lambda^{(b)} = 2 - d' + \frac{1}{2}d$.

(IV.14)

IV.4. Condition for absence of divergences

It is crucial for the consistency of the scheme that there are no ultraviolet (UV) or "catastrophic" (i.e. momentum independent) infrared divergences in the individual skeleton diagrams (at least in a certain range of dimensions d, d'). The importance of this finiteness property comes from the fact that conventional renormalization procedures violate scale (and hence also conformal) invariance (cf.[37]). It is necessary to cover by the argument also vertex functions which appear in the RHS of the bootstrap equation (IV.8) (although no vertex function corrections appear in the skeleton diagrams).

We shall only indicate the main steps of the convergence proof[11].

(i) In order to separate the problem of UV divergences from small momenta singularities, we introduce an infrared cutoff by replacing $(p^2)^{-\lambda}$ in each propagator by $(m^2+p^2)^{-\lambda}$ (with some positive m). Then, we use simple power-counting[38] to show that the contribution of a graph G is convergent, if for any choice of the subgraph $H \subseteq G$ we have

$$\sum_{h \in L(H)} (\lambda_h - 1) > [\tfrac{1}{2}\mu_c]. \qquad \text{(IV.15)}$$

Here λ_h are the exponents associated with individual lines according to (IV.14), L(H) is the set of lines of the subgraph H, $\mu_c = \mu_c(H)$ is the canonical "superficial degree of divergence". If H contains no 3-boson vertices (which is allways the case unless H coincides with the diagram in the RHS of Eq. (IV.8b)), then

$$\mu_c = 4 - B - \frac{3}{2} F \qquad \text{(IV.16)}$$

where B and F are the numbers of external boson and fermion lines of H. Finally, [x] stands for the integer part of x.

In order to verify (IV.15) for all skeleton diagrams it is convenient to express the sum over internal lines on the left-hand side in terms of the parameters of the external lines.

This is achieved by using conservation of dimension law (III.18) which assumes the following form in terms of the parameters λ:

$$\sum_{h \in V} (\lambda_h - 1) = 0 \qquad \text{(IV.17)}$$

where V is the collection of lines incident to a given vertex. It allows to rewrite (IV.15) in the form

$$-\frac{1}{2}\sum_{ext} (\lambda_h - 1) > [\frac{1}{2}\mu_c]. \qquad \text{(IV.18)}$$

The sum on the left is carried over the external lines attached to H. A straightforward but somewhat lengthy analysis[11] of this inequality shows that the infrared-cutoff contribution to any skeleton diagram, as well as to the vertex function graphs in the RHS of (IV.8) is given by an absolutely convergent (Riemann) integral provided that the field dimensions are restricted to the range

$$1 < d < 3 , \qquad d \neq 2, \qquad \frac{3}{2} < d' < \frac{5}{2} . \qquad \text{(IV.19)}$$

Note that this convergence range remains invariant (as expected) under the "shadow transformation" (IV.13).

However, the corresponding x-space integrals are convergent only if we use the set of rules (a) (i.e. Γ, Δ_d and $S_{d'}$) for d, d' < 2 and the set of rules (b) (i.e. s_3, Δ_d^{-1}, $S_{d'}^{-1}$) for d > 2, d' > 2 (see e.g. Appendix A4 of ref. 11 and ref. 33). We could use, however, either set of rules for the whole range (IV.19), defining the x-space integrals, whenever they diverge, through analytic continuation in d and d'.

(ii) The second step consists of removing the infrared cutoff.

Consider an arbitrary connected skeleton diagram G which contributes a term $F_G(p_1,\ldots,p_n;m)\,\delta(p_1+\ldots+p_n)$ to the expansion of $\tilde{s}_n$. The (Euclidean) momenta $p_1,\ldots,p_n$, obeying the conservation law $p_1+\ldots+p_n = 0$, are called exceptional if some partial sum of them vanishes $p_{i_1}+\ldots+p_{i_k} = 0$, $1 \leq k < n$.

One can prove[11] that the limit $m \to 0$ exists for all non-exceptional momenta. The starting point of the proof is the homogeneity condition

$$F_G(p_1,\ldots,p_n;\rho^{-1}m) = \rho^{-\nu}\, F_G(\rho p_1,\ldots,\rho p_n;m) \qquad \text{(IV.20)}$$

where $\nu = \nu(d,d')$ is some function of the field dimensions. We apply Weinberg's power counting theorem[38] in order to estimate the large ρ behaviour of the RHS. From the results of Symanzik[33] even a stronger statement comes out according to which our Green functions are in fact also well defined tempered distributions in the whole momentum space. This is important if we wish to perform back the Fourier transformation of $\tilde{s}_n(p_1,\ldots,p_n)$.

(iii) The last step in our agreement is the proof that in the limit $m \to 0$ in F_G conformal invariance is recovered. This could also be demonstrated by using Weinberg's theorem. However, a more instructive proof is based on the manifestly conformal invariant Feynman rules in six dimensions (see Sec.III.c of the paper of Mack and the author[11]; the manifestly covariant technique goes beyond the scope of the present lecture).

V. RELATED PROBLEMS AND COMMENTS

In this last chapter we shall briefly review several independent from each other topics related to conformal invariant QFT.

V.1. What is conformal invariance good for in massless quantum electrodynamics?

We saw already in Sec. III.3 that if we have a conserved electromagnetic current (with $d_J = 3$) and if its (conformal covariant) 2-point Wightman function

$$w_{\mu\nu}(x,0) = \langle j_\mu(x) j_\nu(0)\rangle_0 =$$

$$= - C \frac{e^2}{\pi^4} \frac{1}{(i0x_0 - x^2)^3} \left(g_{\mu\nu} + 2 \frac{x_\mu x_\nu}{i0x_0 - x^2}\right) \qquad (V.1)$$

($C = 24\pi^2 N_J$, $x^2 = x_0^2 - \underline{x}^2$) does not vanish, then the τ-function of the two currents cannot be defined in a scale invariant way. [The constants are chosen in such a way that in a free quark (or spin 1/2 parton) model in which $j_\mu(x) = e{:}\tilde{\psi}(x)Q\gamma_\mu\psi(x){:}$, where Q is the charge matrix, we have $C = \mathrm{Tr}Q^2$.]

The question arises whether we can avoid such anomalies if we stick to ordinary products (and Wightman functions) and do not use the more singular time ordered products (and τ-functions). The answer is negative: If we make use of the Maxwell equations, in particular

$$\partial^{\mu}F_{\mu\nu}(x) = j_{\mu}(x), \qquad (V.2)$$

and assume conformal invariance of the 2-point Wightman function of the electromagnetic stress tensor $F_{\mu\nu}$, we would have* $d_F = 2$ (because of (V.2)) and using (III.14) we would find that $w_{\mu\nu} = 0$. It follows[39] that $j_{\mu}(x) = 0$.

Nevertheless, the conjecture[40] that the physical value of the fine-structure constant $\alpha = \frac{e^2}{4\pi}$ coincides with the Gell-Mann-Low zero leads to non-trivial consequences. In particular, it turns out that conformal covariant Feynman rules provide a useful tool for calculating single fermion loop contributions to vacuum polarization (see ref. 29). A recent attempt[41] to apply conformal invariance in QED beyond vacuum polarization is rather unconvincing.

It might appear that spontaneously broken conformal symmetry would be a useful concept for gauge theories (see e.g.[42]), but we will not speculate on this here.

---

*A funny way out was suggested to me by K.Wilson (at the Hangl-Bar). One splits the current into two parts. The first one, say j^c_ν, has canonical dimension, $d_c=3$, and is related to the 4-potential, A_μ, by $j^c_\nu = \partial_\nu \partial^\mu A_\mu$. The second part, j^a_ν, is related to the Maxwell Tensor $F_{\mu\nu}$ by (V.2) and is assumed to have anomalous dimension $d_a > 3$. It is automatically conserved (since $F_{\mu\nu}$ is antisymmetric) but is not conformal covariant by itself. Instead, $F_{\mu\nu}$ and j^a_μ form a 10 dimensional non-decomposable representation of the conformal group. The full current j_μ is the sum of j^c_μ and j^a_μ. However, the charge is given as a 3-space integral of the canonical part of the current only:

$$Q = \int j^c_o(x)\, d^3x.$$

V.2. Models

The simplest model of a conformal invariant QFT one can think of is the theory of a generalized free field[34] with 2-point function given by (III.3). The generalized free fields are trivial since all truncated n-point functions for $n \geq 3$ vanish in that case. However, one can obtain non-zero truncated n-point functions by taking the product of two commuting generalized free fields.

Consider as an example the (direct) product of a free zero-mass Dirac field $\psi_o(x)$ with a neutral spinless generalized free field $\phi(x)$ of dimension $d \geq 1$,

$$\psi(x) = \phi(x)\psi_o(x), \qquad \phi = \phi^*, \qquad [\phi(x), \psi_o(y)] = 0 . \tag{V.3}$$

The 2-point function of this field is given by (III.6) with $d' = d+\frac{3}{2}$. All truncated 2n-point functions of $\psi (n \geq 2)$ are nontrivial. For instance, for the truncated 4-point function S^T we can write the following Migdal diagram expansion:

$$S^T_{\psi\psi\tilde{\psi}\tilde{\psi}}(x_1,\dots,x_4) =$$

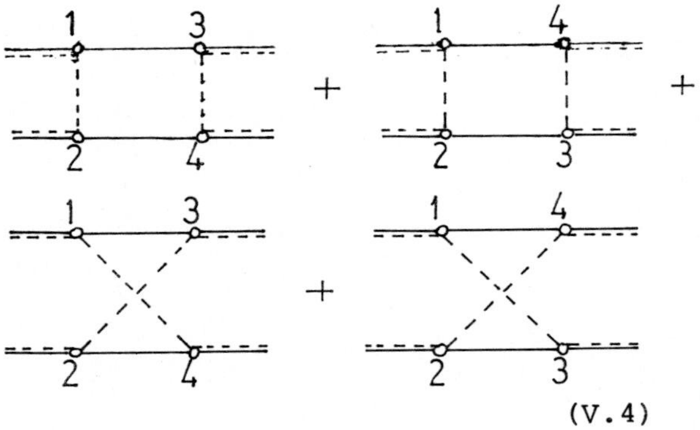

(V.4)

where the dashed line carries the propagator $\Gamma(d+2)\cdot$ $\cdot(\frac{4}{x^2})^d$ of the field ϕ, and the continuous line corresponds to the free Dirac (E-)field propagator

$$S_+(x, m^2 = 0) = \frac{1}{2\pi^2} \frac{\not{x}}{x^4} \, .$$

The electromagnetic current

$$j_\mu(x) = e : \tilde{\psi}_0(x) \gamma_\mu \psi_0(x) : \qquad (V.5)$$

has a non-vanishing 3-point function with ψ:

$$\langle \psi(x_1) \tilde{\psi}(x_2) j_\mu(x_3) \rangle_E =$$ 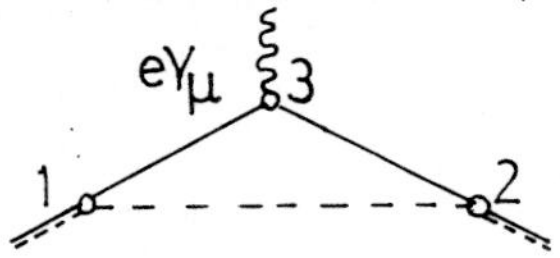

(V.6)

(with the same propagators associated to dashed and continuous lines as in (V.4)).

In this model we also have Bjorken scaling[20] for the virtual Compton scattering. This example indicates that Parisi statement[43] that Bjorken scaling implies free field theory (or, as it is now fashionable to say, parton model) needs some additional assumptions for its validity. (Eq.(4) of ref.43 is violated in our case).

The above model has obviously a trivial nature and it is desirable to have a general requirement on the theory which would automatically discard it. Such is the requirement of local dynamics implemented by a local stress-energy tensor (see Appendix A.2). According to a known result[44] of axiomatic QFT, generalized free fields with not too fast decreasing Källén-Lehmann spectral functions (the power law as in the RHS of (III.4) and (III.7) being included in the admissible class) violate a rather weak requirement of "primitive causality". (The requirement says, that the field algebra generated by smeared fields with test functions with support in a time slice should be irreducible.) This result implies that time translations cannot be generated by a local Hamiltonian, and hence, that no local stress energy tensor exists for our generalized free field $\phi(x)$ (unless $d = 1$). For $d = 1$ we still have conformal invariant Wightman functions (although the 2-point τ-function of ψ is not scale invariant), and we do have an energy momentum tensor $\Theta_{\mu\nu}$. However, we do not have skeleton diagram expansion and generalized unitarity (unless we introduce separately the free constituent fields ϕ and ψ_o which would exhibit the trivial character of the example).

A more sophisticated model with local dynamics was suggested recently by A. Migdal[45]. He defines a scalar field $A(x)$ as a direct product of d commuting free 0-mass fields, $A(x) = \phi_1(x)\ldots\phi_d(x)$ and then performs analytic continuation to non-integer values of d. The model has troubles with spectral conditions and positivity.

One of the most instructive models remains the 2-dimensional Thirring model[46] which served as motivation for the introduction of the important notion of anomalous dimension[7].

V.3. Operator product expansions

There are two ways to look at operator product expansions which lead in fact to quite different problems. The first one is to consider the short distance behaviour of a product of two fields, $A(x_1)\ B(x_2)$. Short distances may mean either that all four coordinates $(12)_\mu = x_{1\mu}-x_{2\mu}$

are small (this is the case of Wilson[7] short distance expansion), or that the Minkowski distance $x_{12}^2 = -(x_1-x_2)^2$ is small (then we have a light-cone expansion of the type considered by Brandt and Preparata and by Frishman[14]). The second approach[15-16, 47] is global. It assumes some (infinite dimensional) algebraic structure of the skeleton theory, valid in the whole 4-dimensional space. Here the assumption of conformal invariance is really essential, since it allows to write down explicit expressions for the "structure functions" of our algebra. Obviously, the second approach is much more restrictive. If we have a global expansion we can go to the limit of short (or near light-like) distances. But we do not need to go into the complications of the global approach if we are only interested in the short distance behaviour.

The basic idea of Wilson short distance expansion can be easily understood from the point of view of conformal invariant n-point functions (cf. Polyakov, ref.18).

Assume, for example, that in the 3-point function (III.17) (III.19) the 4-vector x_1-x_2 goes to 0 while x_3 remains at a finite distance. Then, we will have

$$s(x_1,x_2;x_3) \underset{x_1-x_2\to 0}{\approx} g\,\frac{\Gamma(d'-\frac{1}{2}d)}{\Gamma(4-d'-\frac{1}{2}d)}\,\frac{\Gamma^2(\frac{1+d}{2})}{\Gamma(d)}\left(\frac{4}{x_{12}^2}\right)^{d'-\frac{1}{2}d}\gamma_5\,\Delta_d(x_1-x_3). \qquad \text{(V.7)}$$

We see that at least in this particular matrix element the short distance behaviour of the product of two spinor fields in a theory with a pseudoscalar Yukawa coupling is given by

$$\psi(x_1)\tilde{\bar{\psi}}(x_2) \underset{x_1-x_2\to 0}{\approx} Cx_{12}^{d-2d'} \gamma_5\phi(x_1). \qquad (V.8)$$

It can be argued that the same type of relation remains valid also for other matrix elements. More generally, the RHS of (V.8) should be replaced by a linear combination of operators with different (Lorentz) transformation properties and singular c-number coefficients. Such an expansion was derived[6] under considerably more general premises without assuming conformal invariance. It has found[7, 44] important applications in the analysis of short distance effects (such as the deep inelastic scattering).

The second approach could be viewed as an extension of Lie type models[48] of axiomatic QFT to the case when an infinite number of basic fields is involved. The term "Lie model" is used since the set of fields is assumed to form an (infinite dimensional) Lie algebra under commutation. If the number of basic fields is finite, then it is proven[48] that the model is trivial. (Note that the Lie algebra is still infinite in that case since the space-time coordinates serve as labels of the generators.) There is a hope[47] that for an infinite number of fields a non-trivial algebraic skeleton theory could emerge.

Efremov[48] argues that the Jacobi identity, combined with Wightman axioms, implies the vanishing of the triple commutator of Lie fields. The simple example of the field

$$A(x) = :\phi^*(x)\ \phi(x): \qquad (V.9)$$

where ϕ is a free (charged) 0-mass field, shows that this is not the case for an infinite Lie-field algebra. The infinite Lie algebra, generated by multiple commutators of the form

$$[\ldots[[A(x_1),A(x_2)],\ A(x_3)]\ldots A(x_n)],$$

is spanned by the field A(x), by an infinite set of conserved symmetric tensors:

$$j_\mu(x) = i : \phi^*(x) \overleftrightarrow{\partial}_\mu \phi(x):,$$

$$\Theta_{\mu\nu}(x) = \frac{1}{2}:(\partial_\mu\phi^*(x)\partial_\nu\phi(x) + \partial_\nu\phi^*(x)\partial_\mu\phi(x)): -$$

$$- \frac{1}{2} g_{\mu\nu}:\partial^\lambda\phi^*(x)\partial_\lambda\phi(x): + \frac{1}{6}(\Box g_{\mu\nu} - \partial_\mu\partial_\nu):\phi^*(x)\phi(x):, \text{ etc.} \tag{V.10}$$

and by their derivatives of arbitrary order.

Conventionally, operator product expansions are considered in Minkowski space only. Then, in equalities like (V.8) we have to put $x_{12} = [i(x_{1o}-x_{2o})0 - (x_1-x_2)^2]^{1/2}$ It looks attractive to consider also operator product expansion for E-fields. In that case the image of the light cone is a single point and all fields are strictly commuting or anticommuting (see Appendix B), so that we have to consider only ordinary products. Of course, implications for the real world should be drown only after taking matrix elements from both sides of the expansion and continuing analytically back to Minkowski space.

V.4. Concluding remarks

From a theoretical standpoint the approach reviewed in

these lecture notes offers a new line of attacking the main problem of constructive QFT. Instead of trying to give direct meaning to Lagrangian QFT we start with the following set of axioms.

(i) Wightman axioms[32,34] for a possibly infinite set A of basic fields without the assumption of asymptotic completeness.

(ii) Conformal invariance of n-point functions. This requirement can be stated more precisely as follows.

Each basic field $\psi(x) \varepsilon A$ has a definite scale dimension d (i.e. obeys the transformation law (A.6) under dilatation) and has the property

$$[\psi(0), K_\mu] = 0 , \qquad (V.11)$$

where K_μ are the generators of special conformal transformations (cf. (A.9)). All Wightman functions of basic fields are invariant under dilatations and infinitesimal special conformal transformations (they are also invariant under Poincaré transformations according to (i)).

(iii) The symmetric stress energy tensor $\Theta_{\mu\nu}(x)$ belongs to the set of basic fields. It is conserved and traceless:

$$\partial^\mu \Theta_{\mu\nu}(x) = 0, \qquad \Theta^\mu_\mu(x) = 0, \qquad (V.12)$$

and generates all conformal transformations according to (A.4) (A.7) (A.9) (A.10). (The space integrals in those equations are only required to exist as bilinear forms on the manifold D_0 obtained by acting with polynomials of the smeared fields on the vacuum.)

In addition to these general requirements we add one of the following two alternative hypotheses.

(iva) There exists a skeleton graph expansion for the n-point τ-functions such that the renormalized Schwinger-Dyson equations (IV.8) hold.

(ivb) The basic fields and their derivatives form an (infinite) algebra under commutation of Bose (and Bose and Fermi) fields and anticommutation of Fermi fields.

The two hypotheses look consistent with one another and one can assume both. However, each of them seems sufficiently restrictive so that such an assumption would presumably be superfluous.

The algebraic approach, starting with assumption (ivb), looks particularly attractive. There are however only partial results in this direction[47].

The basic problem in both approaches is the question of self-consistency of the scheme: does a non-trivial example satisfying (i) - (iv) exist? Since the scheme is indeed rather restrictive, if the answer is "no", I hope that we won't have to wait too long to get it.

ACKNOWLEDGEMENTS

I have much learned from our joint work with Gerhard Mack and I take the opportunity to express him my deep gratitude. Stimulating and instructive conversations with S. Ferrara, R. Gatto, A.F. Grillo, A.A.Migdal, G. Parisi, A.M. Polyakov, B. Schroer, D.V.Shirkov and K. Symanzik are also greatfully acknowledged.

It is a pleasure to thank Prof.Dr.L. Streit, Prof.Dr.P. Urban and the University of Graz for their kind hospitality at the 1973 Schladming Winter School.

APPENDIX A

The conformal group of space-time and the field transformation law

A.1. Infinitesimal conformal transformations in Minkowski space.

For the benefit of the reader we summarize here some basic facts about the conformal group in Minkowski space and the fields' transformation laws. For more detail we refer to[9,11].

The conformal group is isomorphic to the 15 parameter group SO(4,2) of pseudorotations in 6 dimension.* As a group in the 4-dimensional Minkowski space M^4 it is compounded by Poincaré transformations, dilatations (of type (I.1)) and special conformal transformations. The special conformal transformations

$$x_\mu \to x'_\mu = \frac{x_\mu - c_\mu x^2}{1 - 2cx + c^2x^2} \qquad \text{(A.1)}$$

do not provide a well defined mapping of M^4 onto itself. Indeed, the denominator in the RHS of (A.1) vanishes on a 3-dimensional hypersurface (a light cone for non-isotropic c, and a light-like hyperplane for $c^2 = 0$). They could be defined as bona fide transformations on a suitable compactification of Minkowski

---

*More precisely, the connected component of the conformal group of space-time is isomorphic to the factor group $SO_0(4,2)/Z_2$, where Z_2 is the centre of $SO_0(4,2)$ (it consists of two elements, 1 and -1, in the 6-dimensional matrix realization of SO(4,2)). I am obliged to L.Castell for this remark.

space[9,11], but we won't go here into this, since these are precisely the transformations which violate the causal order of events, and hence do not seem to have a direct physical significance. (Under transformation of type (A.1) time-like and space-like intervals may exchange places.)

The generators of special conformal transformations, K_μ, and of dilatations, D, obey the following commutation relations among themselves and with the generators $M_{\mu\nu}$ and P_μ of the Poincaré group:

$$[D,M_{\mu\nu}] = 0, \qquad [D,P_\mu] = -iP_\mu, \qquad [D,K_\mu] = iK_\mu,$$
$$[K_\mu,K_\nu] = 0, \qquad [K_\lambda,M_{\mu\nu}] = i(g_{\lambda\mu}K_\nu - g_{\lambda\nu}K_\mu), \tag{A.2}$$
$$[P_\mu,K_\nu] = 2i(D\, g_{\mu\nu} + M_{\mu\nu}) \quad ,$$

where $g_{\mu\nu}$ is the metric tensor in Minkowski space:

$$g_{\mu\nu} = \text{diag}(+---) \, .$$

Let $\psi(x)$ be a Poincaré covariant quantized field, satisfying

$$U(a,\Lambda)\psi^\alpha(x)U^{-1}(a,\Lambda) = V(\Lambda^{-1})^\alpha_\beta \; \psi^\beta(\Lambda x+a) \tag{A.3}$$

where $V(\Lambda)$ is a finite dimensional (in general double valued) representation of the Lorentz group with generators $s_{\mu\nu}$. In infinitesimal form Eq.(A.3) reads

$$[\psi(x),P_\mu] = i\partial_\mu\psi(x),$$
$$[\psi^\alpha(x),M_{\mu\nu}] = [i(x_\mu\partial_\nu - x_\nu\partial_\mu)\delta^\alpha_\beta + (s_{\mu\nu})^\alpha_\beta] \quad \psi^\beta(x); \tag{A.4}$$

$s_{\mu\nu} = 0$ for a scalar field, $s_{\mu\nu} = \frac{i}{4}[\gamma_\mu,\gamma_\nu]$ for a Dirac field, $(s_{\mu\nu})^{\kappa}_{\lambda} = i(\delta^{\kappa}_{\mu} g_{\nu\lambda} - \delta^{\kappa}_{\nu} g_{\mu\lambda})$ for a vector field.

[The γ-matrices in Minkowski space are assumed to satisfy

$$\{\gamma_\mu, \gamma_\nu\} = 2g_{\mu\nu}, \quad \gamma^*_\mu A = A\gamma_\mu \qquad (A = A^*, \ \det A = 1) \tag{A.5}$$

instead of (II.7).]

In general, infinitesimal dilatations and special conformal transformations may also have finite-dimensional matrix counter parts[9] (which is necessarily nilpotent for K_μ). We are assuming throughout that our basic fields transform under such representations for which D and K_μ do not act on the discrete indices. This means that each (basic) field $\psi(x)$ has a definite scale dimension $d = d_\psi$ (in mass units):

$$U(\rho)\psi(x)U^{-1}(\rho) = \rho^d \psi(\rho x), \quad \rho > 0 \tag{A.6}$$

or, in infinitesimal form,

$$[\psi(x), D] = i(d + x^\nu \partial_\nu)\psi(x) \ . \tag{A.7}$$

We shall say that a field ψ, satisfying (A.3) and (A.6), is conformal covariant if it obeys the following infinitesimal conformal transformation law:

$$U(\varepsilon_\mu)\psi(x)U^{-1}(\varepsilon_\mu) \underset{\varepsilon_\mu \to 0}{\sim} \psi(x) - i\varepsilon^\mu[\psi(x), K_\mu], \tag{A.8}$$

where

$$[\psi(x), K_\mu] = i(2dx_\mu + 2x_\mu x_\nu \partial^\nu - x^2 \partial_\mu - 2ix^\nu s_{\mu\nu})\psi(x). \tag{A.9}$$

It was proven in[49] that the infinitesimal operators K_μ, thus defined are essentially self-adjoint. The corresponding global unitary special conformal transformations, however, are given in general by complicated non-local operators (They are implemented by the simple local substitution law (A.1) only in the case of a free field in even space-time dimension, in agreement with the results of ref.25.)

A.2. The stress energy tensor

We shall introduce certain amount of local dynamics in the axiomatic conformal invariant QFT by assuming that the generators of the conformal Lie algebra are given by space integrals of local current densities, related to a conserved (symmetric) stress energy tensor $\Theta_{\mu\nu}(x)$:

$$P_\mu = \int d\sigma^\rho \Theta_{\rho\mu}(x), \qquad M_{\mu\nu} = \int d\sigma^\rho (x_\mu \Theta_{\rho\nu}(x) - x_\nu \Theta_{\rho\mu}(x));$$

$$D = \int d\sigma^\rho\, x^\lambda\, \Theta_{\rho\lambda}(x), \qquad K_\mu = \int d\sigma^\rho\, (2x_\mu\, x^\lambda - x^2 \delta^\lambda_\mu)\Theta_{\rho\lambda}(x) \tag{A.10}$$

where the integration is spread over a space-like (3-dimensional) hypersurface with surface element $d\sigma^\rho$. A canonical choice is given by the plane x_o = const; in this case $d\sigma^\rho = d^3x\delta^\rho_o$. We note that the dilatation current

$$D_\rho(x) = x^\lambda \Theta_{\rho\lambda}(x) \tag{A.11}$$

is conserved only if the stress energy tensor is traceless.

$$\partial^\rho D_\rho(x) = 0 \rightarrow \Theta^\lambda_\lambda(x) = 0 . \tag{A.12}$$

In deducing (A.12) we have used that

$$\partial^\rho \Theta_{\rho\mu}(x) = 0 \tag{A.13}$$

which is the condition of energy momentum conservation.

The integrals in the RHS of (A.10) are only required to converge as bilinear forms on a suitable dense set of state vectors (see, e.g.[50]).

The conformal covariant 2-point Wightman function of $\Theta_{\mu\nu}$ is given by

$$<0|\ \Theta_{\mu_1\mu_2}(x)\Theta_{\nu_1\nu_2}(0)\,|0> = C\,\frac{\Gamma(d_\Theta)}{(4\pi)^2}\left(\frac{4}{i0x_0-x^2}\right)^{d_\Theta} \cdot (V_{\mu_1\nu_1}V_{\mu_2\nu_2}+$$

$$+V_{\mu_1\nu_2}V_{\mu_2\nu_1}-\frac{1}{2}g_{\mu_1\mu_2}g_{\nu_1\nu_2}), \tag{A.14}$$

where in the pseudoeuclidean metric

$$V_{\mu\nu}(x) = g_{\mu\nu} + 2\,\frac{x_\mu\, x_\nu}{i0x_0-x^2} \tag{A.15}$$

(compare with (II.17)). It is consistent with the conservation law (A.13) if $d_\Theta = 4$ (cf.7). This is a special case of a general theorem of Ferrara, Gatto, Grillo and Parisi (ref.15), which states that if a symmetric tensor of rank n is conserved and has a conformal covariant 2-point function, then it has necessarily canonical dimension $d_n = n+2$ $(n = 1,2,\ldots)$.

For canonical dimension ($d_\Theta = 4$) there is no scale invariant time ordered distribution (τ-function), however, which would reduce to (A.14) for $x_0 > 0$. The situation is similar to the case of the s-function of two conserved currents, considered in Sec.III.3 and in Appendix B.1 below.

The 3-point Wightman function of a spinless (neutral) field $\phi(x)$ with $\Theta_{\mu\nu}$ is

$$w_{\mu\nu}(x_1,x_2;x_3) = \langle 0|\phi(x_1)\,\phi(x_2)\,\Theta_{\mu\nu}(x_3)|0\rangle =$$

$$= -\frac{4^d\Gamma(d+1)}{(2\pi)^4}\, x_{12}^{2-2d} x_{13}^{-2}\Big[\frac{(13)_\mu(13)_\nu}{x_{13}^4} + \frac{(23)_\mu(23)_\nu}{x_{23}^4} - \tag{A.16}$$

$$- \frac{(13)_\mu(23)_\nu + (13)_\nu(23)_\mu}{x_{12}^2\, x_{23}^2} + \frac{x_{12}^2}{4x_{13}^2\, x_{23}^2}\, g_{\mu\nu}\Big] x_{23}^{-2}$$

where $x_{jk}^2 = io(x_{jo} - x_{ko}) - (x_j - x_k)^2$.

The normalization constant is chosen in such a way that

$$\delta(x_2^o - x_3^o)\langle 0|\phi(x_1)\,[\phi(x_2),\Theta_{o\nu}(x_3)]\,|0\rangle = \delta(x_2-x_3)\, i\,\partial_{2\nu}\langle 0|\phi(x_1)$$

$$\phi(x_2)\,|0\rangle. \tag{A.17}$$

Because of the singular character of $\Theta_{\mu\nu}$ the corresponding time ordered Green function is not uniquely determined from $w_{\mu\nu}$. Apart from the change of the io prescription in x_{jk}^2 $(io(x_{jo}-x_{ko}) \to i\,o)$ the τ-function may also acquire a quasilocal term. We shall fix this term by demanding the tracelessness condition, $\tau_\mu^\mu = 0$. The result is

$$\tau_{\mu\nu}(x_1,x_2;x_3) = \langle 0|T\phi(x_1)\phi(x_2)\Theta_{\mu\nu}(x_3)|0\rangle =$$

$$= -\frac{4^d\Gamma(d+1)}{(2\pi)^4}\, \{x_{12}^{2-2d}\, x_{13}^{-2}\, [(\overleftarrow{\partial}_3\overrightarrow{\partial}_3 - \overleftarrow{\partial}_3^2 - \overrightarrow{\partial}_3^2)g_{\mu\nu} + \overrightarrow{\partial}_{3\mu}\overleftarrow{\partial}_{3\nu} +$$

$$+ \vec{\partial}_{3\mu}\vec{\partial}_{3\nu} - 2(\overleftarrow{\partial}_{3\mu}\vec{\partial}_{3\nu} + \overleftarrow{\partial}_{3\nu}\vec{\partial}_{3\mu})]\ x_{23}^{-2} -$$

$$- 3i\pi^2\ [\delta(x_1-x_3) + \delta(x_2-x_3)]\ x_{12}^{-2d}\ g_{\mu\nu}\} \tag{A.18}$$

(with $x_{jk}^2 = io - (x_j - x_k)^2$ in this case). This leads to an anomalous Ward identity involving a Schwinger term:

$$\partial_3^\mu \tau_{\mu\nu}(x_1,x_2;x_3) = -i[\delta(x_1-x_3)\vec{\partial}_{1\nu} + \delta(x_2-x_3)\vec{\partial}_{2\nu}]\tau(x_1,x_2) + \\ + \tfrac{i}{4}d\partial_{3\nu}[\delta(x_1-x_3)+\delta(x_2-x_3)]\ \tau(x_1,x_2)\ . \tag{A.19}$$

We emphasize that this anomalous Ward identity is consistent with the correct integrated commutation relations (A.4) (A.7) (A.9) as it is clear from the analysis of the Wightman function (cf.(A.17)).

APPENDIX B

Euclidean free fields

The correspondence between the τ-functions (or the set of all Wightman functions with permuted arguments) and the Euclidean s-functions is unambiguous. However, the relation between fields and states in Minkowski space and their Euclidean counterparts is not so straightforward. It turns out that for a natural realization of the s-functions as expectation values of products of E-space operators the E-fields are not directly related to the conventional fields in space-time but rather to the so-called "hyperquantized" or "quasi" fields[51]. The objective of this Appendix is to illustrate the situation on the simple example of free charged E-fields.

B.1. Spinless charged fields.

Let us start with the spin 0 case.

The only non-vanishing τ-functions of a spinless charged field $\phi(x)$ (in M^4) are those with an equal number of ϕ and ϕ^* :

$$\tau_{2n}(x_1,\ldots,x_n;y_1,\ldots,y_n)=\langle T\phi(x_1)\ldots\phi(x_n)\phi^*(y_1)\ldots\phi^*(y_n)\rangle_o . \tag{B.1}$$

The corresponding s-functions $s_{2n}(x_1,\ldots,x_n;y_1,\ldots y_n)$ satisfy the same type of positivity condition as the Wightman functions in Minkowski (M-)space (see Symanzik, ref.27). Therefore, we can apply a Euclidean version of Wightman reconstruction theorem[52,34]; it enables us to define a Hilbert space H_E of E-state vectors in which there acts a commuting set of E-field operators* $\Phi(x)$, $\Phi^*(y)$ such that

$$S_{2n}(x_1,\ldots,x_n;y_1,\ldots,y_n)=\langle\Phi(x_1)\ldots\Phi(x_n)\Phi^*(y_1)\ldots\Phi^*(y_n)\rangle_E \tag{B.2}$$

($\langle\rangle_E$ stands for Euclidean "vacuum expectation value"; note that there is no time ordering in the RHS).

The fields $\Phi^*(x)$ look at first sight very queer, just because they commute in the whole R^4:

$$[\Phi(x),\ \Phi(y)] = [\Phi(x),\Phi^*(y)] = 0\ . \tag{B.3}$$

---

*To avoid confusion we use different notation for the M-space and for the E-space fields in this Appendix.

The question arises how is it possible to reconcile this exact commutativity with, say, the canonical commutation relations (CCR) of the Minkowski space field $\phi(x)$? Since the question is meaningful already at the level of a free field, we shall look at it in that simple case.

A free charged E-field Φ (with a 2-point function given by (II.5)) can be written in the form

$$\Phi(x) = \Phi_+(x) + \Phi_-^*(x) \tag{B.4}$$

where

$$\Phi_\pm(x) = \int \frac{A_\pm(p)e^{ipx}}{\sqrt{m^2+p^2}}(dp), \quad px = \underline{p}\underline{x} + p_4 x_4, \; (dp) = \frac{d^4p}{(2\pi)^4} \tag{B.5}$$

(and their Hermitian conjugate) play the role of annihilation (and creation) operators in H_E:

$$\Phi_\pm(x)|0>_E = A_\pm(p)|0>_E = 0, \tag{B.6}$$

$$[A_e(p), A_{e'}(p')] = 0,$$

$$[A_e(p), A_{e'}^*(p')] = \delta_{ee'}(2\pi)^4\delta(p-p'), \quad e,e' = \pm . \tag{B.7}$$

It follows from (B.4-7) that

$$[\Phi_e(x), \Phi_{e'}^*(y)] = \delta_{ee'}<\Phi(x)\Phi^*(y)>_E =$$

$$= \delta_{ee'}<\Phi^*(y)\Phi(x)>_E = \delta_{ee'}\Delta_+(x-y) \tag{B.8}$$

where Δ_+ is given by (II.5).

Let us find out how are the CCR reflected onto the Euclidean formalism in this case. A manifestation of the CCR which is appropriate for the 4-dimensional language is provided by the equality

$$K_x T(\phi(x)\phi^*(y)) = \delta(x_o - y_o)[\partial_o\phi(x), \phi^*(y)] = \delta(x-y) \tag{B.9}$$

where K_x is the Klein-Gordon-operator $K_x = m^2 + \Box_x = m^2 + \partial_o^2 - \Delta$.

Now, $\Phi(x)$ and $\Phi^*(y)$ commute identically; hence, so do their derivatives. However, $\Phi(x)$ does not satisfy the Euclidean Klein-Gordon equation: $K^E\Phi(x) \neq 0$, for $K^E = m^2 - \Delta_4$. These two differences between $\phi(x)$ and $\Phi(x)$ sort of compensate each other in the vacuum expectation values. Indeed, using (B.4-7), we obtain

$$\langle K_x^E \Phi(x)\Phi^*(y)\rangle_E = \langle \Phi(x) K_y^E \Phi^*(y)\rangle_E = \delta(x-y) \tag{B.10}$$

in agreement with (B.9).

Another manifestation of the same phenomenon is provided by the current conservation and the Ward identity. The E-picture current $J_\mu(x)$ corresponding to the usual (M-space) current

$$j_\mu(x) = ie{:}\phi^*(x)\overleftrightarrow{\partial}_\mu\phi(x){:} \equiv ie{:}[\phi^*(x)(\partial_\mu\phi(x)) - (\partial_\mu\phi^*(x))\phi(x)]{:} \tag{B.11}$$

is given by

$$J_\mu(x) = e{:}\Phi^*(x)\overleftrightarrow{\partial}_\mu\Phi(x){:} \tag{B.12}$$

(where $:\Phi_1(x)\Phi_2(y): \equiv \Phi_1(x)\Phi_2(y) - <\Phi_1(x)\Phi_2(y)>_E)$.

Unlike j_μ the E-current J_μ is not conserved and commutes with Φ and Φ^*. However, the Ward identity

$$\frac{\partial}{\partial z_\mu} <\Phi(x)\Phi^*(y)J_\mu(z)> = e<\Phi(x)\Phi(y)\cdot$$

$$:((K^E\Phi^*(z))\Phi(z) - \Phi^*(z)K^E\Phi(z)):>_E =$$

$$= e[\delta(x-z) - \delta(y-z)]s(x,y) \qquad (B.13)$$

has the same form as in the M-space picture.

The question arises can we establish some relation between the Euclidean Hilbert space E and the conventional (Minkowskian) Fock space F (before going to the matrix elements). It turns out that F can be identified with a subspace of E generated by polynomials of smeared zero-time creation operators

$$\int \Phi_\pm^*(\underline{x},0)f(\underline{x})d^3x, \qquad f(\underline{x})\varepsilon S(R^3)$$

acting on the Euclidean vacuum $|0>_E$. To see this, we observe that the operators

$$a_\pm(p) = 2p^0 \int A_\pm(p^E) \frac{dp_4^E}{2\pi \sqrt{m^2+(p^E)^2}}$$

and their adjoint, considered as functions on the upper hyperboloid $p^o = \sqrt{m^2 + \underline{p}^2}$ ($\underline{p}^E = \underline{p}$), satisfy the covariant canonical commutation relations

$$[a_\ell(p), a^*_{\ell'}(p')] = (2\pi)^3 2p^o \delta(\underline{p}-\underline{p}') \delta_{\ell\ell'} ,$$

$$p^o = \sqrt{m^2 + \underline{p}^2}$$

(cf. Feldman[28]).

A final remark is due about the singularity structure of s-functions for coinciding arguments. Since the s-functions are an extension of the τ-functions, which are, in general, more singular than the Wightman functions, it is not surprising that in some cases the Wightman functions are well defined (tempered) distributions while the s-functions are not. We shall illustrate the arising complication on the case of the E-space expectation value of the product of two J_μ currents.

According to (B.12) and (B.4-7)

$$S^J_{\mu\nu}(x-y) = \langle J_\mu(x) J_\nu(y) \rangle_E = 2e^2 [\Delta_+(x-y) \partial_\mu \partial_\nu \Delta_+(x-y) -$$

$$- \partial_\mu \Delta_+(x-y) \partial_\nu \Delta_+(x-y)] . \quad \text{(B.14)}$$

The analysis of the small x-y behaviour of the RHS of (B.14) which makes use of the observation that (according to (II.5))

$$\Delta_+(x) \underset{x\to 0}{\sim} \Delta_+(x,m=0) = \frac{1}{(2\pi)^2 x^2} \qquad \text{(B.15)}$$

shows that $S^J_{\mu\nu}(x)$ is not a well defined distribution in the neighbourhood of the origin. It is of course a well defined analytic function away from the origin. However, there is no way to continue it.as a distribution in $S(R^4)$ preserving the homogeneity property

$$S_{\mu\nu}(\rho x,m=0) = \rho^{-6} S_{\mu\nu}(x,0)$$

cf.Chapt.4 of ref.31 and Sec.2 of ref.12).

This reflects the fact that although there exists a scale (and conformal) invariant Wightman function $W_{\mu\nu}$ of two conserved currents the τ-function

$$\tau_{\mu\nu}(x) = \langle Tj_\mu(x) j_\nu(0)\rangle_0$$

cannot be defined in the neighbourhood of the origin in a scale invariant way (provided that $W_{\mu\nu} \neq 0$).

The renormalized expression for $S_{\mu\nu}(x;m)$, satisfying the current conservation condition

$$\partial_\mu S_{\mu\nu}(x,0;m) = 0 \qquad \text{(B.16)}$$

is given by

$$S_{\mu\nu}(x,0;m) = \int (dp) e^{ipx} p^2 (p^2\delta_{\mu\nu} - p_\mu p_\nu) \int_{4m^2}^{\infty} \frac{d\tau}{\tau} \frac{\rho(\tau)}{p^2+\tau} =$$

$$= \frac{e^2}{48\pi^2} \int (p^2\delta_{\mu\nu} - p_\mu p_\nu) e^{+ipx} \Pi(p^2) d^4p \qquad \text{(B.17)}$$

where $\rho(\tau)$ is the Källén–Lehmann spectral function,

$$\rho(\tau) = \frac{e^2}{6\Pi}\left(\frac{\tau-4m^2}{4\tau}\right)^{3/2}$$

and

$$\Pi(p^2) = \left(\frac{4m^2+p^2}{p^2}\right)^{3/2} \ln \frac{2m^2+p^2+\sqrt{(4m^2+p^2)p^2}}{2m^2} - \frac{8}{3} - 8\,\frac{m^2}{p^2} \,. \tag{B.18}$$

B.2. Dirac E-field

For a free Dirac field we set

$$\psi(x) = \int \frac{b_+(p)+b_-^*(-p)}{(m^2+p^2)^{1/4}}\, e^{ipx}(dp) \;, \tag{B.19}$$

$$\tilde{\psi}(x) = \int \frac{b_+^*(-p)-b_-(p)}{(m^2+p^2)^{1/4}}\, \frac{m-i\not{p}}{\sqrt{m^2+p^2}}\, e^{ipx}(dp) \;, \tag{B.20}$$

where $b_\pm^*(p)$ are (4-component!) creation and annihilation operators for an E Fermi field, $b_\pm(p)|0\rangle_E = 0$; they satisfy 4-dimensional anticommutation relations:

$$\{b_e(p), b_{e'}(p')\} = 0, \quad \{b_e(p), b_{e'}^*(p')\} = \delta_{ee'}(2\pi)^4\delta(p-p') . \tag{B.21}$$

In this case $\psi(x)$ and $\tilde{\psi}(x)$ exactly anticommute

$$\{\psi(x), \psi(y)\} = \{\psi(x), \tilde{\psi}(y)\} = 0 \tag{B.22}$$

and

$$\langle\psi(x)\tilde{\psi}(y)\rangle_E = \gamma_5\langle\psi(y)\tilde{\psi}(x)\rangle_E\, \gamma_5 = S_+(x-y) \tag{B.23}$$

where S_+ is given by (II.7). In analogy with (B.10) we have

$$<(m-\not{\partial}_x)\psi(x)\tilde{\psi}(y)>_E = <\psi(x)\tilde{\psi}(y)(m+\overleftarrow{\not{\partial}}_y)>_E = \delta(x-y). \quad (B.24)$$

We note that the hermitean conjugate field ψ^* is not one of the dynamical variables of the Euclidean theory. In fact, it is not even local with respect to $\psi(x)$; we have

$$\{\psi(x),\psi^*(y)\} = \{\tilde{\psi}(x),\tilde{\psi}^*(y)\} = 2\int\frac{e^{ip(x-y)}}{\sqrt{m^2+p^2}}(dp). \quad (B.25)$$

Accordingly, the E-current

$$J_\mu(x) = :\tilde{\psi}(x)\gamma_\mu\psi(x): \quad (B.26)$$

is not Hermitian. It is also not conserved. However, the analytic continuation of its E-matrix elements back to M-space coordinates provides the correct τ-functions (or Wightman functions). In particular, the Ward identity

$$\frac{\partial}{\partial z_\mu}<\psi(x)\tilde{\psi}(y)J_\mu(z)>_E = e[\delta(x-z)-\delta(y-z)]\,S_+(x-y) \quad (B.27)$$

is easily verified (cf. (B.13)).

APPENDIX C

Conformal covariant perturbation theory calculations[33]

In perturbative calculations we encounter integrals of the type

$$\Gamma(x_1,x_2,x_3)=\int\ldots\int S_{d'}^{-1}(x_1-y_1)\,S_{\psi\tilde{\psi}\phi}(y_1,y_2;y_3)\,S_{d'}^{-1}(y_2-x_2)\cdot$$

$$\Delta_d^{-1}(y_3-x_3)d^4y_1 d^4y_2 d^4y_3 = g\frac{(2-d)\sin\pi d}{\pi^3}\sin^2\pi(d'-\tfrac{1}{2})\int\ldots\int$$

$$S_{4-d'}(x_1-y_1)S_{\frac{1}{2}d}(y_1-y_3)\gamma_5 S_{\frac{1}{2}d}(y_3-y_2)\Delta_{d'-\frac{1}{2}d}(y_1-y_2)\cdot$$

$$S_{4-d'}(y_2-x_2)\Delta_{4-d}(y_3-x_3)\prod_{j=1}^{3} d^4y_j . \qquad (C.1)$$

We see that the integrand of each 4-dimensional integral contains three propagators with sum of the indices equal to the number of integrations (four in our case).

C.1. Calculations with spinless propagators in 2D dimensions.

In order to evaluate (C.1) we shall, following[33], first consider the auxiliary integral in 2D dimensions

$$I(x_1\delta_1,x_2\delta_2,x_3\delta_3)=\int\Delta_{\delta_1}(x_1-y)\Delta_{\delta_2}(x_2-y)\Delta_{\delta_3}(x_3-y)d^{2D}y \qquad (C.2)$$

where

$$\delta_1+\delta_2+\delta_3 = 2D . \qquad (C.3)$$

In order to evaluate (C.2) we use the α-representation

$$\Delta_\delta(x) = \frac{4^\delta}{(4\pi)^D}\int_0^\infty e^{-\alpha x^2}\alpha^{\delta-1}d\alpha . \qquad (C.4)$$

Inserting (C.4) in (C.2) and performing the (Gaussian) integration in y we obtain

$$I(x_1\delta_1,x_2\delta_2,x_3\delta_3)=\frac{1}{(2\pi)^{2D}}\int_0^\infty d\alpha_1\int_0^\infty d\alpha_2\int_0^\infty d\alpha_3\frac{\alpha_1^{\delta_1-1}\alpha_2^{\delta_2-1}\alpha_3^{\delta_3-1}}{(\alpha_1+\alpha_2+\alpha_3)^D}.$$

$$\exp\left\{-\frac{\sum_{i<j}\alpha_i\alpha_j x_{ij}^2}{\alpha_1+\alpha_2+\alpha_3}\right\} = \frac{1}{(2\pi)^{2D}}\int_0^\infty\!\!\dots\!\int \prod_{i=1}^{3}\left(\frac{d\alpha_i}{\alpha_i}\,\alpha_i^{\delta_i}\right)\cdot$$

$$\Big(\sum_i \kappa_i\alpha_i\Big)^{-D}\exp\left\{-\frac{\sum_{i<j}\alpha_i\alpha_j\, x_{ij}^2}{\kappa_1\alpha_1+\kappa_2\alpha_2+\kappa_3\alpha_3}\right\} \tag{C.5}$$

where $x_{ij}^2 = (x_i-x_j)^2$ and $\kappa_j > 0$, $\sum\kappa_j > 0$.

The last equality (C.5) follows from the easily verificable identity

$$I = \frac{\Gamma(D)}{(2\pi)^{2D}}\int_0^\infty\!\!\dots\!\int \Big(\prod_{i=1}^{3}\frac{d\alpha_i}{\alpha_i}\,\alpha_i^{\delta_i}\Big)\,\delta\big(1-\sum\lambda_j\alpha_j\big)\Big(\sum_{i<j}\alpha_i\alpha_j x_{ij}^2\Big)^{-D} \tag{C.6}$$

true for every choice of the non-negative numbers λ_j with a positive sum. The independence of the RHS of (C.5) from the choice of κ is a reflection of conformal invariance. It would not be true if the "conservation of dimension law" (C.3) was not fulfilled.

Choosing $\kappa_1 = 1$, $\kappa_2 = \kappa_3 = 0$ we can do the integration in the RHS (C.5) with the result

$$I(x_1\delta_1, x_2\delta_2, x_3\delta_3) = \frac{1}{(2\pi)^{2D}}\Big[\prod_{i=1}^{3}\Gamma(D-\delta_i)\Big](x_{12}^2)^{\delta_3-D}(x_{23}^2)^{\delta_1-D}\cdot$$

$$(x_{13}^2)^{\delta_2-D} = (4\pi)^D\Delta_{D-\delta_1}(x_2-x_3)\,\Delta_{D-\delta_2}(x_1-x_3)\,\Delta_{D-\delta_3}(x_1-x_2)\,. \tag{C.7}$$

C.2. Integrals involving spinor propagators

Going to integrals involving spinor propagators we first note that

$$S_\delta(x) = \not{\partial}\,\Delta_{\delta-\frac{1}{2}}(x) \ . \qquad \text{(C.8)}$$

For a typical integral involving S_δ we can write

$$J(x_1\delta_1, x_2\delta_2, x_3\delta_3) = \int S_{\delta_1}(x_1-y)\,S_{\delta_3}(y-x_3)\,\Delta_{\delta_2}(y-x_2)\,d^{2D}y =$$

$$= \frac{-\not{\partial}_1\not{\partial}_3}{4}\,\frac{1}{(2\pi)^{2D}}\int_0^\infty\!\!\cdots\!\int\Big(\prod_{i=1}^{3}\frac{d\alpha_i}{\alpha_i}\,\alpha_i^{\delta_i}\Big)\,\frac{\exp\{-\frac{\sum\alpha_i\alpha_j x_{ij}^2}{\alpha_1+\alpha_2+\alpha_3}\}}{\sqrt{\alpha_1\alpha_3}\,(\alpha_1+\alpha_2+\alpha_3)^D} \qquad \text{(C.9)}$$

where again the dimensions δ_i satisfy (C.2). Using the identity

$$-\tfrac{1}{4}\not{\partial}_1\not{\partial}_3\exp\{-\frac{\sum_{i<j}\alpha_i\alpha_j x_{ij}^2}{\alpha_1+\alpha_2+\alpha_3}\} = \frac{\alpha_1\alpha_3}{\alpha_1+\alpha_2+\alpha_3}\,[\alpha_2\not{x}_{12}\not{x}_{23}-D-\sum\alpha_i\frac{\partial}{\partial\alpha_i}]\cdot$$

$$\exp\{\ldots\} \quad (\not{x}_{12} = \not{x}_1 - \not{x}_2) \qquad \text{(C.10)}$$

and noticing that the last two terms in the bracket cancel upon partial integration in (C.9) we obtain

$$J(x_1\delta_1, x_2\delta_2, x_3\delta_3) = \not{x}_{12}\not{x}_{23}\; I\,(x_1\delta_1+\tfrac{1}{2}, x_2\delta_2+1, x_3\delta_3+\tfrac{1}{2}) =$$

$$= 4(4\pi)^D\, S_{D-\delta_3}(x_1-x_2)\,S_{D-\delta_1}(x_2-x_3)\,\Delta_{D-\delta_2}(x_1-x_3) \ . \qquad \text{(C.11)}$$

We see at this point that writing down formulae for arbitrary space-time dimension 2D was not just a tribute to mathematical generality: in order to evaluate (C.9) for the physical case 2D = 4 we needed, according to (C.11), to know $I(x,\delta)$ (C.7) for 2D = 6.

Now we are ready to calculate the vertex function (C.1). Integration over y_1 gives

$$\Gamma(x_1,x_2,x_3) = \frac{4g\,\Gamma(\frac{d+1}{2})}{\Gamma^2(\frac{5}{2}-d')\Gamma(d'-\frac{3}{2})\Gamma(d-2)\Gamma(2-d)\Gamma(d'+\frac{1}{2}d-2)} \iint d^4y_2 d^4y_3 \cdot$$

$$S_{2-\frac{1}{2}d}(x_1-y_2)\gamma_5 S_{4-d'}(y_2-x_2)\Delta_{d'+\frac{1}{2}d-2}(y_2-y_3)\Delta_{2-d'+\frac{1}{2}d}(x_1-y_3)\cdot$$

$$\Delta_{4-d}(x_3-y_3)$$

where we have used (C.11) and the identity

$$(4\pi)^2 S_{d'-2}(y_2-y_3)S_{\frac{1}{2}d}(y_2-y_3) = \frac{\Gamma(d'-\frac{3}{2})\Gamma(\frac{d+1}{2})}{\Gamma(d'+\frac{1}{2}d-2)}\Delta_{d'+\frac{1}{2}d-2}(y_2-y_3). \tag{C.12}$$

Applying once more (C.7) and (C.11) we obtain

$$\Gamma(x_1,x_2,x_3) = \frac{g_\Gamma}{\Gamma(d'+\frac{1}{2}d-2)} S_{2-\frac{d}{2}}(x_1-x_3)\gamma_5 S_{2-\frac{d}{2}}(x_3-x_2)\cdot$$

$$\Delta_{2-d'+\frac{1}{2}d}(x_1-x_2) \tag{C.13}$$

with

$$g_\Gamma = \frac{16\,g}{\Gamma^2(\frac{5}{2}-d')\Gamma(2-d)} . \tag{C.14}$$

REFERENCES

1. A. M. Polyakov, ZhETF Pisma Red. 12, 538 (1970) (transl. JETP Lett. 12, 381 (1970)).

2. M. Gell-Mann and F. E. Low, Phys. Rev. 95, 1300 (1954).

3. N. N. Bogolubov and D. V. Shirkov, Introduction to the Theory of Quantized Fields, Moscow, Gostekhisdat, 1957 (transl. Intersc. Publ., New York, 1959).
   I. F. Ginzburg and D. V. Shirkov, ZhETF 49, 335 (1965).
   D. V. Shirkov, Ultraviolet behaviour and finite charge renormalization. Talk at the International Conference on Mathematical problems of Quantum Field Theory and Quantum Statistics, Moscow, Dec., 1972 (to appear).

4. K. Wilson, Phys. Rev. D3, 1818 (1971).
   K. Wilson and J. Kogut, The renormalization group and the ε-expansion, Institute for Advanced Study, Princeton, COO 2220-2 (1972) (to be published in physics Reports).

5. C. G. Callan, Jr., Phys. Rev. D2, 1951 (1970).
   K. Symanzik, Commun. Math. Phys. 18, 227 (1970).

6. W. Zimmermann, Lectures on Elementary Particles and Quantum Field Theory, 1970 Brandeis University Summer Institute in Theoretical Physics, vol. 1 (MIT Press, Cambridge, 1970), pp. 395-589.
   K. G. Wilson, W. Zimmermann, Commun. Math. Phys. 24, 87 (1972).
   B. Schroer, Application of the normal product algorithm to zero mass limits, broken symmetries, and gauge fields, Lecture Notes in Physics, vol. 17, Strong Interaction Physics, International Summer Institute on Theoretical Physics in Kaiserslautern, ed. by W.Rühl and A.Vancura (Springer Verlag, Berlin, 1972) pp.364-4o5.

7. K. Wilson, Phys. Rev. 179, 1499 (1969).

8. K. Symanzik, Commun. Math. Phys. 23, 49 (1971).
B. Schroer, Lett. Nuovo Cimento 2, 867 (1971).

9. G. Mack and Abdus Salam, Ann. Phys. (N.Y.) 53, 174 (1969).
D. G. Boulware, L. S. Brown, R. D. Peccei, Phys.Rev. D2, 293 (1970).

10. A. A. Migdal, Phys. Lett. 37B, 98 and 386 (1971).
G. Parisi and L. Peliti, Lett. Nuovo Cimento 2, 627 (1971).
E. Schreier, Phys. Rev. D3, 980 (1971).
R. Nobili, Conformal covariant Wightman functions, Preprint IFPTH1/72, padova (1972).

11. G. Mack and I. T. Todorov, Preprint IC/71/139, Trieste (1971) and Phys. Rev. D (to be published). This paper contains some additional references. See also I. T. Todorov, Conformal invariant quantum field theory, Dubna report, JINR E2-6642 (1972), Lecture Notes in Physics, Vol. 17, Strong Interaction Physics, International Summer Institute on Theoretical Physics in Kaiserslautern, ed. by W. Rühl and A. Vancura (Springer Verlag Berlin, 1972) pp. 270-299.

12. G. Mack and K. Symanzik, Commun. Math. Phys. 27, 247 (1972).

13. G. Mack, Conformal invariance and short distance behaviour in quantum field theory, Lecture Notes in Physics, Vol. 17, Strong Interaction Physics, International Summer Institute on Theoretical Physics in Kaiserslautern, ed. by W. Rühl and A. Vancura (Springer Verlag Berlin, 1972) pp. 300-374.

14. R. A. Brandt and G. Preparata, Nucl. Phys. B27, 541 (1971).
Y. Frishman, Ann. Phys. (N.Y.) 66, 373 (1971).

15. S. Ferrara, R. Gatto, A.F. Grillo, preprint LNF-71/79. Frascati (1971) (to appear in Springer Tracts in Modern Physics).
S. Ferrara, A. Grillo, R. Gatto, Lett. Nuovo Cim. 2, 1363 (1971).
S. Ferrara, R. Gatto, A. Grillo and G. Parisi, Phys. Lett. 38B, 333 (1972).
V. K. Dobrev, E. H. Hristova, V. B. Petkova, D. B. Stamenov, Symmetric part of the conformal covariant operator product expansion of two spin $\frac{1}{2}$ fields, Lecture at the X-th Winter School of Theoretical Physics, Karpacz, Poland (1973).

16. L. Bonora, Lett. Nuovo Cim. 3, 548 (1972).
S. Ferrara, R. Gatto, A. F. Grillo and G. Parisi, Lett. Nuovo Cim. 4, 115 (1972).

17. B. Bartoli et al., Phys. Rev. D6, 2374 (1972).
About the theoretical prediction of dilatation symmetry for the total cross section of $e^+e^- \rightarrow$ hadrons, see R. Gatto, Riv. Nuovo Cim. 1, 514 (1969).

18. K. Berkelman, Electroproduction and deep inelastic scattering: a look at the final states, Invited talk at the XVI Intern. Conference on High Energy physics, NAL, Batavia, 1972. (Cornell Univ. report CLNS-194). For a theoretical analysis of deep inelastic scattering based on the ideas of a conformal invariant QFT, see G. Mack, Nucl. Phys. B35, 592 (1971) and ref. 13 as well as A. M. Polyakov, Scale invariance of strong interactions and its

application to lepton-hadron reactions, preprint, Landau Institute for Theoretical Physics, Chernogolovka (1972).

19. R. A. Brandt, Phys. Rev. D1, 2808 (1970).
A. Leutwyler and A. Stern, Nucl. Phys. B20, 77 (1970).
R. Jackiw, R. Van Royen and G. B. West, Phys. Rev. D2, 2473 (1971).
N. N. Bogolubov, A. N. Tavkhelidze and V.S.Vladimirov, Dubna reports, JINR P2-6342 and E2-6490 (1972).

20. J. D. Bjorken, Phys. Rev. 179, 1547 (1969).
G. Preparata, Lectures in light cone physics, proceedings on the 7th Finish Summer School in Physics, Loma-Koli, Finland, June-July, 1972, Univ. of Helsinki, 1972, pp. 209-277.

21. V. A. Matveev, R. M. Muradyan and A. N.Tavkhelidze, Particles and Nucleus 2, 7 (1971).
T. D. Lee, Scaling properties in weak and electromagnetic processes, preprint CO-3067 (2-11, New York) (1972).

22. H. Fritzsch and M. Gell-Mann, In: Broken Scale Invariance and the Light Cone, Lectures from the Coral Gables Conference on Fundamental Interactions at High Energy, ed. by M. Dal Gim et al. (Gordon and Breach, New York, 1971), pp. 1-42.
P. Carruthers, Physics Reports, 1C, 1 (1971).
D. J. Gross and S. B. Treiman, Phys. Rev. D4, 2105 (1971).
R. Jackiw, Canonical light cone commutators and their applications, Springer Tracts in Modern Physics, 62, 1, (1972).

23. R. P. Feynman, Phys. Rev. Lett. 23, 1415 (1969);

R. P. Feynman, Photon Hadron Interactions, W. A. Benjamin, New York (to appear);
R. P. Feynman, What neutrinos can tell us about partons, in "Neutrino 72", Europhysics Conference, Balatonfüred, Hungary, June 1972, OMKDK-Technoinform, Budapest (1972), vol. II, pp. 75-100.

24. E. C. Zeeman, J. Math. Phys. 5, 490 (1964).

25. M. Hortacsu, R. Seiler and B. Schroer, Phys. Rev. D5, 2518 (1972).

26. M. Flato and D. Sternheimer, Comptes Rendus Acad. Sc. Paris, 263, A935 (1963).

26a. D. Gross and J. Wess, Phys. Rev. D2, 753 (1970).
M. Flato, J. Simon, D. Sternheimer, Ann. Phys. 61, 78 (1970).

27. J. Schwinger, Phys. Rev. 115, 721 (1959).
T. Nakano, Prog. Theor. Phys. 21, 241 (1959),
E. S. Fradkin, in Quantum Field Theory and Hydrodynamics (Trudy FIAN vol. 29, Moscow, 1965, in Russian), pp. 7-138.
Later development in Euclidean QFT is reviewed in K. Symanzik, Euclidean quantum field theory, in Coral Gables Conference on Fundamental Interactions at High Energy, ed. by T. Gudehus et al. (Gordon and Breach, New York, 1969), pp. 19-31.

28. E. Nelson, Quantum fields and Markoff fields, Summer Institute on Partial Differential Equations, Berkeley (1971).
E. Nelson, Time-ordered operator products of sharp — time quadratic forms. J. Functional Analysis (1972).
E. Nelson, Construction of quantum fields from Markoff fields, J. Functional Analysis (to be

published).
E. Nelson, The free Markoff field, Princeton University preprint (1972).
K. Osterwalder and R. Schrader, Axioms for Euclidean Green's functions, Havard University preprint (1972).
J. Feldman, A relativistic Feynman-Kac formula, Havard University preprint (1972).
K. Osterwalder and R. Schrader, Euclidean Fermi fields and a Feynman-Kac formula for boson-fermion models, Havard University preprint (1972).
The author is indebted to A. Jaffe and B. Simon for acquainting him with the results of this work.

29. S. Adler, Massless, Euclidean quantum electrodynamics on the 5-dimensional unit hypersphere. Preprint NAL-THY-58, Institute for Advanced Study, Princeton (1972).

30. B. Schroer, Fortschr. d. Physik, 11, 1 (1963).

31. I. M. Gel'fand and G. E. Shilov, Generalized Functions, vol. 1 (Academic Press, New York, 1964), (see in particular Sec. 3 of Chapt. 1 and Chapt. 4).

32. A. S. Wightman and L. Garding, Ark. f. Phys. 28, 129 (1965).

33. M. D'Eramo, L. Peliti and G. Parisi, Lett. Nuovo Cim. 2, 878 (1971).
K. Symanzik, Lett. Nuovo Cim. 3, 734 (1972).

34. R. Jost, The General Theory of Quantized Fields (American Math. Soc. Providence, Rhode Island, 1965).
N. N. Bogolubov, A. A. Logunov, and I. T. Todorov, Introduction to Axiomatic Quantum Field Theory (Nauka, Moscow, 1969; Benjamin, New York, to be

published).

35. V. Glaser, A. Lehmann, W. Zimmermann, Nuovo Cim. 6, 1122 (1957).

36. S. F. Edwards, Phys. Rev. 90, 284 (1953).
E. S. Fradkin, Zh. Exp. Teor. Fiz. 29, 121 (1955). transl. JETP 2, 148 (1956).
I. Bialinicki-Birula, Bull. Acad. Pol. Sci. 13, 499 (1965).
In the present context the bootstrap equation was considered by A. A. Migdal, ref. 10.

37. S. Coleman and R. Jackiw, Ann. Phys. (N.Y.) 67, 552 (1971).

38. F. J. Dyson, Phys. Rev. 75, 1736 (1949).
S. Weinberg, Phys. Rev. 118, 838 (1960).

39. F. Strocchi, Phys. Rev. D6, 1193 (1972).

40. S. Adler, Phys. Rev. D5, 3021 (1972).

41. H. J. Schnitzer, Constraints and anomalies in finite quantum electrodynamics, Brandeis University preprint (Oct. 1972).

42. C. J. Isham, Abdus Salam and J. Strathdee, Ann. Phys. (N. Y.), 62, 98 (1971).
B. Zumino, Effective Lagrangians and broken symmetries, In: Lectures on Elementary particles and Quantum Field Theory, 1970 Brandeis University Summer Institute in Theoretical Physics, vol. 2 (MIT Press, Cambridge, 1970), pp. 437-500.

43. G. Parisi, Does Bjorken scaling imply the parton model? Nota Interna 391, Università di Roma (1972).

44. R. Haag and B. Schroer, J. Math. Phys. 3, 248 (1962).

45. A. A. Migdal, 4-dimensional soluble models of conformal field theory, Preprint, Landau-Institute for Theoretical Physics, Chernogolovka (1972).

46. K. Wilson, Phys. Rev. D2, 1473 (1970), J.H.Lowenstein, Commun. Math. Phys. 16, 265 (1970), J. H. Lowenstein and B. Schroer, Phys. Rev. D3, 1981 (1971); G. F. Dell'Antonio, Y. Frishman and D. Zwanziger, Phys. Rev. D6, 988 (1972).
S. Ferrara, R. Gatto, A. F. Grillo, Nota Interna 385, Università di Roma (1972).
M. Gomes, J. H. Lowenstein, Nuclear Phys., B45, 252 (1972).
J. Geicke and S. Meyer, The Callan-Symanzik coefficient in a massive Thirring model in $2+\varepsilon$ dimensions. Preprint Freie Univ. Berlin (1972).
G. Dell'Antonio, Model field theories (see Lectures at this School).

47. A. M. Polyakov, Duality in conformal field theory (to be published).

48. O. W. Greenberg, Ann. Phys. (N.Y.) 16, 158 (1961).
D. W. Robinson, Phys. Lett. 9, 189 (1964).
A. S. Wightman, Introduction to some aspects of relativistic dynamics of quantum fields, Cargèse lectures, 1964.
A. V. Efremov, A model of Lie fields, preprint JINR, P2-3731 Dubna (1968).

49. I. A. Swieca and A. H. Völkl, Remarks on conformal invariance, Preprint Dept. de Fisica, PUC, Rio de Janeiro, Brazil (1972).

50. M. O. Hortacsu, Conformal group in some model field theories; University of Pittsburgh dissertation (1971). See in particular Sec. 5 of Part I.

51. S. Coester, Phys. Rev. 95, 1318 (1954). Yu. A. Gol'fand, Zh. Exp. Teor. Fiz., 28, 140 (1955). Yu. V. Novozhilov, Zh. Exp. Teor. Fiz. 31, 493 (1956).

52. A. S. Wightman, Phys. Rev. 101, 860 (1956).

Acta Physica Austriaca, Suppl.XI, 317–340 (1973)

GAUGE THEORIES AND SUPERSELECTION RULES*

BY

R. STREATER
Bedford College, London

I. There are four absolutely conserved quantum numbers in elementary particle physics; these are Q, the charge, B, the baryon number, L, the lepton number and M, the muon number. The following table shows the values these quantum numbers take in various states.

Particle	Q	B	L	M	Other states in the same sector
$e^{\pm}$	± 1	0	∓ 1	0	$\vert e^{\pm}\gamma\rangle, \vert 2e^{\pm}, e^{\pm}\rangle, \vert e^{\pm}, p, \bar{p}\rangle \ldots$
p	+1	+1	0	0	$\vert p, \gamma\rangle, \vert p \cdot \pi^{0}\rangle, \vert p\Lambda^{0}\bar{\Sigma}^{0}\rangle \ldots$
n	0	+1	0	0	$\vert n, \gamma\rangle, \vert n, K^{0}\rangle, \vert n, \bar{K}^{0}\rangle \ldots$
$\bar{p}$	-1	-1	0	0	$\vert \bar{p}, \gamma\rangle, \vert \bar{p}\ p\ \bar{p}\rangle, \vert \bar{p}\Omega^{-}\ \bar{\Sigma}^{-}\rangle \ldots$
ν_e	0	0	1	0	$\vert e^{-}\ p\ \bar{n}\rangle, \vert e^{-}, K^{0}\rangle, \vert e^{-}, \nu_e, e^{+}\rangle \ldots$
$\bar{\nu}_e$	0	0	-1	0	$\vert e^{+}\bar{p}n\rangle, \vert \bar{\nu}_e\gamma\rangle, \vert e^{+}K^{-}\rangle, \ldots$
$\mu^{\pm}$	± 1	0	∓ 1	∓ 1	$\vert \nu_e e^{+}\nu_{\mu}\rangle$ or $\vert \bar{\nu}_e\ e^{-}\bar{\nu}_{\mu}\rangle$
π^{+}	+1	0	0	0	$\vert \Sigma^{+}\bar{\Lambda}^{0}\rangle, \vert K^{+}\rangle, \vert p\ \bar{n}\rangle$

*Lecture Notes given at XII. Internationale Universitätswochen für Kernphysik, Schladming, February 5 - 17, 1973.

π^-	-1	0	0	0	$\|\pi^-,\pi^0>, \|\pi^-\gamma>, \ldots$
D	1	2	0	0	$\|p,n>$
H_1	0	1	1	0	$\|p,e^->$
H_2	0	2	1	0	$\|p,n,e^->$
U_{238}	0	238	92	0	|state of 92 protons, 92 electrons, 146 neutrons>
....	...	...	...	...	...
Ω	0	0	0	0	$\|e^+e^->, \|p,\bar{p}>, \|\pi^+\pi^->, \|K^0>, \ldots$

Notice that I have only written the absolutely conserved laws, and have omitted the broken symmetry groups such as SU_3. In this way the K^0 and the $\bar{K}^0$, for instance, both have the quantum numbers of the vacuum. All states with the same values of Q,B,L and M are said to lie in the same sector (originally called coherent subspace by Wightman [1]). Let $H_{q,b,\ell,m}$ be such a subspace, where q, b, ℓ and m label eigenvalues of Q, B, L and M respectively. Then $H = \bigoplus_{q,b,\ell,m} H_{q,b,\ell,m}$ is the space of all states. Since Q, B, L and M can be simultaneously measured, they commute and so do the unitary groups $e^{i\alpha Q}$, $e^{i\alpha B}$, $e^{i\alpha L}$, $e^{i\alpha M}$ that they generate. These are called gauge groups (the charge-gauge, the baryon-gauge etc.). The main hypothesis in the theory of superselection rules is that for an operator on H to represent an observable, it must commute with all the gauge groups, that is, must be gauge invariant. Wightman and Strocchi [2] have shown that this follows, at least for the charge gauge, from the usual laws of quantum electrodynamics. From now on, the label q will be used to denote the set of quantum numbers q, b, ℓ, m,... (or whatever numbers occur in the theory) rather than just the electric charge. Any observable must be neutral, that is, must map each H_q into itself. [In this connection it must be said that most of the "currents" discussed in the theories of the algebra of currents are not observable.] For convenience, let us discuss bounded observables, and let A denote the C^*-algebra of operators on H generated by the observables. Then gauge invariance implies that A is reduced by the direct sum $\bigoplus_q H_q$, so that if $\psi = \bigoplus_q \psi_q$, then $A \in A$ can be written in block-diagonal form:

$$A\psi = \begin{bmatrix} A_1 & 0 & & & \\ 0 & A_2 & \ddots & & 0 \\ & & & A_q & \\ & & & & \ddots \\ & 0 & & & \end{bmatrix} \cdot \begin{bmatrix} \psi_1 \\ \psi_2 \\ \vdots \\ \psi_q \\ \vdots \end{bmatrix}$$

Let A' denote the set of bounded operators on H that commute with A; A' is called the commutant of A. In elementary particle physics, A' is generated by $e^{i\alpha Q}$, $e^{i\alpha B}$, $e^{i\alpha L}$, $e^{i\alpha M}$ and is therefore abelian. This led Wightman [3] to his hypothesis of commutative superselection rules, which states that in any quantum theory with superselection rules (so that not every self-adjoint operator is observable), the commutant of the observables should be abelian.

The next step was taken by Haag and Kastler [4]. Let A_q denote the C^*-algebra (on the coherent subspace H_q) generated by the operators A_q, as A runs over A. Haag and Kastler postulate that

1) each A_q is irreducible on H_q.
2) Each A_q is a representation of the <u>same</u> abstract C^*-algebra, which can be taken to be A; the representation can be denoted Π_q, so that $A_q = \Pi_q(A)$ and $A_q = \Pi_q(A)$.
3) For different values of q, the representations are inequivalent, i.e. the reduction of A into irreducibles is <u>multiplicity free</u>.

Given 1) and 2) it can be shown that A' is abelian if and only if 3) holds. Of course, this does not constitute a proof of the hypothesis of commutative superselection rules, but has simply re-expressed it.

Let us define the terms; an (abstract) C^*-algebra is an algebra with a $*$-operation and a norm, $||\cdot||$, such that $||AB|| < ||A||\ ||B||$ and $||A^*A|| = ||A||^2$ for all A, B ε A, and such that A is topologically complete in the norm; the field of numbers used is, of course, the complex field. It can easily be shown that a <u>"concrete" C^*-algebra,</u> namely a norm closed $*$-algebra of bounded operators on a Hilbert space, is an abstract C^*-algebra in this sense. Not every "abstract" algebra is concrete however. (These terms should not be taken literally, since all mathematics is concrete.) A <u>representation</u> of a C^*-algebra A, denoted (Π, H), is a map Π from A to bounded operators on H, such that $\Pi(A^*) = \Pi(A)^*$; $\Pi(AB) = \Pi(A)\Pi(B)$; $\Pi(\lambda A) = \lambda\Pi(A)$ and $\Pi(A+B) = \Pi(A) + \Pi(B)$, for all $\lambda \varepsilon C$ and A,B ε A. Two representations, say Π_q and $\Pi_{q'}$, on spaces H_q and $H_{q'}$, are said to be equivalent if there exists a unitary operator, V, taking H_q to $H_{q'}$, such that $V\Pi_q(A)V^{-1} = \Pi_{q'}(A)$ for all elements A ε A.

The hypotheses of Haag and Kastler provide us with a method for predicting the remarkable superselection structure of the elementary particles. Indeed, just as the Casimir operators of a Lie group provide labels for the different irreducible inequivalent represenatations (the multiplets), so Q, B, L, M provide such labels for different representations of A: they take constant values on each irreducible, and so are the analogues of the Casimir operators of a group. We need "only" to choose the right algebra A, find its representations, reject those with unphysical properties, and we have explained the world!

A few remarks about states will enable us to clarify the concept of "physically realizable" as used in [1] . A <u>state</u> on a C^*-algebra A is an "expectation value" ρ, that is, to each $A \in$ A is assigned its expectation value, $\rho(A)$, which satisfies $\rho(A) \in C$ and

1. $\rho(\lambda A) = \lambda\rho(A)$, $\rho(A+B) = \rho(A) + \rho(B)$
2. $\rho(A^*A) \geq 0$
3. $\rho(1) = 1$.

Following Segal, we shall take it that all the physical properties of a state are known if the expectation value ρ is given for every observable A. A representation (Π, H) defines a set of states, namely those given by vectors $\Psi \in H$: $\rho_\Psi(A) = \langle\Psi, A\Psi\rangle$. Such states are called <u>vector states</u> relative to the representation Π. If A is a concrete C^*-algebra (on a Hilbert space H) then the vectors of H are often said to define vector states of A. Since two vectors Ψ and $e^{i\alpha}\Psi$ that differ by a phase define the same expectation values, they define the same state.

A state ρ is said to be impure if there exist states, ρ_1 and ρ_2, not equal, such that $\rho = \lambda\rho_1 + (1-\lambda)\rho_2$, where $0 < \lambda < 1$. Otherwise it is said to be pure. An interesting construction is the way (classical) stochastic mixtures of states can be made. In the above example, ρ is a mixture of ρ_1 with probability λ and ρ_2 with probability $1-\lambda$. More generally, if (Π, H) is a representation, let σ denote a density matrix on H, that is, an operator with complete set of eigenvectors belonging to non-negative eigenvalues, σ_i, such that $\Sigma\sigma_i < \infty$. Then if $A \in \mathcal{A}$, σA also has finite trace, and the map $A \to \mathrm{tr}(\sigma A)$ defines a state on $\mathcal{A}$, said to be normal relative to Π. If Π is irreducible, then a normal state is pure if and only if it is a vector state (a special case, where σ is a 1-dimensional projection).

The famous Gelfand-Naimark-Segal theorem says that every state ρ is a vector state ρ_Ψ in some representation Π, and that this representation is irreducible if and only if ρ is pure; the connection is given by $\rho_\Psi(A) = \langle\Psi, \Pi(A)\Psi\rangle$.

Let us now return to the theory of superselection rules, and consider the vacuum representation (Π_0, H_0). Any state in a specific sector, H_q, defines a pure state which is not normal relative to Π_0 if $q \neq 0$. Similarly, the states of H_q are not normal relative to $\Pi_{q'}$, if $q' \neq q$. However, if we consider a vector ψ in the total space, $H = \bigoplus_q H_q$, which has compenents in more than one sector H_q, then ψ defines an impure state on $\mathcal{A}$, rather than a pure state. In Wightman's original version, such a ψ was called physically unrealizable. Clearly, a coherent mixture is impossible, and ψ does not represent a coherent mixture, of states of different charge; I think it is better to regard Ψ as possible but mixed state, rather than as not realizable; this point of view allows one to understand approximate symmetries such as SU_2 in terms of gauge groups. If we ignore SU_2 breaking terms, the proton and neutron become identical. Observables must then be invariant under the nonabelian gauge group SU_2 as well as the usual gauges. Thus the observable algebra becomes smaller if we cannot use the electromagnetic and weak interactions. In this way one can, at least approximately, define the strong algebra of observables as the SU_2 invariant subalgebra. This concept will be slightly time-dependent. A state, that is pure relative to the full algebra of observables, might become mixed when restricted to the strong observable subalgebra. Thus the K_{01} and K_{02} states, which are fully coherent with respect to the full algebra of observables, become mixed states relative to the strong algebra alone. This explains why full coherent mixing is possible, even though the observables concern only the weak interaction, and so might be thought to allow only very tiny coherent mixing. Haag, Doplicher and Roberts note that a non-abelian gauge group is associated with the existence of paraparticles. For instance, the nucleon doublet behaves as a parafermion of order 2. Clearly in this case one chooses not to have a multiplicity free representation of A, but some states, e.g. nucleon, are repeated twice. Doplicher and Roberts explain why this is a natural thing to do if the gauge group is not abelian.

II. The Physical Representations

Our programme, then, is to choose A and then find its representations. Put not all representations are going to occur in practice; there are far too many of them. We must stay close to the actual mechanism used in practice. We assume that the algebra A is derived from a field theory, so it will have the usual Haag axioms;

that is, to each open region O of space-time, is associated an algebra $A(O)$ of observables that can be measured in O. A is the C^*-algebra generated by all the $A(O)$. We assume that if O_1 and O_2 are spacelike separated, then $A(O_1)$ and $A(O_2)$ commute, and that P (the Poincaré group) acts on A as transformations preserving the algebraic structure, that is, as <u>automorphisms.</u> An automorphism is a 1:1 map $A \overset{\tau}{\to} A$ such that $\tau(A) = A$ and $\tau(A+B) = \tau(A)+\tau(B)$; $\tau(\lambda A) = \lambda\tau(A)$; $\tau(A^*) = \tau(A)^*$; $\tau(AB) = \tau(A)\tau(B)$. Thus to each $L = (\Lambda,a) \varepsilon P$, there is a τ_L such that it transforms $A(O)$ to $A(O_L)$, where O_L is the L-transform of O. We also require that $\tau_{L_1} \tau_{L_2} = \tau_{L_1 \ L_2}$. The physical interpretation requires that if $O_1 \subset O_2$ then $A(O_1) \subset A(O_2)$. A few further axioms are sometimes put, which we shall not mention.

If A is a C^*-algebra, τ an automorphism and (Π,H) a representation, then we say that τ is <u>implemented in the representation Π</u> if there exists a unitary operator $U: H \to H$, such that $\tau(A) = U A U^{-1}$ for all $A \ \varepsilon \ A$. In the case of a one-parameter family of automorphisms, such as time-evolution, τ_t, it is usual to require strong continuity of the implementing unitaries, U_t, before we consider the representation useful. Thus, we can select, from the myriad of representations of the observables, only those in which the Poincaré automorphisms are implementable continuously. In the others, this symmetry will be spontaneously broken. Moreover, we can restrict the representations even further by requiring that the energy-momentum spectrum lie in the forward light-cone. Such representations are called <u>covariant.</u> The implementability of time evolution in a representation is often related to the question of whether the representation Π contains states of finite energy (relative to the vacuum).

A further restriction on the representations comes from the idea of "the particle behind the moon", due to Haag and Kastler [4]. Suppose we have a theory of observables as a concrete algebra in the vacuum sector (Π_o, H_o) containing the vacuum Ω. This state is used to represent a far more complicated state, represented below:

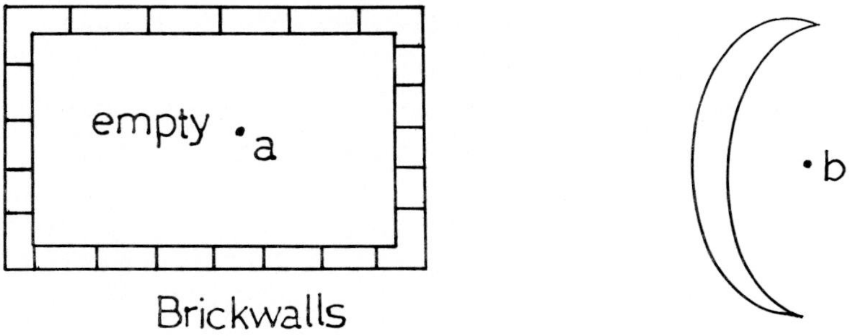

While Ω is a good approximation for the empty space, it is not supposed to describe the walls or the moon. Suppose now, to the Ω we apply an observable which creates an electron in the region of a and a positron in the region b. This would be done by photon-photon collisions with a subsequent separation of charge. Let us call this state $A_o\Omega$. This state should be interpreted as a mere electron state at a, since the positron would never have reached b but for the fact that Ω omitted the walls. Thus we see that the state $A_o\Omega$ is interpreted correctly everywhere except in the neighbourhood of b; here we must change $A_o\Omega$ locally back to the vacuum. This changing is done by a relabelling of the observables in the neighbourhood of b, say by an automorphism γ of A that actually is the identity outside the region of b. Thus, the observer assigns the state $\langle A_o\Omega, \gamma(A) A_o\Omega\rangle = \rho(A)$ to this set up, and the representation $\Pi_\gamma(A) \equiv \Pi_o(\gamma(A))$. If ρ is a state of a C^*-

algebra A, and γ is an automorphism, then ρ_γ, defined as $\rho_\gamma(A) = \rho(\gamma(A))$, is also a state. If ρ is a pure state, and $(\Pi_\rho, H_\rho, \Omega_\rho)$ the canonical representation defined by ρ, then it can be seen that ρ_γ is a vector state relative to Π_ρ if and only if γ is implementable in Π_ρ. Thus, to get from the vacuum sector Π_0 to one of the charged sectors, that is inequivalent, one must apply a non-implementable automorphism γ to A, and transform the vacuum sector states with it. In the above example, γ destroys a positron behind the moon. Applied to the vacuum, it creates an electron. Thus γ is, really, the quantized electron field, in a new guise.

The role of the field operators carrying charge has been clearly explained by Borchers [5] . In a Wightman field theory, the set of all fields form an irreducible set of operators, whereas the observables will be reducible if there are any superselection rules. It is usually assumed that the fields commute with the observables at space-like separation, so that if $A \in A(O)$ and f is a test-f$\underline{\text{n}}$ with support in O_1, space-like relative to O, then $\psi^*(f)A = A\psi^*(f)$. If ψ^* creates one unit of charge, then the states on the left, on which A acts, have one unit less of charge than the states on which A acts on the right. So this equation could be written

$$\psi^*(f)\Pi_q(A) = \Pi_{q+1}(A)\psi^*(f), \quad A \quad \varepsilon \quad A\ (O).$$

Now if ψ is a Fermi field, $\psi^*(f)$ is bounded, but in any case will have a unitary or isometric part. Assuming that the same equation holds for this unitary or isometric part of ψ, we see that $\psi^*(f)$ acts as an intertwining operator between two inequivalent representations, Π_q and Π_{q+1}, provided that A is localized space-like to f. Thus Π_q restricted to any A(O), where O has an open space-like complement, is equivalent to Π_o, (if there are field operators with unitary parts satisfying local commutativity). If this holds, we say Π_q and Π_o are strongly locally equivalent representations of A. These ideas led Haag, Doplicher and Roberts [6] to the following axioms. We are given a local algebra system $O \to A(O)$ with the usual properties, and a vacuum representation (Π_o, H_o) with a Lorentz invariant vector Ω; the representation Π_o is to satisfy the spectrum condition. Then the physical representations $\{\Pi_q\}$ are those that are obtained from Π_o from an automorphism γ_q, so that $\Pi_q(A) = \Pi_o(\gamma_q(A))$. γ_q is to reduce to the identity outside a bounded set, and is to be strongly locally implementable. Finally, we require γ to be such that P is implementable in Π_q.

The main result is that these properties single out representations that fall into representations of a compact abelian group, and that local intertwiners ψ^* exist, which may be chosen to satisfy the usual connection between spin and statistics. The dynamics and mass spectrum in a charged representation are entirely determined by the dynamics of the observables and the choice of automorphism γ.

Remarks (a) This provides the first real idea of why particles are either Bosons or Fermions.

(b) By allowing γ to be a morphism and not an automorphism, HDR obtain parastatistics for their particles, and conjecture that a compact non-abelian gauge group can always be chosen to give exactly this charge spectrum.

(c) In two dimensional space-time, particles other than Bose and Fermi are possible [7].

We now illustrate these ideas with models from statistica mechanics, and solid state physics, as is compulsory at this school.

III. Non-relativistic models

In statistical mechanics one is interested in representations of Bose or Fermi systems, or spin systems defined by states satisfying the KMS boundary condition. These representations are labelled by the temperature $\frac{1}{\beta}$, chemical potential μ, magnetization $\vec{M}$ or other classical parameters. We would like to suggest that these parameters are Casimir operators for the algebras that enter. Indeed, Takesaki [8] has shown that for states obeying the KMS condition, representations belonging to different representations are <u>disjoint</u> (have no non-zero intertwiner). These representations are, of course, reducible, unless the temperature is zero. The closest analogy with field theory comes from quantum spin systems. For more details see my Karpacz School lectures, 1968, and my Inaugural Lecture, (available from Bedford College).

A lattice in s dimensions is denoted Z^s, and has a natural translation group Z^s: $\underline{n} \to \underline{n} + \underline{m}$ is the translation by $\underline{m}$. At each site $\underline{n}$, let $C^2_{\underline{n}}$ be a copy of C^2 (spinor space), and let $A_{\underline{n}}$ be the algebra of all linear operators on $C^2_{\underline{n}}$. The "local algebras", $A(\Lambda)$, are associated with finite subsets $\Lambda \subset Z^s$, and is defined by $A(\Lambda) = \bigotimes_{\underline{n}\in\Lambda} A_{\underline{n}}$. The algebra of observables, A, is that generated by all the $A(\Lambda)$. The algebra $A_{\underline{n}}$ is generated by the spin operators σ^1, σ^2, σ^3 acting on C^2. If we choose the Heisenberg Hamiltonian

$$H = \sum_{\underline{n}\,\underline{m}}^{n\,m} f_{\underline{n},\underline{m}}\ \sigma_{\underline{n}}\ \sigma_{\underline{m}}$$

where $f_{\underline{m}\,\underline{n}} < 0$, then the automorphisms

$$\sigma_n \to e^{iHt}\ \sigma^i_n\ e^{-iHt} = \tau_t\ \sigma^i$$

are implemented in the "vacuum representation". All spins are pointing in the same direction in the ground state Ω. In one dimension we get this picture for Ω:

$$\ldots\uparrow\otimes\uparrow\otimes\uparrow\otimes\uparrow\otimes\uparrow\otimes\ldots$$

Actually, the direction of the spins is not determined; there is a spontaneous breakdown of rotation symmetry in this representation. Let us choose one vacuum and call the corresponding representation Π_o. A basis in the representation space H_o of Π_o, can be obtained by choosing any A, finite, and then all spin configurations in Λ, keeping the values of the spin outside Λ to be the same as before. The mean magnetization,

$$\lim_{\Lambda\to Z^s} \frac{1}{N(\Lambda)} \sum_{\underline{n}\in\Lambda} \langle \psi\ \underline{\sigma}_n \psi \rangle = \vec{M}$$

is then the same for all states ψ in H_o, and is just the asymptotic value. It can be shown that different values of $\vec{M}$ correspond to different representations. Thus $\vec{M}$ is one of our superselection quantum numbers, when the representation is such that it exists.

There is another type of representation representing two and not one magnetic domains. In one dimension, we may form the product state which looks like this:

$$\ldots \otimes \uparrow \otimes \uparrow \otimes \uparrow \otimes \downarrow \otimes \downarrow \otimes \downarrow \ldots$$

It can be easily proved that space-time translation automorphisms are implemented in this representation if $s = 1$. There is no invariant state (under time) in this representation. The energy is locked localized in the region A, the "Bloch wall" separating the two magnetic domains. In fact, this state moves and carries its energy and momentum locally around with it - it behaves as a particle. The motion of this particle is completely determined by the automorphism τ_t. The representation $\Pi_{\vec{q}}$ containing this state has not get a definite value for $\vec{M}$, but the limit

$$\vec{q} = \lim_{\Lambda \to \infty} \Big(\sum_{n<0} \langle \psi, \vec{\sigma}_n \psi \rangle - \sum_{n \geq 0} \langle \psi, \vec{\sigma}_n \psi \rangle \Big)$$

exists (the difference in magnetization to the left and the right). This is our Casimir operator in this case.

We obtain $\Pi_{\vec{q}}$ from Π_0 by the moon argument. H_0 contains states of finitely many but many spin waves (a spin wave is a spin pointing oppositely to $\vec{M}$), for instance:

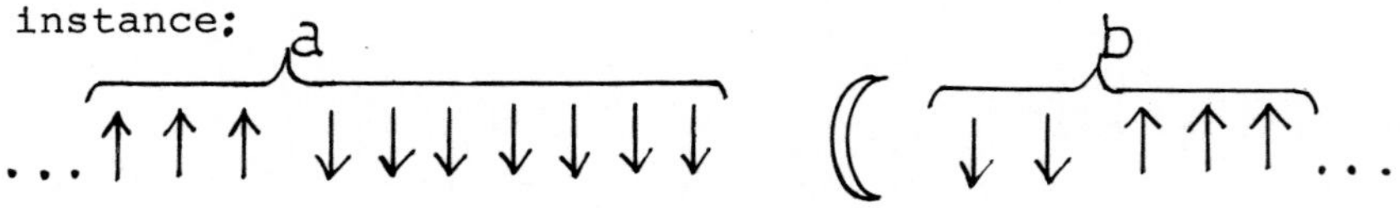

Again, in region (a) we see, locally, a particle with positive $\vec{q}_z$, and in region (b), locally one with negative $\vec{q}_z$. We obtain a state in a new representation by rotating the spins down, at (b), and to the right of (b). This can be achieved by an automorphism of the algebra, unfortunately not localized.

In the sector $H_{\vec{q}}$, one can create additional spin waves, to either side of the break. They interact with the Bloch wall, bleeding off more spin waves. This dynamical production of spin waves is entirely controlled by the dynamics in Π_o.

In higher dimensions, a Bloch wall is not so nicely localized, and space translations are not implementable. However, there is a useful physical interpretation of some of these states in terms of dislocations. Consider the two-dimensional spin system, and the representation containing the state below:

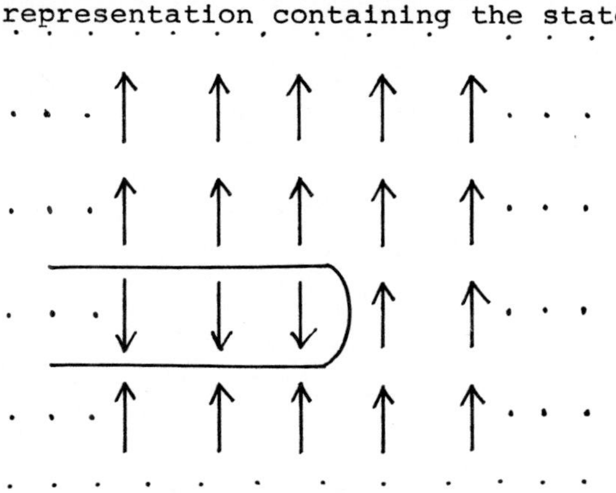

The circled region represents a dislocation. Space translations across the page but not up the page are implemented in this representation. This means that momentum exists across the page, and that we can send a wave along the dislocation (which is tied to ∞ at a fixed point) so that a dislocation spectrum results. Aigain, all the dynamics will be derivable from the (Π_o, H_o) dynamics. These models suggest that elementary particles carrying quantum numbers should be twists or knots in some basic nuclear field; they should be localized and contain finite energy. Dislocations of the nuclear field would not represent particles as such, but would be interesting states to prepare. They would be of macroscopic size and energy, which would be quantum mechanically coherent. An infinite plane electromagnetic wave is already such a state of the electromagnetic field.

IV. Relativistic Models [7], [9]

1957 Model
(Thirring)

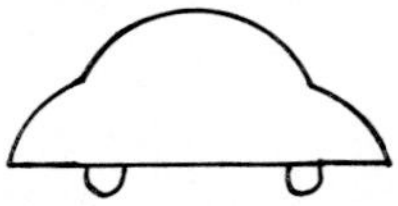

1958 Model
(Berezin)

1960 Model
(Johnson)

1960 Model
(Skyrme)

1964 Model
(Klaiber)

1970 Model
(Streater & Wilde)

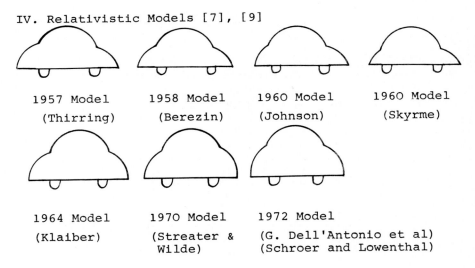

1972 Model
(G. Dell'Antonio et al)
(Schroer and Lowenthal)

Let ϕ be the free field of mass 0 in two dimensional space-time, so that

$$\frac{\partial^2\phi}{\partial t^2} - \frac{\partial^2\phi}{\partial x^2} = 0 \quad .$$

It is well known that this model is not a Wightman theory, since the two-point function is not a distribution. This difficulty is purely a matter of which class of test-functions is chosen, and is easily avoided, as suggested by Schroer, by using a special class of functions. This is preferable to introducing an indefinite metric. In any case, the model falls within the abstract Weyl quantization of Segal.

Let $K = L^2(R, \frac{dp}{|p|})$ be the one-particle space, and H = Fock space over K. Then

$$\phi(x,t) = \frac{1}{(2\pi)^{1/2}} \int (a^*(p)+a(-p))e^{ipx} \frac{dp}{\sqrt{2}\sqrt{|p|}}$$

is the field. There is a 1:1 correspondence between vectors in K and real solutions η of $\Box\eta = 0$, that have finite norm. See Segal [10]. We shall not use the whole of K as testfunctions, but only those corresponding to C^∞ functions $\eta \in M_0 = \{\eta; \eta(x,0), \dot\eta(x,0) \in D$ and $\int\dot\eta dx = 0\}$. Then $\{\phi,\eta\} = \int\{\phi(x,0)\dot\eta(x,0) - \dot\phi(x,0)\eta(x,0)\}dx$ is self-adjoint. Since $\int\dot\eta(x,0)dx = 0$, the first term has no infrared divergence (Schroer's remark). We note that the Wronskian $\{\phi,\eta\}$ is invariant under the simultaneous actions of the Poincaré group $\phi(x) \to \phi(\Lambda x+a)$ and $\eta(x) \to \eta(\Lambda x+a)$. We define the Weyl operator $W(\eta) = e^{i\{\phi,\eta\}}$, $\eta \in M_0$. To any interval $I \subset R$, connected and finite, is defined the causal square of the interval I at time t:

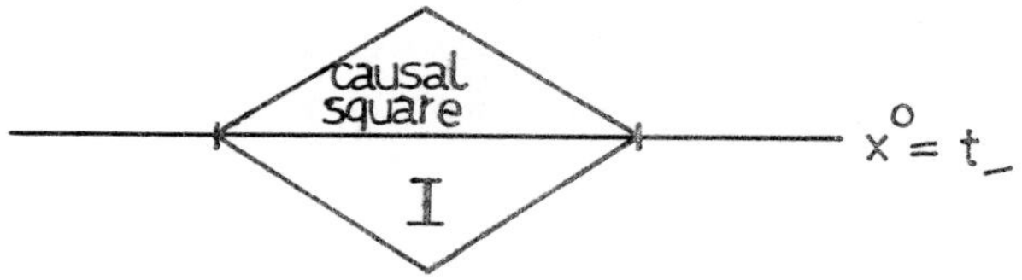

To this region we associate the W^*-algebra $A(I,t)$ generated by all the Weyl operators $e^{i\{\phi,\eta\}}$ with $\eta \in M_0(I) = \{\eta \in M_0; \eta(x,t) = 0$ and $\dot{\eta}(x,t) = 0$ outside $(x \in I)\}$, that is, we find the weak closure of the $*$-algebra generated by the Weyl operators of the region. We should note that M_0 is Lorentz invariant; indeed, $\int \dot{\eta}(x,0)dx$ is the Wronskian of the two solutions 1 and $\eta(x,t)$ of the wave equation.

The quasi-local algebra A is the norm closure of the union of all the $A(I,t)$. The algebra A is smaller than the "Weyl algebra" as defined by Segal [10], because not every vector in K corresponds to a solution in M_0. By restricting ourselves to test functions, for $\phi(x,0)$, that integrate to zero, we are really saying that it is $\nabla\phi$, and not ϕ, that generate the observables.

The usual free field of zero mass is "invariant" under the gauge transformation $\phi(x,t) \to \phi(x,t)+\eta$; $\eta \in R$. This defines a non-implementable automorphism τ_η (in 4 dimensions) of a quasi-local algebra, which transforms the vacuum to a continuum of others, namely the Bose condensate of strength η, $\eta \in R$, leading to the model of spontaneously broken symmetry [11]. Our situation is different. Because the observables are taken to be generated by the field $\partial\mu\phi$, and not by the "potential" ϕ, we see that all these states obtained from the vacuum by applying τ_η to A, coincide as states on A. So our model has a unique vacuum (as far as we know).

The use of the term "gauge group" for $\phi \to \phi+\eta$ is usual in physics, but does not exactly coincide with the full gauge group of the theory in the sense of Haag, Doplicher and Roberts.

Having obtained our algebra satisfying the Haag

axioms, and found the vacuum representation, the charge and superselection properties of the theory are, in principle, determined - we just need to find all the automorphisms of A that have the properties listed; we can find some, using the ideas of gauge transformations of the second kind. Let $\eta_o(x,t)$ a solution of $\Box\, \eta = 0$, and consider the transformation $\phi(x,t) \to \phi+\eta_o$. This leads to the transformation of the Weyl operators $W(\eta) \overset{\tau_{\eta_o}}{\to} W(\eta)e^{i\{\eta_o,\eta\}}$. Remember, we are looking for automorphisms that are not implemented. A lemma of Manuceau [13] completely solves this question for the transformations τ_{η_o}: it is implemented if and only if the functional $\eta \to \{\eta_o,\eta\}$ is continuous in the Hilbert space K, for $\eta \in M_o$. Applying this, we see that if $\eta_o \in M_o$, then τ_{η_o} is implementable. Indeed, the implementing operator is just $W(\eta_o)$. Thus, τ_{η_o} applied to the vacuum is a coherent state in H_o. Physically, it is a Bose condensate of varying strength, $\eta_o(x,t)$. Let us choose a solution η_o with the shape:

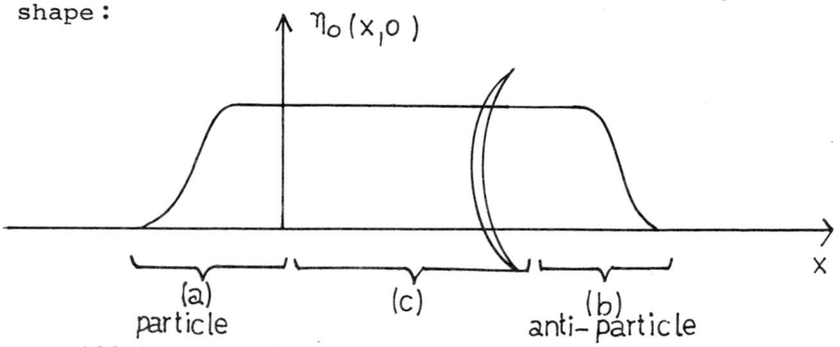

Energy will be stored around (a) and (b) in the classical wave, and also in the coherent state $W(\eta_o)\Omega$. Now, an observer will not notice the Bose condensate in the region (c), because only $\nabla\phi$ and not ϕ is observable. Thus we have the typical particle-antiparticle situation, and the automorphism which leads to a charged state is obtained by making region (b) look like the vacuum, that is, to choose $\eta_0(x,0)$ like this:

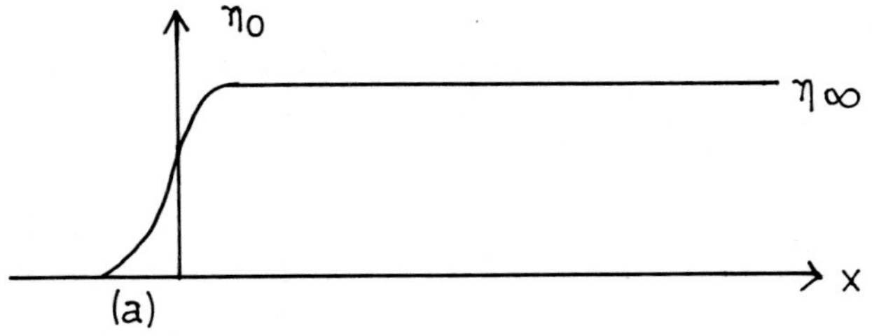

It can be shown that this leads to an automorphism τ_{η_0} that is not implementable but is the identity outside a finite set (namely (a)), and it leads to a covariant representation. The Casimir operator, or charge, is $\int \nabla\eta \, dx = \eta_\infty - \eta_{-\infty}$ in this case.

Another case of non-implementable τ_{η_0} is if $\int \dot{\eta}(x,0)dx \neq 0$. Such a function is obtained from M_0 by the usual trick: choose η_0 first in M_0, so that $\int \dot{\eta}_0(x,0)dx = 0$:

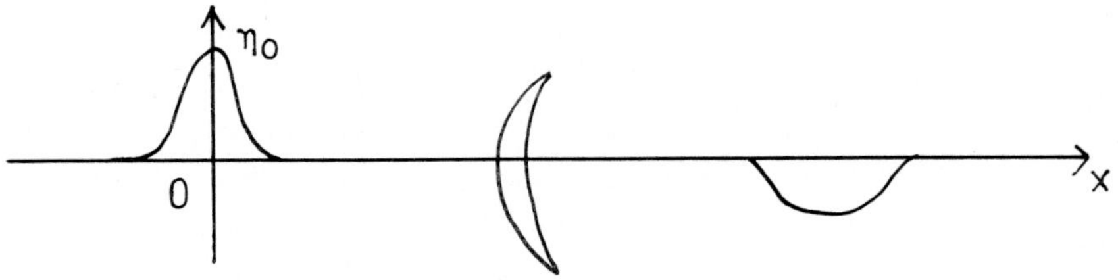

Then, removing the piece behind the moon, we get a τ_{η_0} that is not implemented. The "charge" in this case is

$$\int \dot{\eta}(x,0)dx \quad .$$

We can see that these automorphisms commute and lead to

two branches of sectors. They are not implementable as can be seen heuristically by the remark that $\int_{\infty}\dot{\phi}(x,0)\cdot \eta_0(x,0)dx$ has no meaning, if $\eta_{0\infty} - \eta_{0-\infty} \neq 0$, and also $\int \phi \dot{\eta} dx$ has no meaning if $\int \dot{\eta} dx \neq 0$. Hence the putative operator $W(\eta_0)$, which would show unitary equivalence, do not exist.

Both these sets of representations are strongly locally equivalent to the Π_0. Thus, by the general theory [6] , we would expect to be able to construct charged fields. This is how it is done. We first form the double stack of all the charged sectors we have: a direct <u>sum</u> of spaces $H_{m,n}$ labelled by two real numbers, $\int \dot{\eta} dx$ and $\int \nabla \eta dx$. Then we construct a map from $H_{o,o}$ to $H_{m,n}$, creating a <u>standard shape</u> $\hat{\eta}_0(x,0)$, and another creating a standard shape $\eta_0(x,0)$, of various sizes. These maps we could call U_0 and V. These are the charged fields localized where $\dot{\eta}_0$ and $\nabla \hat{\eta}_0$ are non zero. To get charged fields localized elsewhere, we just translate these by using the Weyl operators to destroy and create Bose condensate locally.

Thus our model of a charged state is a wave of Bose condensate, parametrized globally by charges $m = \int \dot{\eta} dx$ and $n = \int \nabla \eta\, dx$, or, equally well, by the parts that move to the left or to the right, $k > 0$ or $k < 0$. In order to obtain the commutation relations (Bose or Fermi) for the charged field, we may write $W(\eta_0)$ for the "creation operator" for the state η_0; even though this does not exist on Fock space, it can map into $H_{m,n}$. This is neater than writing $W(\eta_0 - \hat{\eta}_0)UV$, which does exist and has the same commutation relations; we find that $W(\eta_0)$ and $W(\eta_0')$ do not always obey the Bose-Fermi alternative;

$$W(\eta_o)W(\eta_o') = W(\eta_o')W(\eta_o)e^{i\{\eta_o,\eta_o'\}}$$

and is Bose-Fermi only if $\{\eta_o,\eta_o'\}$ is $\eta\Pi$. Indeed, the commutation relations between $W(\eta_o)$ and the observables, ϕ' and $\nabla\phi$, are <u>exactly</u> those postulated by Dell'Antonio in his lectures on the Thirring model, where $W(\eta_o)$ is one of his "cut-off" Fermi fields. Thus we have derived the Thirring model intrinsically as the set of charged sectors of the Bose field with Bose or Fermi statistics. There would appear to be no need in the algebraic approach to define point Fermi fields, although this is the non-trivial part of the Thirring model. This also explains why we must change the momentum as well as the energy: in a charged sector the energy is $\int|k|a^*(k)a(k)dk$ and the momentum $\int ka^*(k)a(k)dk$, where a^*, a are in the displaced Fock representations.

Either the Thirring model has no interaction, or we can generate interactions by these gauge transformations - the latter would be an interesting new approach to dynamics. But I rather fear the former is the case, and that if we can gauge the interaction here we can gauge it away again.

Bonnard and I studied [9] free massive particles with U(1) or O(2) symmetry, and again find, in 2 dimensions, that local gauge transformations lead to charged sectors. It might be interesting to ask what solvable model they correspond to.

Barut [12] has attempted a model in the 4 dimensional photon field theory, but it is too early to report on his results.

REFERENCES

1. R. F. Streater and A. S. Wightman, PCT, Spin and Statistics and All That, Benjamin, 1964.

2. F. Strocchi and A. S. Wightman, Princeton preprint.

3. Ref. [1] p. 5.

4. D. Kastler and R. Haag, An Algebraic Approach to Quantum Field Theory, J. Mathematical Phys. 5, 848 (1964).

5. H. J. Borchers, Trieste Lectures in Theoretical Physics (A. Salam Ed), Vienna 1966.

6. R. Haag, S. Doplicher and J. E. Roberts, Fields, Observables and Gauge Transformations, Commun. Math. Phys. 13, 1-23 (1969) and 15 (73-2oo) (1969) and Local Observables and Particle Statistics, Commun. Math. Phys. 23, 199-230 (1971).

7. R. F. Streater & I. F. Wilde, Fermion States of a Boson Field, Nucl. Phys. B24, 561 (1970) and R. F. Streater, The C^*-Algebra Approach to Quantum Field Theory, Phys. Scripta 3, 5 (1971).

8. M. Takesaki, Disjointness of the KMS-States of Different Temperature, Commun. Math. Phys. 17, 33-41 (1970).

9. R. F. Streater, Local Gauge Models Predicting Their own Superselection Rules, Report to the Moscow Conference, On Mathematical Problems of Quantum Field Theory and Quantum Statistical Mechanics, Dec. 1972.

10. I. E. Segal, Mathematical Problems of Relativistic Physics, A. M. S. 1963.

11. R. F. Streater, Spontaneous Breakdown of Symmetry in Axiomatic Theory, Proc. Roy. Soc. (1965).

12. A. O. Barut, Fermi States of a Boson Field, Trieste Preprint 1972.

13. J. Manuceau, Ann. Inst. Henri Poincaré, 8, 117 (1968).

Acta Physica Austriaca, Suppl.XI, 341–387 (1973)

FIELD STRUCTURE THROUGH MODEL STUDIES: ASPECTS OF NONRENORMALIZABLE THEORIES*

BY

John R. KLAUDER
Bell Laboratories
Murray Hill, New Jersey 07974

ABSTRACT

Qualitative features of quantum theories are examined with the aid of models. Attention is focussed on the fact that for sufficiently singular interaction potentials turning off the coupling need not restore the theory to a free theory. This effect is illustrated in a simple, single degree of freedom system. Analysis of the causes leads us to suggest that this effect may arise for covariant nonrenormalizable quantum field theories, and heuristic arguments are developed to support this proposal. For certain soluble noncovariant nonrenormalizable quantum field theories we verify the reality of the effect and comment on its significance, especially for perturbation analyses.

* Lecture given at XII. Internationale Universitätswochen für Kernphysik, Schladming, February 5 - 17, 1973.

1. INTRODUCTION

One can imagine perturbations so strong or so singular that once turned on they cannot be completely turned off but instead leave a residual imprint on the system. In order to illustrate this effect we first study a class of single degree of freedom systems. With such examples as motivation, it is suggestive that such behavior may occur for many degree of freedom systems, and for infinite degree of freedom systems, i.e., for quantum fields. While such examples may be built by repeating single degree of freedom cases one after another, we are more interested in examples of greater physical interest. In particular, we should like to suggest that this phenomenon arises for nonrenormalizable quantum fields-specifically, that once a nonrenormalizable interaction is turned on it cannot be fully turned off, and instead of returning to the free field theory it passes continuously to a "pseudo-free" theory which we shall characterize in a formal way [1].

In order to make it at least plausible that nonrenormalizable field theories have this behavior, we appeal to the formal techniques of functional integration (path integrals) for Euclidean quantum field theory [2]. To set the stage for our field analysis, we first discuss the single degree of freedom examples in a similar language, which is related to the Feynman-Kac formula, etc., [3]. While we are the first to admit that our arguments by no means prove that nonrenormalizable fields have such effects, they nevertheless seem to provide a natural class of models if such effects are to occur at all.

Although we are unable to give more than formal

arguments for covariant theories, the effects we describe do arise in certain model field theories, namely, ultralocal models [4,5]. In the final section we discuss these models and show that the known solutions for these models converge as the interaction is turned off to a pseudo-free theory, which is fundamentally different from a free theory. This pseudo-free theory is covariant under scale transformations of the field (with noncanonical dimensions, etc.), and in its massless form is actually scale invariant. Moreover, since it is continuously related to the interacting theory, it provides a far more natural starting point for a (generally asymptotic) perturbation theory than the usual free theory to which it has no continuous connection.

2. ELEMENTARY EXAMPLE:
PERTURBATION OF THE HARMONIC OSCILLATOR

Our simple example is based on perturbations of the free harmonic oscillator Hamiltonian

$$H_o = P^2 + Q^2$$

with standard eigenstates denoted by $|n\rangle$ and eigenvalues $E_n = 2n + 1$, for $n = 0,1,2,\ldots$. The perturbation is taken as

$$V = |Q|^{-\alpha}$$

and we ask what is the limit of

$$H = P^2 + Q^2 + \lambda V$$

as $\lambda \downarrow 0$ (i.e., $\lambda \to 0^+$). By convergence we mean (strong) convergence of the eigenfunctions and convergence of the eigenvalues. For $\alpha < 2$ (including $\alpha < 0$) $H \to H_o$, while for $\alpha \geq 2$, $H \to H_o'$ a pseudo-free Hamiltonian the properties of which we shall describe.

Among all perturbations we first distinguish the class of analytic perturbations (the classic source for operator perturbations is Kato [6]). For these there exist real constants a and $b < 1$ such that (as forms)

$$-(a+bH_o) \leq \lambda V \leq (a+bH_o) \quad . \tag{1}$$

The familiar resolvent equation

$$(x+H)^{-1} = (x+H_o)^{-1} - (x+H_o)^{-1}\lambda V(x+H)^{-1}$$

may (for $x > 0$) be rewritten in terms of

$$S = (x+H_o)^{\frac{1}{2}}(x+H)^{-1}$$

as

$$S = (x+H_o)^{-\frac{1}{2}} - (x+H_o)^{-\frac{1}{2}}\lambda V(x+H_o)^{-\frac{1}{2}} S \; .$$

If we let

$$B \equiv (x+H_o)^{-\frac{1}{2}}\lambda V(x+H_o)^{-\frac{1}{2}}$$

then

$$(1+B)S = (x+H_o)^{-\frac{1}{2}} ,$$

which has a convergent series solution

$$S = \sum_{n=0}^{\infty} (-B)^n (x+H_o)^{-\frac{1}{2}}$$

provided $||B|| \equiv c < 1$. The bound (1) on the potential leads to $c = \max(a/x, b)$ which for sufficiently large x can be made less than 1. It follows, therefore, that

$$(x+H)^{-1} = \sum_{n=0}^{\infty} (x+H_o)^{-1}[-\lambda V(x+H_o)^{-1}]^n .$$

As $\lambda \to 0$, $(x+H)^{-1}$ converges strongly to $(x+H_o)^{-1}$, which since H_o is semibounded and has discrete states implies that the eigenfunctions $|n,\lambda>$ and eigenvalues $E_n(\lambda)$ of H converge, as desired, to $|n>$ and E_n.

The analytic perturbation covers the cases $-2 \leq \alpha < 1$. For $\alpha = -2$, i.e., for $V = Q^2$, the resolvent series converges for λ near zero (which is adequate for our purposes), while for $-2 < \alpha < 1$ the resolvent series converges for all finite λ. For $V = |Q|^p$, $0 \leq p < 2$ these properties are clear.

The more interesting examples are those for which $0 < \alpha < 1$, and we consider them further. In a Schrödinger representation for Q and P we learn that

$$\lambda\int|x|^{-\alpha}|\psi(x)|^2dx \leq \lambda \int_{-1}^{1}|x|^{-\alpha}|\psi(x)|^2dx + \lambda\int|\psi(x)|^2dx$$

$$\leq \lambda \sup_x |\psi(x)|^2 \int_{-1}^{1}|x|^{-\alpha}dx + \lambda||\psi||^2 .$$

Since

$$|\psi(x)| = |\int e^{ikx}\tilde{\psi}(k)dk| \leq \int|\tilde{\psi}(k)|\, dk$$

it follows that

$$\sup_x |\psi(x)|^2 \leq (\int|\tilde{\psi}(k)|dk)^2$$
$$= (\int \sqrt{\varepsilon k^2+1}\ |\tilde{\psi}(k)|dk/\ \sqrt{\varepsilon k^2+1})^2$$
$$\leq \frac{\pi}{\sqrt{\varepsilon}} \int(\varepsilon k^2+1)|\tilde{\psi}(k)|^2 dk$$
$$= \pi\sqrt{\varepsilon}\,||P\psi||^2 + (\pi/\sqrt{\varepsilon})||\psi||^2 . \tag{2}$$

Combining these facts, we find that

$$\lambda|Q|^{-\alpha} \leq a + bH_o$$

where $b = 2\pi\sqrt{\varepsilon}\ \lambda/(1-\alpha)$, and given λ and α, one may always choose $\varepsilon > 0$ so that $b < 1$. It is interesting to note the trade-off in such bounds as (2) that arises when ε varies (consistent with $0 < \varepsilon < \infty$).

We next consider perturbations $V = |Q|^p$, $2 < p < \infty$. As a typical case we treat $V = Q^4$. A rather complete study of these cases has been given by Simon [7]. It is reasonable that some $k < \infty$ exists such that

$$Q^4 \leq k(P^2+Q^2)^2 .$$

This is one case of the more general problem: Find K_{2m} such that

$$Q^{2m} \leq K_{2m}(P^2+Q^2)^m$$

is satisfied. It is clear that $K_{2m} \neq 1$ since the ground state expectation value of each side yields

$$\langle 0|Q^{2m}|0\rangle = \frac{1}{\sqrt{\pi}} \int x^{2m} e^{-x^2} dx = \frac{1}{\sqrt{\pi}} \Gamma(m+\tfrac{1}{2}) \leq K_{2m} .$$

A crude estimate shows that

$$K_{2m} = (2m+1)(3m)^m .$$

With the existence of k ($=K_4$) assured we note for $V = Q^4$ that

$$H_o \leq H_o + \lambda V \leq H_o + \lambda k H_o^2 ,$$

and it seems reasonable that as $\lambda \downarrow 0$ the operator H is forced to become H_o. Indeed, since for $\tau \geq 0$

$$e^{-\tau H_o} - e^{-\tau H} = \int_0^\tau e^{-(\tau-s)H} \lambda V e^{-sH_o} ds ,$$

we have (for $V = Q^4$) the bound

$$\|(e^{-\tau H_o} - e^{-\tau H})|n\rangle\| \leq \tau \|\lambda V|n\rangle\|$$

$$\leq \tau \lambda K_8^{\frac{1}{2}} \|H_o^2|n\rangle\| = \tau\lambda K_8^{\frac{1}{2}} E_n^2 .$$

Consequently, for all $\tau \geq 0$ and arbitrary n,

$$\lim_{\lambda \downarrow 0} e^{-\tau H} |n\rangle = e^{-\tau E_n} |n\rangle ,$$

which ensures strong convergence of the eigenvectors and convergence of the eigenvalues to their free values. Similar remarks apply to $V = |Q|^p$, $2 < p < \infty$.

Although the perturbation Q^4 is not analytic,

there is a close reformulation of the problem that makes use of the analytic perturbations and that has heuristic value.

If A, B, p and p' are positive numbers with $p^{-1} + p'^{-1} = 1$, then one has the inequality

$$AB \leq \frac{A^p}{p} + \frac{B^{p'}}{p'} \quad .$$

From the spectral resolution for H_o, this relation implies that

$$\lambda k H_o^2 = \delta^{-1}(\delta\lambda k H_o^2) \leq a + \frac{1}{2}\, \varepsilon\, H_o^3 \; ,$$

where we have set $A = \delta^{-1}$ and $p' = \frac{3}{2}$. We can always choose ε as small as desired by adjusting δ. From this inequality it follows that

$$0 \leq \lambda V \leq a + \frac{1}{2} H_{o\varepsilon} \quad ,$$

where $H_{o\varepsilon} \equiv H_o + \varepsilon H_o^3$. For this relation we can apply analytic perturbation theory and find, for sufficiently large x, that

$$(x+H_\varepsilon)^{-1} = \sum_{n=0}^{\infty} (x+H_{o\varepsilon})^{-1}[-\lambda Q^4(x+H_{o\varepsilon})^{-1}]^n \; ,$$

where $H_\varepsilon \equiv H_{o\varepsilon} + \lambda Q^4$. Note that in this convergent expansion the eigenvectors of $H_{o\varepsilon}$ are the same as those of H_o (and thus matrix elements of the potential are unchanged); only the energy eigenvalues are modified - and even that can be made arbitrarily small below any pregiven energy level. The resolvents for each $\varepsilon > 0$

extend to expressions $(z+H_\varepsilon)^{-1}$ that are analytic everywhere in z except perhaps for Rez < 0. It is plausible that the resolvent of interest can be obtained from

$$(z+H)^{-1} = \lim_{\varepsilon\downarrow 0} (z+H_\varepsilon)^{-1} ,$$

and this is actually the case in the example we have considered.

Although perturbation theory is not literally true for $V = Q^4$, it can be made applicable by modifying the energy denominators appropriately as discussed above. Clearly similar remarks apply to other "power-bounded" potentials such as $|Q|^p$, $2 < p < \infty$, for which

$$0 \leq V \leq kH_0^{p/2}$$

for some finite k.

Next we turn to the cases where $\alpha \geq 1$. Since domain changes arise*, we appeal directly to the form of the solutions of the Schrödinger equation, namely, to

$$(-\frac{d^2}{dx^2} + x^2 + \frac{\lambda}{|x|^\alpha})\psi = E\psi \quad .$$

The term x^2 controls the behavior at $\pm\infty$ and leads to discrete energy levels. The singularity at the origin, on the other hand, must be dealt with by the derivative term. The principal behavior is already apparent in the equation

---

* This analysis is one without ambiguity. A self-adjoint extension of a symmetric operator is at issue, and this may be done in more than one way. Our argument is openly influenced by scale transformation and dimensional considerations (see below).

$$(-\frac{d^2}{dx^2} + \frac{\lambda}{|x|^\alpha})\psi = 0, \tag{3}$$

and the behavior of its solutions near $x = 0$. For $(2-\alpha)^{-1}$ non-integer this equation has two independent solutions [8] given by

$$\psi = \sqrt{x}\, I_{\pm\frac{1}{2-\alpha}}\left(\frac{2\sqrt{\lambda}\, x^{1-\frac{1}{2}\alpha}}{2-\alpha}\right),$$

where I_ν is a standard Bessel function. If $(2-\alpha)^{-1} = m$ is an integral the two solutions involve I_m and K_m. For any $0 < \alpha < 2$ both solutions are locally bounded and locally L^2, and, as $\lambda \downarrow 0$, they pass continuously to the form const. or const. x appropriate to the equation $(d^2/dx^2)\psi=0$. Although, for all $\lambda > 0$, the eigenstates in cases where $1 \leq \alpha < 2$ are not in the form domain of V (and therefore not in the form domain of H_o) it is entirely consistent that the eigenvectors and eigenvalues should converge to those of the free Hamiltonian H_o.

For $\alpha = 2$, the two solutions to (3) have the form

$$\psi \sim x^{1+\gamma}, \quad \psi \sim x^{-\gamma}$$

where $\gamma \geq 0$ and $\lambda \equiv \gamma(\gamma+1)$. The first solution is bounded and locally L^2, while the second solution is unbounded and locally L^2 only for $\gamma < \frac{1}{2}$. Closer inspection reveals that the second solution is related to an infinite attractive δ-function potential at the origin the coefficient of which is divergent for any $\gamma > 0$. That explains how an (approximate) eigenfunction could <u>in-</u>

crease on tunnelling into a potential barrier. We reject such solutions on physical grounds, and are left with the only possible form $\psi \sim x^{1+\gamma}$. This means that for even and odd solutions we have, respectively,

$$\psi_{even} \sim |x|^{1+\gamma} ; \psi_{odd} \sim x|x|^{\gamma} .$$

Now, in fact, for $\alpha = 2$ the eigensolutions to the equation

$$(-\frac{d^2}{dx^2} + x^2 + \frac{\lambda}{x^2})u_n(x) = \varepsilon_n u_n(x)$$

are explicitly known in terms of Laguerre polynomials [9]. As $\lambda \downarrow 0$ these solutions tend to those of the pseudo-free Hamiltonian H_0'. If $\psi_n(x)$ and $E_n = 2n + 1$, $n = 0,1,2,...,$ denote the eigenfunctions and eigenvalues for H_0, then those for H_0' are given by

$$\psi_{2n}'(x) = (\text{sign } x)\psi_{2n+1}(x)$$

$$\psi_{2n+1}'(x) = \psi_{2n+1}(x)$$

$$E_{2n}' = E_{2n+1}' = E_{2n+1} = 4n + 3.$$

In particular the ground state of H_0' is

$$\psi_0'(x) = \frac{\sqrt{2}}{\pi^{\frac{1}{4}}} |x| e^{-\frac{1}{2}x^2}$$

and has energy eigenvalue $E_0' = 3$. All levels of H_0' are doubly degenerate, and if desired the eigenfunctions could alternately be chosen to vanish for $x < 0$ or for

$x > 0$. Quite clearly the limiting form of $P^2 + Q^2 + \lambda Q^{-2}$ as $\lambda\downarrow 0$ is <u>not</u> the free Hamiltonian.

For $\alpha > 2$ only one solution of (3) is locally L^2 at $x = 0$ and it has the form $(x > 0)$

$$\psi \sim x \exp\left[-\frac{\gamma}{(\frac{1}{2}\alpha-1)}\left(\frac{1}{x^{\frac{1}{2}\alpha-1}} - 1\right)\right] .$$

This relation is normalized such that as $\alpha\downarrow 2$, $\psi\to x^{1+\gamma}$, while for $\gamma\downarrow 0$ (i.e., $\lambda\downarrow 0$), $\psi\to x$. That is, for $\alpha > 2$, as $\lambda\downarrow 0$ all acceptable solutions are of the form $\psi \sim x$ (i.e., $\psi_{even} \sim |x|$, $\psi_{odd} \sim x$). This is just the form encountered in the pseudo-free case, and we conclude that for $\alpha > 2$ the limiting form for H is also given by the pseudo-free form H_0'.

A convenient pictorial representation of our results may now be given in the form of Fig. 1.

Scale Transformations

A few remarks on scale covariance are in order. Let us reintroduce the usual factors of ω so that $H_0 = P^2 + \omega^2 Q^2$, and $E_n = \omega(2n+1)$. This example is scale covariant under the following scale transformation

$$P\to S^{-1}P; \qquad Q\to SQ; \qquad \omega\to S^{-2}\omega$$

where $S > 0$. Under this transformation

$$H_0\to S^{-2}H_0; \qquad E_n\to S^{-2}E_n,$$

while the eigenstates are invariant, $|n\rangle\to|n\rangle$.

Besides H_o, the pseudo-free case H_o' is also scale covariant under the same transformation. Consider first $\alpha = 2$ for which it is clear that

$$H = P^2 + \omega^2 Q^2 + \lambda Q^{-2} \rightarrow S^{-2} H$$

under the above transformation along with $\lambda \rightarrow \lambda$. Since this holds for all λ, it holds in the limit $\lambda \downarrow 0$, and therefore

$$H_o' \rightarrow S^{-2} H_o' ; \qquad E_n' \rightarrow S^{-2} E_n' \ .$$

This fact is also quite clear from the close relationship between the eigenfunctions and eigenvalues of H_o and H_o'.

For $\alpha \neq 2$, H is not scale covariant but rather

$$H = P^2 + \omega^2 Q^2 + \lambda |Q|^{-\alpha} \rightarrow S^{-2} [P^2 + \omega^2 Q^2 + \lambda S^{2-\alpha} |Q|^{-\alpha}] \ .$$

However, as $\lambda \downarrow 0$ the limiting form of the expression in the brackets is the same whether $S = 1$ or $S \neq 1$ and it only depends on whether $\alpha > 2$ or $\alpha < 2$. Scale covariance arises in both the free and pseudo-free theories even though it may or may not be present when the coupling constant $\lambda > 0$.

A useful way to exhibit scale covariance is with the expectation function, e.g.,

$$E_\omega(s) = \langle 0 | e^{isQ} | 0 \rangle = \sqrt{\frac{\omega}{\pi}} \int e^{isx - \omega x^2} dx$$

appropriate to the free case. Since $|0\rangle$ is invariant, we have the relation

$$E_{S^{-2}\omega}(S^{-1}s) = E_\omega(s)$$

for all $S > 0$. In similar fashion, the expectation function for the pseudo-free case, given by

$$E'_\omega(s) = 2\sqrt{\frac{\omega^3}{\pi}} \int e^{isx} x^2 e^{-\omega x^2} dx \ ,$$

fullfills the same symmetry, i.e.,

$$E'_{S^{-2}\omega}(S^{-1}s) = E'_\omega(s) \ .$$

We shall find this a convenient way to investigate scale covariance in our model field solutions.

Coupling Constant Dimensionality

It is interesting to relate the change in behavior at $\alpha = 2$ to the dimensionality of the coupling constant λ. If we assign dim $\omega = 1$, then it is clear that

$$\dim \lambda = 1 - \tfrac{1}{2}\alpha \ .$$

Thus for dim $\lambda > 0$ (i.e., $\alpha < 2$) we have convergence to H_0, while for dim $\lambda \leq 0$ (i.e., $\alpha \geq 2$) we have convergence to H'_0. A standard separation in covariant quantum field theory is also made on the basis of coupling constant dimensionality. Thus it is suggestive that $\alpha < 2$ corresponds to the super renormalizable interactions, $\alpha = 2$ to the renormalizable interactions and $\alpha > 2$ to the non-renormalizable interactions. Of course, such a connection should be viewed heuristically since field theory is far more complex than "potential theory", and also coupling constant dimensionality is not invariably a correct guide to such behavior. Nevertheless, this is an important viewpoint, and there is a considerable literature dealing with singular power-law potentials for the Schrödinger and Bethe-Salpeter equations where the potentials are regarded as analogues of super renormalizable, renormalizable and nonrenormalizable interactions, and are suggested, in part, by coupling constant dimensionality [10].

Other Perturbation Examples

Before closing this section it is worthwhile noting that analogous behavior arises for other perturbations, and in particular we wish to remark briefly about the case

$$H = P^2 + Q^2 + \lambda |P|^{-\alpha} \quad .$$

Entirely similar results to those derived above hold in this case as well. One need only employ momentum space ($P = k$, $Q = i \frac{\partial}{\partial k}$) rather than configuration space ($P = -i \frac{\partial}{\partial x}$, $Q = x$). We specifically note the following: For $\alpha < 0$, i.e., for $p = -\alpha > 0$, all cases of the form

$$H = P^2 + Q^2 + \lambda |P|^p$$

pass continuously to the free Hamiltonian H_0 as $\lambda \downarrow 0$. The interested reader may work out the present form of the pseudo-free Hamiltonian H_0' that arises in the limit $\lambda \downarrow 0$ when $\alpha \geq 2$.

3. FUNCTIONAL FORMULATION

If operator solutions to nonrenormalizable covariant quantum theories were available, one could simply check whether or not the effects described in Sec. 2 held. For example, one could study a sequence of Wightman functions for such theories and see to what limit they approached as $\lambda \downarrow 0$. Unfortunately such a procedure is not open to us at present, and we are forced to argue otherwise.

Single Particle Formulation

The functional formulation of quantum theory has been shown to be equivalent to the operator formulation under very wide circumstances. One result, known as the Feynman-Kac formulation, relates the solution of the differential equation

$$\frac{\partial\psi}{\partial t} = \frac{\partial^2}{\partial x^2}\psi - V(x)\psi \quad , \qquad (4)$$

with $\psi = \delta(x-y)$ at $t = 0$, to the functional integral

$$\int e^{-\int_0^t V(x(t))dt} d\mu_W(x) \quad , \qquad (5)$$

where μ_W is the measure for a conditional Wiener process defined on the space C of continuous paths starting at y and ending at x. As is well known, Wiener measure is characterized in an essential way by the elementary solution of the diffusion equation ($\partial\psi/\partial t = \partial^2\psi/\partial x^2$) given by*

*Strictly speaking our normalization here is such that $\dot{x} = 2p$, rather than $\dot{x} = p$, but we shall ignore this inessential fact.

$$\psi(x,t;y) = \frac{1}{\sqrt{\pi t}} e^{-\frac{(x-y)^2}{t}} .$$

Equivalence of the two forms of the solutions (4) and (5) has not been demonstrated for all potentials $V(x)$, but is initially confined to those potentials that are locally bounded [3]. Unfortunately, we need to consider more general potentials.

Rather than Wiener-process averages, we appeal to the closely related Ornstein-Uhlenbeck process (stationary Markov-Gaussian process [11]) which incorporates into the measure the x^2 (or $\omega^2 x^2$) term we invariably include. Specifically, we let μ_F (F for free) denote a Gaussian measure on the path space of continuous functions defined for $-T \leq t \leq T$, $T < \infty$, and characterized by the fact that

$$C_F(s) \equiv \int e^{i\int s(t)x(t)dt} d\mu_F(x) = e^{-\frac{1}{4}\int s(t)D(t-t')s(t')dtdt'} \tag{6a}$$

where the support of $s(t)$ lies within $|t| < T$. Here the kernel D fulfills

$$\left(-\frac{d^2}{dt^2} + 1\right)D(t) = \delta(t) ; \tag{6b}$$

namely

$$D(t) = \frac{1}{2} e^{-|t|} . \tag{6c}$$

In addition, we consider the following average in μ_F:

$$C(s) \equiv \int e^{i\int s(t)x(t)dt-\lambda\int V(x)dt} d\mu_F(x) . \tag{7}$$

Although this quantity is not identical to the expression given by the Feynman-Kac formula, it contains that information within it. For if one set $s(t) = -a\delta(t-\tau) + b\delta(t)$ and formed $(2\pi)^{-1}\int e^{i(xa-yb)}C(s)\,da\,db \equiv \psi(x,\tau,y)$, the result would be a solution of the equation

$$\frac{\partial\psi}{\partial\tau} = \frac{\partial^2\psi}{\partial x^2} - x^2\psi - \lambda V(x)\psi \tag{8}$$

which (up to a scale factor) fulfills the boundary condition $\psi = \delta(x-y)$ at $\tau = 0$. Consequently, we focus our study on the expression (7) and we ask what is the limit of $C(s)$ as $\lambda\downarrow 0$.

Consider first $V = x^2$. It follows from (6) that

$$\int\int x^2(t)\,dt\,d\mu_F(x) = \tfrac{1}{4}\int dt = \tfrac{1}{2}T$$

and therefore that $\int x^2(t)\,dt < \infty$, μ_F a.e.. Consequently for $V = x^2$ it follows that

$$\lim_{\lambda\downarrow 0} C(s) \to C_F(s) \ .$$

A similar argument for $p \geq 0$ shows that $\int|x(t)|^p dt < \infty$, μ_F a.e., and for $0 \leq \alpha < 1$ that $\int|x|^{-\alpha}dt < \infty$, μ_F a.e.. Consequently for $V = |x|^{-\alpha}$, $\alpha < 1$,

$$\lim_{\lambda\downarrow 0} C(s) \to C_F(s) \ ,$$

which corresponds to the fact that $H \to H_0$ under the same circumstances.

Let us consider the changes that arise if we choose $T = \infty$, i.e., if we consider paths defined for all t. In that case it is not too difficult to be persuaded that $\int |x(t)|^p dt = \infty$, μ_F a.e. for all $p \geq 0$ (or $\alpha < 1$ also). This includes $p = 2$ which simply corresponds to a "mass shift," a change of the coefficient of x^2 in (8). Clearly, under these circumstances, $C(s) \equiv 0$ in (7) for $\lambda > 0$ and we do not recover $C_F(s)$ as $\lambda \downarrow 0$. This type of behavior is not solely confined to an infinite time interval. Suppose, for example, we consider perturbations of the form $V = |P|^p$, $p > 0$, rather than $V = |Q|^p$. Such questions involve expressions like

$$\int e^{i\int s(t)x(t) - \int F_\lambda(|\dot{x}(t)|)dt} d\mu_F(x) \equiv F$$

for suitable functions F_λ with the property that $F_\lambda(\infty) = \infty$. But (6) leads quite simply to the fact that for almost all t

$$\int e^{-\dot{x}^2(t)} d\mu_F(x) = 0 ;$$

namely, that $|\dot{x}(t)| = \infty$, μ_F a.e., leading to $F \equiv 0$. This behavior seems to conflict with the symmetry between P and Q we know to be present (see previous section), but it simply reflects the fact that μ_F is inevitably connected with the configuration space formulation, a fact already emphasized by (8). Even for $p = 2$, where we simply deal with a change of the coefficient of P^2, the problem cannot be approached directly through the measure μ_F we have introduced. One is confronted here with examples of orthogonal measures in path space, a well known but troublesome property. Let us consider such questions more generally.

Digression on Measures on Function Space

Many of the properties of functional space measures can be seen more clearly in quite simple examples. Let us employ a countable number of variables $x_r \in R$, $r = 1,2,\ldots,$ and examine several simple measures on them. Consider first the characteristic functional

$$e^{-\frac{1}{2}\alpha^{-1}\Sigma s_r^2} = \int e^{i\Sigma s_r x_r} \, d\mu_\alpha(x)$$

where formally it is suggestive that

$$d\mu_\alpha(x) = \prod_r \left(\sqrt{\frac{\alpha}{2\pi}} \, e^{-\frac{1}{2}\alpha x_r^2} \, dx_r \right) .$$

It is instructive to examine the support of such measures and their relationship for different α, where $0 < \alpha < \infty$. From the formal relation for $d\mu_\alpha(x)$ it would seem that the support is characterized by $\Sigma x_r^2 < \infty$, but this is incorrect. To investigate this question, let us consider the expression

$$R \equiv \int e^{-\frac{1}{2}\varepsilon\Sigma\beta_r x_r^2} d\mu_\alpha(x) = \prod_r \left(1 + \frac{\varepsilon\beta_r}{\alpha}\right)^{-\frac{1}{2}} ,$$

which holds provided $\alpha > -\varepsilon\beta_r$. Assume first that $\varepsilon = 1$, $\beta_r \geq 0$ and that $\Sigma\beta_r = \infty$. Then $R \equiv 0$ which implies that $\Sigma\beta_r x_r^2 = \infty$, μ_α a.e.; or stated otherwise that the set of points for which $\Sigma\beta_r x_r^2 < \infty$ has μ_α measure zero. Next take $\beta_r \geq 0$, $\Sigma\beta_r < \infty$ (excluding the case $\beta_r \equiv 0$) and set $b = \sup(\beta_r/\alpha)$. Then $0 < R < \infty$ for $-b^{-1} < \varepsilon < b^{-1}$, and it follows that $\Sigma\beta_r x_r^2 < \infty$, μ_α a.e., in the case where $\Sigma\beta_r < \infty$. These relations show that the measure is concentrated on the Hilbert space $H_{\{\beta\}}$ formed of sequences for which $\Sigma\beta_r x_r^2 < \infty$ for any $\{\beta_r\}$ provided $\Sigma\beta_r < \infty$, and therefore on the space V that is the intersection of such spaces for all $\{\beta_r\}$ for which $\Sigma\beta_r < \infty$. This still yields a support that is larger than ℓ^2, as it must be, and also a support that is independent of the parameter α. On the other hand, within that support the measures for each α are mutually orthogonal, $\mu_\alpha \perp \mu_{\alpha'}$, $\alpha \neq \alpha'$, and therefore each measure "lives" on a different subset. Orthogonality means that $\int\sqrt{d\mu_\alpha(x)}\ \sqrt{d\mu_{\alpha'}(x)} = 0$, an expression which is calculated here as the $N\to\infty$ limit of A^N, where

$$A \equiv \sqrt{\frac{\sqrt{\alpha\alpha'}}{2\pi}} \int e^{-\frac{1}{4}(\alpha+\alpha')x^2} dx = \sqrt{\frac{\sqrt{2\alpha\alpha'}}{(\alpha+\alpha')}} \ .$$

Since $A < 1$, if $\alpha \neq \alpha'$, $A^N \to 0$ as $N\to\infty$. While each μ_α has support on V, as discussed above, further refinement should be possible since within that support measures for different α's are mutually orthogonal. Further deliniation of the support for μ_α may be given as follows [12]. For each sequence $\{x_r\} \in V$ define

$$r(x) \equiv \overline{\lim_{N\to\infty}}\ N^{-1}\Sigma_1^N x_r^2,$$

which exists and respects $0 \leq r(x) \leq \infty$. Let us introduce the spaces $V_\alpha \equiv \{\{x_r\} \in V : r(x) = \alpha^{-1}\}$ for all α, $0 \leq \alpha \leq \infty$. Clearly such spaces are disjoint and $V = \bigcup_\alpha V_\alpha$. On the other hand, for $0 < \alpha < \infty$, the random variable $N^{-1}\Sigma_1^N x_r^2$ has mean α^{-1} and variance $O(N^{-1})$ in the measure μ_α. Consequently, when $N\to\infty$ it follows that

$$\lim_{N\to\infty} \frac{1}{N}\ \Sigma_1^N\ x_r^2 = \alpha^{-1},\ \mu_\alpha\ ,\ \text{a.e.}\quad .$$

This implies that μ_α is concentrated on V_α, namely on just those sequences $\{x_r\} \varepsilon V$ for which $\lim_{N\to\infty} N^{-1}\Sigma_1^N x_r^2 = \alpha^{-1}$. One can even carry this argument further to see that

$$\lim_{N\to\infty} \frac{1}{N} \Sigma_1^N e^{ipx_n} = e^{-\frac{p^2}{2\alpha}} , \quad \mu_\alpha , \text{ a.e.},$$

which holds for all p. Stated otherwise, the measure μ_α is concentrated on sequences $\{x_n\}$ for which this relation holds for all p. Expanding in powers of p we recover the variance rule given previously along with other relations such as $\lim_{N\to\infty} N^{-1}\Sigma_1^N x_r^m = 0$, m odd, and $\lim_{N\to\infty} N^{-1}\Sigma_1^N x_r^4 = 3\alpha^{-2}$, etc. Such rules deliniate rather closely the sequences $\{x_r\}$ on which the support of μ_α is concentrated.

With these facts clear let us examine the suggestive formal expression

$$d\mu_\alpha(x) = e^{-\frac{1}{2}\alpha\Sigma x_r^2} \, Dx ,$$

where Dx symbolically denotes a translation invariant measure (but one which is nonexistent!). While the integrand in this formal representation is an incorrect guide to the support properties of μ_α, there is no question that the integrand provides a qualitative way to compare support properties of different measures and thereby gives a kind of "zeroth approximation" to the relative support of μ_α. What we have in mind here can be best explained by an example. Consider the two cases formally defined by

$$d\mu_{\{\alpha\}}(x) = e^{-\frac{1}{2}\Sigma\alpha_r x_r^2} \, Dx$$

and by

$$d\mu_1(x) = e^{-\frac{1}{2}\Sigma x_r^2}\, Dx \quad .$$

The measure $\mu_{\{\alpha\}}$ is supported on the space V' formed by the intersection of those spaces for which $\Sigma \beta_r \alpha_r x_r^2 < \infty$, for all $\{\beta_r\}$, $\beta_r \geq 0$, such that $\Sigma \beta_r < \infty$. If $\alpha_r \to \infty$ or $\alpha_r \to 0$, V' is different than the space V which is appropriate for the support of μ_1. Such a difference is already correctly signalled by the fact that under these circumstances the spaces for which $\Sigma \alpha_r x_r^2 < \infty$ and $\Sigma x_r^2 < \infty$ are also different. If $\alpha_r \to \alpha \neq 1$ (with $0 < \alpha < \infty$), then the spaces V' and V are the same, although we know that the support of the two measures is disjoint. Clearly this kind of refinement cannot be dealt with by our zeroth approximation based on the finiteness of the formal integrand.

It is convenient to introduce one additional formal notion. We have seen how different α parameters label orthogonal measures, and we wish to introduce a "liberalized viewpoint" adapted to such possibilities. Let

$$K_N = e^{-\frac{1}{2}\beta\Sigma_1^N x_r^2}$$

which for $N < \infty$ satisfies $0 < K_N < \infty$, μ_1 a.e., while for $N = \infty$, $K_\infty = 0$, μ_1 a.e. . On the other hand, we can clearly find numbers N_N such that

$$\lim_{N\to\infty} N_N \int e^{i\Sigma s_r x_r} K_N d\mu_1(x) = e^{-\frac{1}{2(1+\beta)}\Sigma s_r^2} =$$

$$= \int e^{i\Sigma s_r x_r}\, d\mu_{1+\beta}(x) \quad . \qquad (9)$$

Moreover, this result is so defined that as $\beta\to 0$ we recover the measure μ_1. In the liberalized viewpoint $K_\infty = 0$ (almost everywhere) is taken as a signal to allow for possible alternative normalization prescriptions. To schematically represent this, we let σ denote unnormalized measures. That is, we would interpret the expression

$$\int e^{i\Sigma s_r x_r - \frac{1}{2}\beta\Sigma x_r^2}\, d\sigma_1(x) = e^{-\frac{1}{2(1+\beta)}\Sigma s_r^2}$$

recognizing that σ is to be normalized along with an appropriate weight function. Actually, this latter view ("the σ-standpoint") is properly understood as working with the characteristic functions and not with the measures. In the other view ("the μ-standpoint"), in which μ is normalized, there is no justification for the "reinterpretation" of $K_\infty = 0$, μ a.e..

Of course, there is nothing special about our use of Gaussian examples above and all the same remarks apply, i.e., the μ- and σ-standpoints exist, etc., if we had chosen quite general integrands - a simple example being

$$K_N = e^{-\beta\Sigma_1^N x_r^4} \quad ,$$

or even more interesting if

$$K_N = e^{-\beta(\Sigma_1^N x_r^2)^\tau}$$

for some $\tau > 1$.

Significance for Single Particle Formulation

With the aid of the preceding discussion we have attempted to alert the reader to the occurrence of measures becoming orthogonal and the corresponding limitations for quantum applications of "the μ-standpoint." Although in the case $T = \infty$ it is true that $\exp(-\lambda\int|x(t)|^p dt) = 0$, μ_F a.e., a suitable normalization can be found such that $C(s)$ exists, fulfills $C(0) = 1$, and converges to $C_F(s)$ as $\lambda \downarrow 0$. Similar statements apply to a perturbation by P^2, for example, even when $T < \infty$. It is important to note that even though orthogonal measures on path space arise this in no way precludes (in "the σ-standpoint") the convergence of the characteristic functionals to the free one as $\lambda \downarrow 0$.

We have previously noted the ambiguity in the operator treatment that arises in the range $1 \leq \alpha < 2$. On the other hand, it seems that the functional integral approach effectively singles out a particular self-adjoint extension; specifically, the functional integral already exhibits the anomalous behavior of the pseudo-free theory whenever $\alpha \geq 1$. It is not surprising that the functional integral technique leads to the pseudo-free solution since, from its very construction, the functional integral does not seem able to contend with competing domain requirements between V and $P^2 + Q^2$, the latter being buried inside the measure. If further study confirms that $\alpha = 2$ is the proper dividing line between free and pseudo-free behavior, then the selection of $\alpha = 1$ by the functional integral technique must be regarded as spurious. On the other hand, if the operator analysis is revised so that $\alpha = 1$ becomes the dividing line, then this would clearly be in greater harmony with the functional integral results. These questions are under further study at the present time. Let us proceed to discuss $\alpha \geq 2$ and subsequently relate the cases $1 \leq \alpha < 2$ to that discussion.

For $\alpha \geq 2$ the basic stochastic process is modified in a fundamental way that can be best understood if we temporarily return to the Wiener process, i.e., Brownian motion. Imagine we had an "absorbing wall" at $x = 0$ that absorbs all particles that reach or would have crossed $x = 0$. Then the modified set of paths $x(t)$ passing between $a > 0$ and $b > 0$ would be the same as those entering the Wiener process with the exception that all paths that reach or cross zero are dropped. If $a > 0$ and $b < 0$, then <u>all</u> connecting paths would have to cross zero at least once, and so they are all discarded. Such processes are known in the literature of stochastic processes [13] as diffusion with absorbing boundaries. In our case such diffusion is based on the elementary solution of the diffusion equation given by

$$\frac{1}{\sqrt{\pi t}}\,[e^{-\frac{(x-y)^2}{t}} - e^{-\frac{(x+y)^2}{t}}]$$

for $xy > 0$ rather than just the first term alone. When $xy < 0$ then one chooses the solution zero. If we extend this result to the Ornstein-Uhlenbeck process, it means for $xy > 0$ that we consider the elementary solution

$$\sum_n \psi_n(x)\, e^{-tE_n}\,[\psi_n(y) - \psi_n(-y)] \qquad (10)$$

of Eq. (8) for $V = 0$ rather than the usual solution

$$\Sigma\ \psi_n(x)\ e^{-tE_n}\ \psi_n(y) \quad .$$

Here $\psi_n(x)$ and $E_n = 2n + 1$ are the eigenfunctions and eigenvalues of the usual harmonic oscillator. The even solutions drop out of (10) and we are left with

$$2 \sum_{odd} \psi_n(x)\ e^{-tE_n}\ \psi_n(y) \quad .$$

For $xy < 0$ one chooses the solution zero. Such an expression may be clearly rewritten as

$$\Sigma\ \psi'_n(x)\ e^{-tE'_n}\ \psi'_n(y)$$

in terms of the eigenfunctions and eigenvalues of the pseudo-free solution, and this automatically incorporates the correct boundary condition for $xy < 0$. The very presence of the potential, when $\alpha \geq 2$, drastically changes the fundamentals of the stochastic process, essentially projecting out certain paths, namely those that reach or cross zero. Similar conclusions also hold for $1 \leq \alpha < 2$.

We can summarize these findings, in a schematic way, by the expression

$$C(s) = \int e^{i\int s(t)x(t)dt-\lambda\int|x|^{-\alpha}dt}\ \chi(x)\,d\sigma_F(x) ,$$

where $\chi(x)$ denotes a formal characteristic function. For $\alpha < 1$, $\chi(x) \equiv 1$ (i.e., no paths are dropped) and the normalization is chosen so that $C(0) \equiv 1$ for all λ. In this case $C(s) \to C_F(s)$ as $\lambda \downarrow 0$. For $\alpha \geq 2$, $\chi(x) \not\equiv 1$, and we are instructed to drop paths that reach or cross zero, and this holds for $1 \leq \alpha < 2$ as well. Clearly this fundamentally changes the basic process, and we may introduce the (unnormalized) measure appropriate to the new basic process,

$$d\sigma_{PF}(x) \equiv \chi(x)\ d\sigma_F(x) \ .$$

Again the normalization is chosen so that $C(0) = 1$ for all λ, and we observe that as $\lambda \downarrow 0$

$$C(s) \to \int e^{i\int s(t)x(t)dt}\ d\sigma_{PF}(x) \equiv C_{PF}(s)$$

appropriate to the pseudo-free case. We do not recover $C_F(s)$ simply because the paths projected out can never be restored.

What seems basic to the path space formulation is the appearance of some formal projection onto a subset of paths denoted by $\chi(x)$. Whether $\chi \equiv 1$ or $\chi \not\equiv 1$ is unrelated to the measures being orthogonal or not, but has rather to do with whether certain paths are given distinctly zero weight with respect to other paths. In this context it is appropriate to recall for Wiener process theory (and most certainly for the Ornstein-Uhlenbeck process as well) that the set of paths $x(t)$, $|t| \leq T$, for which there exists some finite h such that

$$|x(t)-x(t')| \leq h|t-t'|^{\beta}$$

has probability zero if $\beta > \frac{1}{2}$ and probability one if $\beta < \frac{1}{2}$ [3]. If a path between $a > 0$ and $b > 0$ never reaches zero, then $\int|x|^{-\alpha}dt < \infty$ no matter what is the value of α. If a path crosses zero and $\alpha > 2$, then already the first crossing leads with probability one to the divergence of $\int|x|^{-\alpha}dt$, and a refined argument shows that to be true for $\alpha = 2$. When $1 \leq \alpha < 2$, on the other hand, the divergence of $\int|x|^{-\alpha}dt$ comes not from the first crossing but from the fact that once a path reaches zero it spends, with probability one, sufficient time in the neighborhood of zero, zigzagging back and forth, to lead to an infinite integral. It is interesting to observe that the path space formulation also makes some distinction in the character of the divergence that arises in the cases $1 \leq \alpha < 2$ and $\alpha \geq 2$*.

Finally, let us indicate how our zeroth approximation viewpoint signals the fundamental change of the basic process that arises when $\alpha \geq 2$. We may formally rewrite the characteristic functional as

$$C(s) = \int e^{i\int s(t)x(t)dt-\lambda\int|x|^{-\alpha}dt-\int(\dot{x}^2+x^2)dt}\, \mathcal{D}x$$

where again $\mathcal{D}x$ is a (nonexistent) translation invariant measure. In the spirit of our zeroth approximation, we consider paths for which $\int(\dot{x}^2+x^2)dt < \infty$ - which of course are a set of μ_F measure zero since almost all paths are nowhere differentiable [3]. Be that as it may, we examine further the properties of the functions $x(t)$ in that subset. For such functions we note that

$$x(t) - x(t') = \int_{t'}^{t} \dot{x}(\bar{t})d\bar{t} ,$$

*It is a pleasure to acknowledge discussions with L.Shepp that have helped clarify this situation.

and thus by Schwarz's inequality

$$|x(t)-x(t')| \leq |t-t'|^{\frac{1}{2}}(\int_{t'}^{t}\dot{x}^2 d\bar{t})^{\frac{1}{2}}$$

$$\leq |t-t'|^{\frac{1}{2}}(\int(\dot{x}^2+x^2)d\bar{t})^{\frac{1}{2}} ,$$

where in the last expression we have extended the integration limits to $\pm T$. Now consider paths that reach zero for the first time at t', and thus $x(t') = 0$. Then there is a constant $K < \infty$ such that

$$\int |x(t)|^{-\alpha}dt \geq \int_{-T}^{t'}|x(t)|^{-\alpha}dt \geq K \int_{-T}^{t'}|t-t'|^{-\alpha/2}dt ,$$

which diverges whenever $\alpha \geq 2$ even though $\int(\dot{x}^2+x^2)dt < \infty$. Consequently, the space of functions having finite formal integrand is reduced from that suggested by the free term alone, and this is in full accord with the fundamental change that we have previously noted in the underlying process. This result strengthens our conviction that the zeroth approximation can imply meaningful properties about the complete measure.

When $0 < \alpha < 2$, the preceding argument is inconclusive. There surely are some paths $x(t)$, e.g., $x(t) \equiv \exp[-(t-t')^{-2}]$, for which $\int|x|^{-\alpha}dt = \infty$; however, these are only a special subclass of paths that reach zero and their relative importance cannot be decided a priori in the zeroth approximation. Nonetheless, it may be suspected that such special paths occur rarely, and for $0 < \alpha < 1$ this conviction is solidly borne out since in that case $\int|x|^{-\alpha}dt < \infty$, μ_F a.e..

When $2 \geq p \equiv -\alpha \geq 0$, Hölder's inequality gives $[K' = (2T)^{1-\frac{1}{2}p}]$

$$\int |x(t)|^p dt \leq K' (\int x^2 dt)^{\frac{1}{2}p} \leq K' \{\int (\dot{x}^2+x^2) dt\}^{p/2} .$$

When $p \geq 2$ we can use the facts

$$\int |x(t)|^p dt \leq \sup |x(t)|^{p-2} \int x^2(t) dt$$

and [just as in (2) with $\varepsilon = 1$]

$$\sup |x(t)|^2 \leq \pi \int (\dot{x}^2+x^2) dt ,$$

to learn that (with $K = \pi^{\frac{1}{2}p-1}$)

$$\int |x(t)|^p dt \leq K \{ \int (\dot{x}^2+x^2) dt\}^{p/2} .$$

Evidently the interaction is power bounded for any $p \geq 0$. These relations suggest, correctly, that for $p \geq 0$ there is no fundamental change in the measure such as takes place for $\alpha \geq 2$. If $p \geq 2$, this conclusion even holds when $T = \infty$, in which case a perturbation of the form $\int |x|^p dt$ leads to orthogonal measures.

Quantum Field Formulation

We next turn our attention to quantum field theory. To attempt to make the basic effect plausible, we employ the functional integration form of Euclidean quantum field theory. Specifically, if $h(x)$ denotes a test function on $x \in R^n$ (n = space-time dimension), then the formal analogue to our previous expressions is given by

$$S(h) = \int e^{i(h,\Phi)-\lambda V(\Phi)-W_o(\Phi)} D\Phi$$

where $\Phi(x)$ are c-number "fields",

$$(h,\Phi) = \int h(x)\Phi(x)dx \ ,$$

$$W_o(\Phi) = \tfrac{1}{2}\int (|\nabla\Phi|^2+m^2\Phi^2)dx \ ,$$

where $dx = d^n x$, and $V(\Phi)$ denotes the interaction potential. In standard language this is the T-product generating functional in Euclidean space time [2]. In particular

$$|\nabla\Phi|^2 \equiv \sum_{j=1}^{n} (\partial\Phi/\partial x^j)^2 \ ,$$

which is evidently expressed with a Euclidean metric.

The free field theory in n space-time dimensions is characterized by the fact that

$$S_F(h) \equiv \int e^{i(h,\Phi)} d\mu_F(\Phi) = e^{-\frac{1}{2}\int \frac{|\tilde{h}(k)|^2}{k^2+m^2} dk} \ , \qquad (11)$$

and the formal identification is such that

$$d\mu_F(\Phi) = e^{-W_o(\Phi)} D\Phi \ .$$

As in the Brownian motion problem it is suggestive to regard $W_o(\Phi) < \infty$ as the zeroth approximation of the relative support of the measure μ_F. Since our treatment of the potential term is restricted to this level of analysis, we shall not go beyond the heuristic physical ideas implicit in this zeroth approximation.

We now study the interacting case and consider

$$\int e^{i(h,\Phi)-\lambda V(\Phi)} \, d\sigma_F(\Phi) \quad . \tag{12}$$

Observe that we have adopted "the σ-standpoint" which is mandatory for even the simplest of potentials. Suppose V just represents a mass shift, $m^2 \rightarrow \bar{m}^2$, the new functional for which would have the form indicated in (11) with m^2 replaced by $\bar{m}^2$. Clearly this expression is continuous in the mass [as was the example in Eq. (9) continuous in β], in spite of the fact that the measures μ_F are mutually perpendicular for different masses. Indeed "relatively few" expressions admit "the μ-standpoint", but notable among them are the normal ordered $P(\Phi)_2$ theories defined in a two space-time dimensional field theory [14]. However, that is not our principal concern - instead we are potentially interested in higher space-time dimensions (like four) and in singular potentials.

Suppose we assume that $V(\Phi) \geq 0$, and we let $\chi(\Phi)$ denote the formal characteristic function defined by

$$\chi(\Phi) = 1; \qquad W_o(\Phi) < \infty, \qquad V(\Phi) < \infty,$$

$$\chi(\Phi) = 0; \qquad W_o(\Phi) < \infty, \qquad V(\Phi) = \infty.$$

Than we may formally write ($\lambda > 0$)

$$\int e^{i(h,\Phi)-\lambda V(\Phi)} \, \chi(\Phi) d\sigma_F(\Phi)$$

without changing the expression (12). As $\lambda\downarrow 0$ this expression becomes

$$\int e^{i(h,\Phi)}\ \chi(\Phi)d\sigma_F(\Phi)\ ,$$

which if $\chi(\Phi) \not\equiv 1$ formally defines the pseudo-free theory and suggests we introduce

$$d\sigma_{PF}(\Phi)\ \equiv\ \chi(\Phi)d\sigma_F(\Phi)$$

and

$$S_{PF}(h)\ \equiv\ \int e^{i(h,\Phi)}\ d\sigma_{PF}(\Phi)\ .$$

Whenever $\chi(\Phi) \not\equiv 1$ we have definitely changed the set of fields having finite formal integrand, and if the zeroth approximation provides any clue about the true support of the measure then it is plausible that a pseudo-free theory arises that is distinct from a free theory.

How do we reconcile the condition $V(\Phi) \geq 0$ with the standard picture that normal ordering is required? An expression such as (12) is largely of heuristic value. In general, to carry out such an integral, many renormalizations, such as an infinite mass renormalization, Wick ordering, etc., are often required. On the other hand, the nature of these modifications is largely determined by the form of W_o and V. It is the spirit of the zeroth approximation that qualitative, relative support properties of the measure may be correctly inferred from the finiteness of the formal integrand. This conviction is of course based on the assumption that the proper measure is in some sense an extension of the finite integrand (i.e., of the classical action) beyond its finite domain of applicability. The validity of this viewpoint can hardly be tested for covariant theories until their solutions are available. (However, model fields exhibiting the expected qualitative features will be discussed in the next section.)

One might approach the whole problem by introducing cutoffs, say both in momentum and configuration space. The so-regularized interaction is now finite, as are any Wick subtractions and necessary renormalizations. As the cutoffs are removed, and the measure approaches an orthogonal one, it is not inconceivable that certain fields receive ever and ever smaller relative weight and in the limit do not contribute at all. This would be one possible way of constructing the formal characteristic function χ belonging to a pseudo-free theory, but that is beyond our ability at the present time.

We adopt, instead, our physically motivated criterion and study the relation between

$$V(\Phi) = \int |\Phi(x)|^p \, dx$$

and the free term $W_o(\Phi)$. The properties of $\chi(\Phi)$ may then be essentially reduced to arguments involving classical Sobolev embedding theorems and multiplicative inequalities. In this context we follow closely the results presented in Ref. [15] to which we refer the reader for proofs. Our notation is as follows:

$$||\Phi||_p \equiv (\int |\Phi(x)|^p dx)^{1/p} ,$$

$$||\nabla\Phi||_2 \equiv (\int |\nabla\Phi(x)|^2 dx)^{\frac{1}{2}} ,$$

$dx = d^n x$, and n = space-time dimension.

Case A: $n = 2, \quad 2 \leq p < \infty$.

For $\Phi(x) \in C_o^\infty(R^2)$ one has

$$||\Phi||_p^2 \leq 2\pi\{k||\nabla\Phi||_2^2 + (\frac{p}{k})^{\frac{p}{2}-1}||\Phi||_2^2\}$$

for any $0 < k < \infty$. Another relation is given by

$$||\Phi||_p \leq 5\sqrt{p}\,||\nabla\Phi||_2^{1-2/p}\,||\Phi||_2^{2/p} .$$

(Although proved for C_o^∞ these inequalities extend to fields for which the right side is finite.) As a consequence we observe that finite constants exist such that

$$\int|\Phi|^p dx \leq KW_o(\Phi)^{p/2} , \tag{13a}$$

$$\int|\Phi|^p dx \leq (\int[\varepsilon|\nabla\Phi|^2 + m^2(\varepsilon)\Phi^2]dx)^{p/2} , \tag{13b}$$

$$\int|\Phi|^p dx \leq \varepsilon(\int|\nabla\Phi|^2 dx)^{p/2} + M^2(\varepsilon)(\int|\Phi|^2 dx)^{p/2}, \tag{13c}$$

the latter two holding for arbitrary $\varepsilon > 0$. Traditionally such interactions are associated with super renormalizable theories.

Case B: $n \geq 3, \quad 2 \leq p \leq 2n/(n-2)$.

For $\Phi(x) \in C_o^\infty(R^n)$, the basic inequality (as we shall need it) reads

$$||\Phi||_p \leq 4||\nabla\Phi||_2^\alpha\,||\Phi||_2^{1-\alpha} \tag{14}$$

where

$$\alpha = \frac{n(p-2)}{2p} ,$$

which as p varies satisfies $0 \leq \alpha \leq 1$.

Subcase B1: $n \geq 3$, $0 \leq \alpha < 1$. With $\alpha \neq 1$ we deal with

$$2 \leq p < \frac{2n}{n-2}$$

and the multiplicative inequality (14) leads to

$$||\Phi||_p \leq \varepsilon||\nabla\Phi||_2 + c(\varepsilon)||\Phi||_2$$

for any $\varepsilon > 0$. From this relation we again determine three expressions like (13). Again these interactions are traditionally associated with super renormalizable theories.

Subcase B2: $n \geq 3$, $\alpha = 1$. With $\alpha = 1$ we deal with

$$p = \frac{2n}{n-2}$$

and we learn that (14) becomes

$$||\Phi||_p \leq 4||\nabla\Phi||_2 + \delta||\Phi||_2$$

for any $\delta > 0$. Note that we necessarily have a minimum coefficient of the gradient term required to obtain bounding. Hence we obtain a bound of the type

$$\int|\Phi|^p dx \leq KW_o(\Phi)^{p/2}$$

but not of the other types with arbitrarily small coefficient of the gradient term. Such interactions are traditionally associated with renormalizable interactions (e.g., for $n = 4$, $p = 4$, while for $n = 3$, $p = 6$).

Case C: $n \geq 3$, $p > 2n/(n-2)$.

These interactions are not power bounded. Let $\Phi = r^{-\gamma}$, and assume Φ falls smoothly to zero outside a compact region. Then $\int|\Phi|^p dx = \infty$ provided $p\gamma \geq n$, while $W_o(\Phi) < \infty$ provided $\gamma < \frac{1}{2}n - 1$. Hence for a given $p > 2n/(n-2)$ we can always choose a range of γ fulfilling these two criteria, and as a consequence we have fields for which $W_o < \infty$ while $V = \infty$, i.e., a characteristic function $\chi(\Phi) \not\equiv 1$. Traditionally such interactions are associated with nonrenormalizable interactions. We leave it as an exercise for the reader to confirm that the introduction of suitable momentum space cutoffs restores power bounding to the nonrenormalizable interactions.

In summary we are led to the conjecture that renormalizable covariant field theories are continuously connected to the free theories while nonrenormalizable covariant field theories are not. Instead, as $\lambda \downarrow 0$ these cases tend to pseudo-free theories formally characterized by

$$\int_{"L^p"} e^{i(h,\Phi)}\, d\sigma_F(\Phi) \tag{15}$$

which in the zeroth approximation is restricted to fields for which

$$\int|\Phi|^p dx < \infty \; .$$

Observe that such a criterion is scale invariant; namely, if

$$\Phi'(x) \equiv S^d \Phi(Sx)$$

denotes a scale transformed field, then $||\Phi||_p$ and

$||\Phi'||_p$ are either both finite or both infinite. Consequently, one would expect the pseudo-free theory to be scale covariant, i.e., changing the mass term in σ_F as well. Indeed, if the mass in σ_F were set to zero, one would even expect the pseudo-free theory to be scale invariant. In either case, one suspects that the pseudo-free theory is a better starting point for the interacting theory than the free theory. [Whether or not the pseudo-free theories formally given by (15) are different for different p is not known.]

Finally, we should like to remark that although our zeroth approximation classical inequalities have put the renormalizable theories into the free category (i.e., continuously connected to the free theory) it is not impossible that better arguments may put some of the renormalizable theories into the pseudo-free category. (Recall that in the simple model of Sec. 2 the case $\alpha = 2$ was related to the pseudo-free cases.)

4. ULTRALOCAL MODELS

In this last section we consider ultralocal model field theories and show that they possess the basic properties ascribed to the nonrenormalizable interactions. Such models have been treated elsewhere [4,5] and we shall not review them here in any detail. The defining characteristic of such models is the dropping of the spacial derivatives (but not the time derivative) from the Lagrangian. In the language of Sec. 3 the free action $W_o(\Phi)$ is changed to become

$$W_o'(\Phi) = \frac{1}{2} \int [(\frac{\partial \Phi}{\partial \tau})^2 + m^2\Phi^2]dx \quad ,$$

where τ is the Euclidean coordinate distinguished as the imaginary time. Quite clearly

$$W'_o(\Phi) \leq W_o(\Phi)$$

for all $\Phi(x)$, and any theory for which $W_o(\Phi) < \infty$ does not imply $V(\Phi) < \infty$ will have the same property for $W'_o(\Phi)$. Indeed, it is not difficult to see that for $n \equiv s + 1 \geq 2$ the potential $\int|\Phi|^p dx$ is not power bounded by W'_o for <u>any</u> $p > 2$. One need only choose $\Phi(x) = h(\tau)u(\underset{\sim}{x})$, or a sum of such functions, where $h \in C_o^\infty$ and $u(\underset{\sim}{x}) \in L^2(R^s)$ but $u(\underset{\sim}{x}) \notin L^p(R^s)$. Then

$$W'_o(\Phi) = \tfrac{1}{2}||u||_2^2 \int[\dot{h}^2 + m^2h^2]d\tau < \infty$$

while

$$V(\Phi) = ||h||_p^p \;\; ||u||_p^p = \infty \;.$$

The free field measure associated with $W'_o(\Phi)$ is well defined, and in fact

$$S'_F(h) = \int e^{i(h,\Phi)} \, d\mu'_F(\Phi)$$

$$= e^{-\frac{1}{2}\int\frac{|\tilde{h}(k)|^2 dk}{k_o^2+m^2}} \;.$$

On the other hand, we suspect for any $p > 2$ that the interacting theories are not continuously connected with the free theory but with the pseudo-free theory formally given by

$$S'_{PF}(h) = \int_{"L^p"} e^{i(h,\Phi)}\, d\sigma'_F(\Phi) \quad .$$

Proper Solutions

We now discuss the true solutions of the ultralocal models and examine them for the expected features. We temporarily specialize to the time-zero field $\phi(\underset{\sim}{x})$ to introduce the basic ideas. In that case dynamical independence of the field at distinct spacial points and simple symmetries of the model lead to the expectation functional

$$E(f) = e^{-\int d\underset{\sim}{x} \int [1-\cos\lambda f(\underset{\sim}{x})] c^2(\lambda) d\lambda} \quad . \qquad (16)$$

Here $c(\lambda)$, $\lambda \in R$, is related to the particular potential V in a certain way. The analogous expression for the free ultralocal field is given by

$$E_F(f) = e^{-\frac{1}{4m}\int f^2(\underset{\sim}{x}) d\underset{\sim}{x}} \quad .$$

All interacting models with symmetric, semi-bounded potentials V fit into the form of the functional (16). The function $c(\lambda)$ may be further delineated in the form

$$c(\lambda) = \frac{1}{|\lambda|^\gamma} e^{-\frac{1}{2}m\lambda^2 - y(\lambda,\kappa)} \quad ,$$

where γ is a parameter to be determined and y is related to the potential. This choice corresponds [4] to the formal Hamiltonian

$$\int \{\tfrac{1}{2}[\pi^2(\underset{\sim}{x}) + m^2\phi^2(\underset{\sim}{x})] + \kappa V[\phi(\underset{\sim}{x})]\}d\underset{\sim}{x} ,$$

where we have denoted the coupling constant by κ in this section. As $\kappa\downarrow 0$, $y(\lambda,\kappa)\to 0$ smoothly, and the pseudo-free theory is characterized by

$$E_{PF}(f) = \exp \{-\int d\underset{\sim}{x}\int[1-\cos\lambda f(\underset{\sim}{x})] \frac{e^{-m\lambda^2}}{|\lambda|^{2\gamma}} d\lambda\} . \qquad (17)$$

Observe that this expression depends on just the parameter m (γ will be fixed). If $m\to 0$, then we find

$$E_{PF}(f) = \exp \{-\int d\underset{\sim}{x}\int[1-\cos\lambda f(\underset{\sim}{x})] \frac{d\lambda}{|\lambda|^{2\gamma}} \}$$

$$= \exp \{-k\int d\underset{\sim}{x}|f(\underset{\sim}{x})|^{2\gamma-1}\} . \qquad (18)$$

Scale covariance of (17) [and scale invariance of (18)] under the transformation

$$\phi(\underset{\sim}{x}) \to S^d\phi(S\underset{\sim}{x}) ,$$
$$m \to S^{-\beta}m$$

is ensured [5] provided that

$$d = s - \tfrac{1}{2}\beta ,$$
$$\gamma = \frac{s}{\beta} + \frac{1}{2} .$$

This relates the parameter γ to the space dimension s and to the assumed scaling law of the mass m. As it stands (17) applies only for $\frac{1}{2} \leq \gamma < \frac{3}{2}$, while (18) requires $\frac{1}{2} < \gamma < \frac{3}{2}$. However, the covariant examples that motivate our study are those for which $m \to S^{-1}m$, i.e., $\beta = 1$, and for those $\gamma = s + \frac{1}{2}$, which lies outside the range of (17) and (18). This simply means that the sharp time fields do not exist for such fields and time smearing is required to obtain meaningful operators. However, suitable renormalized operator products [4,5] do exist as sharp time fields and since each of these (!) is maximally abelian they will suffice for our purposes. Consequently, we consider the expectation functional for the renormalized field power $\phi_r^\theta(\underset{\sim}{x})$ [4,5] given by

$$E_{PF}^{\theta}(f) = \exp\{-\int d\underset{\sim}{x} \int [1-\cos\lambda^{\theta} f(\underset{\sim}{x})] \frac{e^{-m\lambda^2}}{|\lambda|^{2\gamma}} d\lambda\} ,$$

where $\theta > \gamma - \frac{1}{2} = s$. This field (for $\beta = 1$) has scale dimension $d_\theta = s - \frac{1}{2}\theta$, and we can always choose θ such that $d_\theta > d_c \equiv \frac{1}{2}(s-1)$.

At the very least the two expectation functionals illustrate a genuine distinction that exists between the free theory and the pseudo-free theory demonstrating the basic effect we propose. To emphasize the fundamental distinction in these two cases, let us examine the measures that lead to the free case and to the pseudo-free case, which is meaningful since the appropriate field is maximal abelian in each case. Thus we consider the measures given by means of

$$E_F(f) = \int e^{i(f,\Lambda)} d\hat{\mu}_F(\Lambda) ,$$

and

$$E_{PF}^{\theta}(f) = \int e^{i(f,\Lambda)} d\hat{\mu}_{PF}^{\theta}(\Lambda) .$$

These measures are orthogonal ($\perp$) to one another; but we also know they are fundamentally different from one another in the following sense. If $\hat{\mu}^{(g)}(A) \equiv \hat{\mu}(A+g)$ for arbitrary (measurable) sets A defines a translated measure, then it is known for the free measure that

$$\hat{\mu}_F^{(g)} \approx \hat{\mu}_F \, , \qquad \text{for all } g \in L^2$$
$$\hat{\mu}_F^{(g)} \perp \hat{\mu}_F \, , \qquad \text{for all } g \notin L^2 \, ,$$

where $\approx$ denote equivalence of measures (roughly, the same support). On the other hand [16], for the pseudo-free measure

$$\hat{\mu}_{PF}^{\theta\,(g)} \perp \hat{\mu}_{PF}^{\theta} \, , \text{ for } \underline{\text{all}} \; g \neq 0 \, .$$

Thus the character of the two measures is seen to be fundamentally different, and it is tempting to formally interpret this fact as the consequence of projecting out certain functions by an appropriate $\chi(\Phi)$. In any case, we strongly feel that the basic concepts underlying the pseudo-free models have been demonstrated and that they are worthy of further development for application to covariant problems.

ACKNOWLEDGMENTS

It is a pleasure to thank Professors H. Ezawa and L. Faddeev for their comments on some of the ideas presented herein while they were in a formative stage, and to Professors H.Ezawa for a critical reading of the final text. This written version of the lectures has measurably benefited from remarks and comments those lectures engendered, especially from Professors G. Dell'Antonio, B. Simon, L. Streit, K. Symanzik and W. Thirring.

REFERENCES

1. J. R. Klauder, "Remarks on Nonrenormalizable Interactions", (unpublished).
2. K. Symanzik, Proc. of the International School of Physics "Enrico Fermi", Varenna Course XLV, ed. R. Jost, Academic Press (1969).
3. See, e.g., I. M. Koval'chik, Russian Mathematical Surveys 18, 97 (1963).
4. J. R. Klauder, Acta Physica Austr., Suppl. VIII, 227 (1971).
5. J. R. Klauder, "Functional Techniques and Their Application in Quantum Field Theory," Boulder Summer School (1971) (to be published).
6. T. Kato, Perturbation Theory of Linear Operators, Springer-Verlag (1966).
7. B. Simon, Annals of Physics (New York) 58, 76 (1970).
8. E. Jahnke and F. Emde, Tables of Functions, Dover (1945).
9. G. Szego, Orthogonal Polynomials, Amer. Math. Soc. (1959).
   See also G. Parisi and F. Zirilli, "Anomolous Dimensions in One-Dimensional Quantum Field Theory", J. Math. Phys. (to be published).

10. See, e.g., the review article by W. M. Frank, D. J. Land and R. M. Spector, Rev. Modern Phys. 43, 36 (1971), which also contains numerous references.

11. M. C. Wang and G. E. Uhlenbeck, Rev. Modern Phys. 17, 323 (1945), reprinted in Selected Papers on Noise and Stochastic Processes, edited by N. Wax, Dover (1954).

12. T. Hida, Theory of Prob. Appl. (Moscow) 15, 119 (1970).

13. S. Chandrasekhar, Rev. Modern Phys. 15, 1 (1943), reprinted in Selected Papers on Noise and Stochastic Processes, edited by N. Wax, Dover (1954); M. S. Bartlett, An Introduction to Stochastic Processes, Cambridge (1966).

14. F. Guerra, L. Rosen and B. Simon, "The $P(\phi)_2$ Euclidean Quantum Field Theory as Classical Statistical Mechanics" (in preparation).

15. O. A. Ladyzenskaja, V. A. Solonnikov and N. N. Ural'ceva, Linear and Quasi-Linear Equations of Parabolic Type, Trans. of Math. Mono., Vol. 23, American Math. Society (1968).

16. G. C. Hegerfeldt and J. R. Klauder, Il Nuovo Cimento 10A, 723 (1972).

FIGURE CAPTION

Fig.1 Schematic representation of "theories" based on Hamiltonians $H = P^2 + Q^2 + \lambda|Q|^{-\alpha}$, $\lambda \geq 0$. For $\alpha < 2$ such theories are continuously related to the free theory (F), while for $\alpha \geq 2$ they are continuously related to the pseudo-free theory (PF), as $\lambda\downarrow 0$. If one starts with (F), i.e., $H_0=P^2+Q^2$, and turns on $\lambda|Q|^{-\alpha}$, $\alpha \geq 2$, one immediately goes from (F) to the branch connected with (PF). Reducing λ to zero takes one back to (PF) and not to (F).

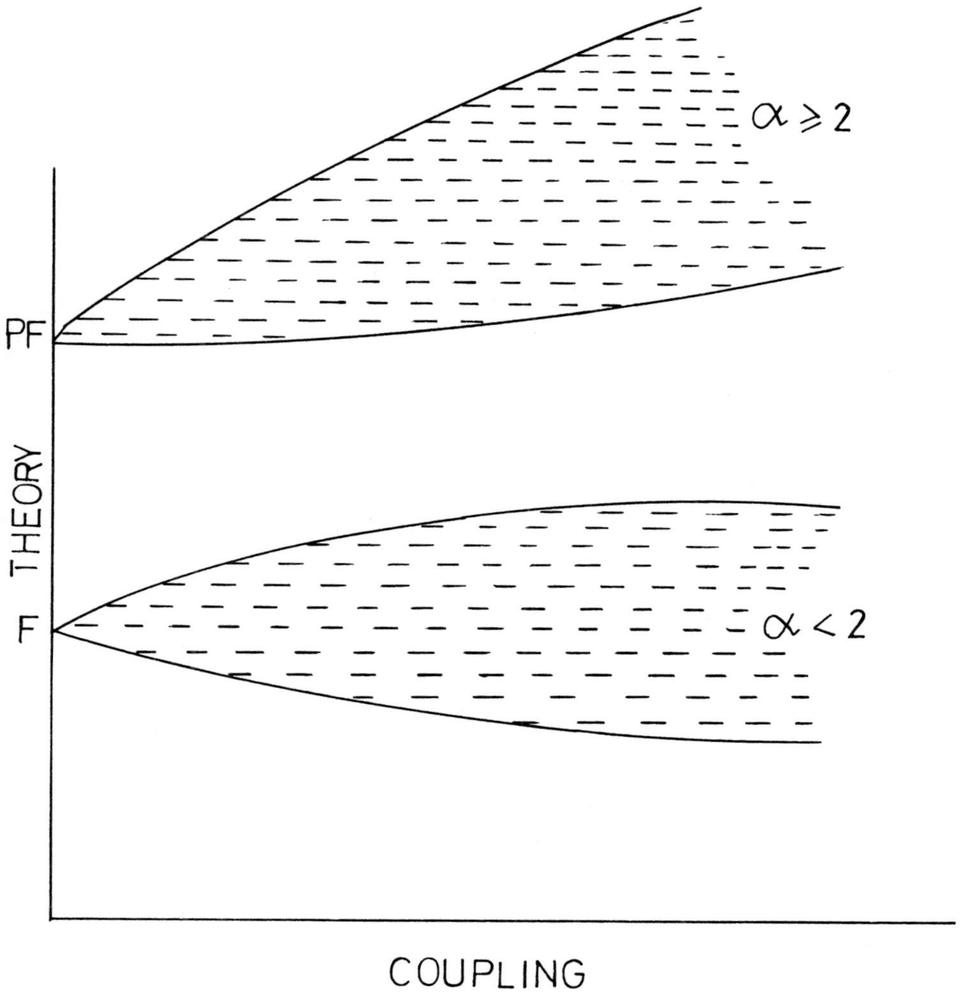

Fig. 1

Acta Physica Austriaca, Suppl.XI, 389–473 (1973)

RENORMALIZED LOCAL QUANTUM FIELD THEORY AND CRITICAL BEHAVIOUR*

BY

F. JEGERLEHNER and B. SCHROER
Institut für Theoretische Physik
Freie Universität Berlin

INTRODUCTORY REMARKS

In these lectures we discuss critical phenomena using methods of renormalized local quantum field theory. Such methods have already been used in the following articles:

1. G. Mack, Kaiserslautern lectures 1972, Lecture Notes in Physics, Vol. 17, Springer Verlag
2. C. Di Castro, Lettere Nuovo Cim. 5, 69 (1972).
3. B. Schroer, "A Theory of Critical Phenomena based on the Normal Product Formalism", FU Berlin preprint 1972.

---

*Lecture given at XII. Internationale Universitätswochen für Kernphysik, Schladming, February 5 - 17, 1973.

Supported in part by the Deutsche Forschungsgemeinschaft Contract No.: 160/1

4. E. Brezin, I. C. Guillou and J. Zinn-Justin, "Wilson Theory of Critical Phenomena and Callan-Symanzik Equations in 4-ε Dimensions". Saclay preprint.

5. P. K. Mitter, "Callan-Symanzik Equations and ε-Expansion". University of Maryland Technical Report.

Dedicated to K. G. Wilson, whose ideas have enriched Statistical Mechanics and rejuvenated Quantum Field Theory.

TABLE OF CONTENT

1. SIMILARITIES AND DIFFERENCES BETWEEN RELATIVISTIC QFT AND THE STATISTICAL MECHANICS OF CRITICAL PHENOMENA

Before we go into the details of use of renormalized field theory to the description of critical behaviour some physically qualitative and mathematically formal remarks are in order. The simplest, and for a long time only description of phase transitions near the Curie point is the famous molecular field theory. In the ferromagnetic language, the thermal correlation functions of the spinvariable $\sigma(x)$ (for simplicity we talk about scalar variables corresponding to a Lenz-Ising system)

$$< \sigma(\vec{x}_1) \ldots \quad \sigma(\vec{x}_N) >_c , \quad c = \text{connected} \tag{1.1}$$

furnish a complete description of the system. For the molecular field theory we have:

$$< \sigma(\vec{x}) > \; = \; M \; = \; \text{density of magnetization} \tag{1.2}$$

$$< \sigma(\vec{x})\sigma(\vec{y}) >_c = \frac{1}{(2\pi)^3} \int \frac{e^{-i\vec{k}(\vec{x}-\vec{y})}}{\vec{k}^2+\xi^{-2}} d^3k \approx \frac{e^{\frac{|\vec{x}-\vec{y}|}{\xi}}}{|\vec{x}-\vec{y}|} .$$

The correlation length ξ is related to the temperature $t = T - T_c$: (for magnetic field $H = o$)

$$\xi \sim |t|^{-\frac{1}{2}} \tag{1.3}$$

and the mean field theory leads to an equation of state of the form:

$$H = c_1 t \; M + c_2 \; M^3 \; . \tag{1.4}$$

A simple derivation of this statement in a field theoretical language can be found in the review article by Kogut and Wilson[1]. In our derivation of equation of state it will appear as a special case. These correlation functions appear as analytically continued vacuumexpectation values of a 3 dimensional free relativistic theory with a mass $m^2 = \xi^{-2}$. This transition from Minkowski points to euclidean points may also be raised to an algebraic level by introducing the commuting euclidean fields[2]:

$$\sigma(\vec{x}) = \frac{1}{(2\pi)^{3/2}} \int \frac{e^{-i\vec{k}\vec{x}}}{\sqrt{\vec{k}^2+m^2}} (A^+(\vec{k})+A(-\vec{k}))d^3k + M \tag{1.5}$$

with $$[\ A(\vec{k}),\ A^+(\vec{k}')\ ] = \sigma\ (\vec{k}-\vec{k}')\ . \tag{1.6}$$

In this euclidean Fock-space the σ act cyclically on the vacuum, but because of

$$[\sigma(\vec{x}),\ \ \sigma(\vec{y})] = 0 \tag{1.7}$$

the thermal correlation functions lead to a highly reducible euclidean field. This is of course related to the fact that thermal states are always impure. In the discrete formalism of the lattice gas it corresponds to the well known statement that in the space in which the transfer matrix acts on the order variable, σ only generate a subalgebra of the total algebra of this space. (The transfer-matrix space for example also contains disorder variables.) This situation which we discussed in trivial case of the molecular field approximation also pertains to the general case. As Wilson[3] pointed out, there is a perfect analogy between the generating functional of a relativistic theory

$$Z(J)=N^{-1}\int d[A(x)]\exp\{i\int A(x)J(x)d^4x+i\int L(A)d^4x\} \qquad (1.8)$$

with $L(x) = \frac{1}{2}\partial_\mu A\partial^\mu A - \frac{m^2}{2}A^2 - \frac{\lambda}{4!}A^4 - cA(x)$

and that of the continous version of the Lenz-Ising model ($\beta = \frac{1}{kT}$ has been absorbed into the field and the constants):

$$Z(J)=N^{-1}\int d[\sigma(x)]\exp\{\int\sigma(x)J(x) - V(\sigma)\} \qquad (1.9)$$

with $V(\sigma) = \int d^3x\{\frac{1}{2}\partial_i\sigma\ \partial_i\sigma - \frac{m^2}{2}\sigma^2 - \frac{\lambda}{4!}\sigma^4 + H\sigma\}$

H = external magnetic field.

The physical origin of the quadrilinear term is (as Wilson explained) the wanted peaking of the probability at $\sigma = \pm 1$.

It should be stressed that the idealization from the lattice system to the continous version (1.9) is useful only near the critical point. In this description the lattice distance has been summarily taken into account by the introduction of a cut-off in momentum space. The cut-off therefore has a fundamental physical interpretation in the euclidean field theory (EFT) defined by (1.8) whereas in the relativistic quantum field theory corresponding to (1.8) the cut-off is at best a phenomenological concept (expressing perhaps the omission of other local couplings) and therefore should be eliminated by renormalization. If we consider the statistical correlation function right at the critical point where $\xi \to \infty$, the influence of the cut-off should disappear and the elimination by renormalization is expected to be a legitimate procedure. In this way the close analogy between the two theories is reestablished. Then the only difference remaining is a physical one: relativistic QFT is expected to give us interesting insight into the physics of elementary particles for short distances, whereas statistical E F T should tell us something about the long range fluctuations which are responsible for the critical behaviour. As will become clear in section 4 the mass is insignificant if we are interested in short-distance scaling whereas for the discussion of long distance saling we first have to perform the limit $\xi \to \infty$. The two renormalized theories corresponding to (1.8) and (1.9) are just related by analytic continuation from Minkowski points to euclidean points. From the mathematical point of view they are identical[4]. The selection between the long distance asymptote which is different from the short distance asymptote originates from the physical distinction between statistical mechanics and elementary particle physics.

In the next section we will formulate our general framework of renormalization for the relativistic case. The transscription to the euclidean form is always straightforward.

2. THE ALGEBRAIZATION OF THE DYNAMICAL CONTENT

The limitation of the formal canonical version of QFT led to the LSZ- and the Wightman-framework. Mainly from the work of Wilson[5], Brandt[6] and Zimmermann[7] it became clear that canonical arguments have no place even in Lagrangian field theory. The new formulation of Lagrangian field theory which emerged during recent years can be best understood by looking first at classical fields. The set of point-like classical observables is generated by all monomials $\Theta(A)$ of the field $A(x)$ including derivations (examples: $\Theta(A(x)) = A^2(x)$, $\partial_\mu A \partial_\nu A$, $\partial_\mu \partial_\nu A$ etc.). The equation of motion

$$(\partial^2 + m^2) \quad A(x) = \frac{\lambda}{3!} A^3(x) \tag{2.1}$$

gives a "texture" to the formal classical algebra: it defines a (two-sided) ideal in this algebra. The corresponding quantum fields ("composite fields") are called normal products $N[\Theta]$. To one classical Θ there correspond many $N[\Theta]$'s which will be distinguished by indices. The equation of motion for $A(x)$ as well as the equation of motion for an A inside a normal product again lead to an ideal in the Normal-product algebra. However in contradistinction to the classical case and in correspondence with the nonuniqueness of normal products belonging to a classical Θ, we have other algebraic relations between normal products. These are called Zimmermann identities[7]. Together with the algebraic relations due to the equation of motion they furnish a <u>complete</u> set of relations between normal products. Since this completeness can be rigorously established one is sure to have a set of algebraic relations from which all <u>structural</u> (i.e. non numerical) properties <u>must</u> follow. The power of this approach has been amply demonstrated by deriving the Adler-Bardeen statement about absence of radiative correction in QED anomalies[8], the Callan-Symanzik and Renormalization group equations[9], the vanishing of the Callan-Symanzik function β in the Thirring model[29], the Wilson short distance expansion[10] etc.

In the following we will adopt the point of view that the algebraic relations between composite fields are the fundamental structural properties of a field theory defined in terms of a Lagrangian. For discussions of critical behaviour one then derives the set of "prescaling equations" (this set contains the Callan-Symanzik equations and the Renormalization-group equations). The pre-scaling equations are then used to construct the perturbative pre-asymptotic theory. The scale invariant short- and long-distance behaviour emerges if one sums up the pre-asymptotic theory.

Let us look at these problems in the simplest case of a relativistic A^4 coupling without symmetry-breaking (i.e. with c = 0). The unrenormalized (formal) Feynman rules for the Green-functions are given in terms of the Gell-Mann Low formula (which may for example be derived from formula

$$< TX>_{unr} = N^{-1} <\Phi_o \,| TX_o \, e^{i\int :L_{int}(A_o):d^4x} | \Phi_o > \tag{2.2}$$

$$N = <\Phi_o | \, e^{i\int : L_{int}(A_o):d^4x} \, | \Phi_o >$$

with: Φ_o = Fockvacuum, A_o = free field

$$X = \prod_{i=1}^{N} A(x_i)$$

The renormalized Greens function with the normalization properties (prop = 1 particle irreducible part)

$$\Gamma^{(2)}(-p,p)\Big|_{p^2=m^2} = <TA(o)\tilde{A}(p)>^{prop}\Big|_{p^2=m^2} = 0 \qquad (2.3a)$$

$$\Gamma^{(2)}(-p,p)\Big|_{p^2=-\mu^2} = <TA(o)\tilde{A}(p)>^{prop}\Big|_{p^2=-\mu^2} = i(-\mu^2-m^2) \qquad (2.3b)$$

$$\Gamma^{(4)}(-\sum_1^3 p_i,p_1,p_2,p_3)\Big|_{s.p.} = <T\ (A(o)\tilde{A}(p_1)\tilde{A}(p_2)\tilde{A}(p_3)>^{prop}\Big|_{s.p.} = -i\lambda \qquad (2.3c)$$

(where s.p. = symmetry point = $p_i p_j = -\frac{1}{3}(4\delta_{ij} - 1)\mu^2$
μ^2 = spot of normalization $\neq -m^2$)

can be obtained by the finite B P H Z version of formula (2.2):

$$<TX> = \text{"Finite Part"} <\Phi_o\Big|TX_o e^{i\int :L^{eff}_{int}(A_o(x):d^4x}\Big|\Phi^o>_x \qquad (2.4)$$

x = omission of vacuum loops.

$$L^{int}_{eff} = \frac{b}{2}\partial_\mu A\partial^\mu A + \frac{a}{2}A^2 - \frac{\lambda-c}{4!}A^4 \qquad (2.5)$$

The innocent looking "Finite Part" is a complicated operation expressed in terms of Taylor-subtractions which acts on the Feynman-integrands in momentum space:

$$I_\Gamma \; (p,k) \xrightarrow{\text{F.P.}} R_\Gamma (p,k) \qquad (2.6)$$

$\Gamma = n^{th}$ order Graph.

This R-process in the form given by Zimmermann[7] solves the problem of arbitrary overlapping divergencies and leads to an absolutely convergent integrand R_Γ. The integral over the loop momenta k is the contribution of Γ to the renormalized Greens function in momentum space. The F. P. acts on all one particle irreducible sub-graphs $\gamma \subset \Gamma$ which are "renormalization parts" i.e. for which the degree of superficial divergence is non-negative

$$d(\gamma) = 4 - N(\gamma) \geq 0 \qquad (2.7)$$

$N(\gamma)$ = number of external lines of γ.

The combinatorical aspects of the overlapping divergencies are solved by the Zimmermann "forest" formula:

$$R_\Gamma (p,k) = \sum_{U \varepsilon F_\Gamma} \prod_{\gamma \varepsilon U} (-t^{d(\gamma)}) \; I_\Gamma \qquad (2.8)$$

$t^{d(\gamma)}$ are the Taylor operators up to derivations of degree $d(\gamma)$ in the external momenta of the renormalization parts γ. The product goes over a set U (called forest) of nonoverlapping γ's (γ_1 nonoverlapping to γ_2 if either $\gamma_1 \cap \gamma_2 = \emptyset$ or $\gamma_1 \subset \gamma_2$ resp. $\gamma_2 \subset \gamma_1$) including the empty set (for which the product gives I_Γ itself) and finally we sum over all forests, the symbol F_Γ denotes the family of forests. The novice is not expected to grasp the full content of (2.8). It should however be clear to everybody that an explicit way of doing renormalization on the Feynman integrand without introducing any cut-off or regularization as proposed by Zimmermann, makes algebraic relations between different sets of renormalized Feynman-graphs completely visible. There are also other ways of renormalization (for example the BPH method which does the Taylor-operations on suitably regularized Feynman integrals) which lead to the causal unitary Greens functions and they are even more economical from a computational point of view, however, in those other approaches one would have a hard time to derive the renormalized field equation, Ward identities etc.

Before we go over to the construction of composite fields we should comment on the form of (2.5). As far as the Zimmermann operation in (2.4) is concerned the a,b and c are free parameters. So in addition to the old Feynman rules one has new 2- and 4-vertices

$i(a+bp^2)$

and  i c. By playing around with a,b and c one can change the normalization of Greens functions. The adoption of the normalization conditions (2.3a-c) allows to fix these in a recursive fashion as power series in λ and $\frac{m^2}{\mu^2}$: $a = m^2 f(\lambda, \frac{m^2}{\mu^2})$

$$b = b(\lambda, \frac{m^2}{\mu^2}) , \quad c = c(\lambda, \frac{m^2}{\mu^2}) .$$

$$(2.3a) \quad \rightarrow \quad i(a + b\, m^2 - \hat{\Pi}(m^2)) = 0 \qquad (2.9a)$$

$$(2.3b) \rightarrow \quad i(-\mu^2-m^2+a-b\mu^2-\hat{\Pi}(-\mu^2)) = i(-\mu^2-m^2) \qquad (2.9b)$$

$$(2.3c) \rightarrow \quad -i(\lambda-c+\hat{\Lambda})\Big|_{s.p.} = -i\lambda \qquad (2.9c)$$

with $-i\hat{\Pi}(p^2) = R$ {  + ...} = nontrivial BPHZ 2-vertex contribution

and $-i\ \hat{\Lambda} = R$ { 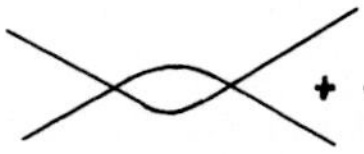 + ...} = nontrivial BPHZ 4-vertex contribution.

The a,b and c have now become genuine "conterterm constants" i.e. constants which start in 2nd order perturbation theory.

Remark: R { 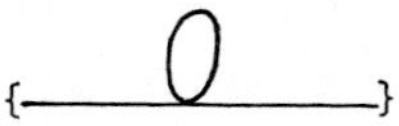 } $= 0$ (2.10)

The introduction of Greens functions involving composite fields procedesin a similar manner. For the simple case of one composite field Θ (monomial of A and its derivations) we define

$$< T\ N\ [\Theta]\ (x)\ X > = F.P. <\Phi_o \Big| T : \Theta_o(x) : X_o \; e^{i\int :L^{eff}_{int}(A_o):d^4x} \Big| \Phi_o > \qquad (2.11)$$

The general Feynman graph for this object has the form:

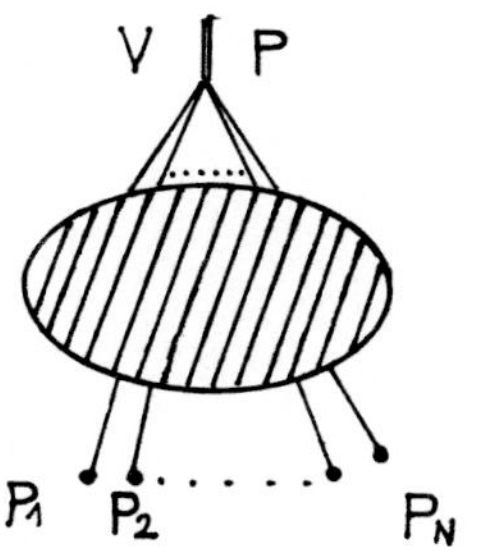

The N external legs correspond to $X = \prod_{1}^{N} A(x_i)$, whereas the composite field corresponds to the above vertex V with ℓ lines (ℓ = degree of monomial Θ in A). The F.P. has to be newly defined at least for those subgraphs which contain the new vertex V. We choose the superficial divergence function:

$$d(\gamma) = 4 - N(\gamma) - \begin{cases} 4 - \dim \Theta & \text{if } V \in \gamma \\ 0 & \text{otherwise} \end{cases} \qquad (2.12)$$

here dim Θ = can. dim. of Θ (i.e. the dimension of the corresponding free field expression Θ_o). If we oversubtract by choosing dim. $\Theta \longrightarrow \delta_\Theta >$ dim. Θ, we will denote the so constructed object by $N_\delta[\Theta]$; the N in (2.11) without a subscript is always minimally subtracted according to (2.12).

Instead of a simple "isotropic" oversubtraction one can also introduce anisotropically subtracted normal products. In the following we give an example. Let $\Theta(A)$ be a product of two monomials $P_1(A)$ and $P_2(A)$ with canonical dimensions d_1 and d_2. Then $N_{d_1+d_2+\alpha}[P_1|P_2]$ denotes a normal product where the Feynman-Integrand I_Γ^Θ (p,k) has been treated with the F.P. operation corresponding to

$$\delta(\gamma) = 4 - N(\gamma) \qquad \text{if } V_\Theta \not\in \gamma \tag{2.13}$$

$$\delta(\gamma) = 4 - N(\gamma) - (4-d_1-d_2) + \begin{cases} \alpha \text{ if } \gamma \text{ contains lines from } P_1 \\ \text{as well as from } P_2 \\ 0 \text{ otherwise} \end{cases}$$

The important point is that the introduction of these anisotropic objects is not only an interesting mental exercise, but they automatically enter if one wants to prove the equivalence theorem[11] or derive equations of motions for field inside ordinary products[12] (see later remark in connection with the equation of motions). For the later discussion of critical phenomena we will have no need for anisotropic normal products.

In addition to trivial properties (following directly from the presence of Taylor-operators in (2.8))

as:
$$\partial_\mu N_\delta [\Theta] = N_{\delta+1} [\partial_\mu \Theta] \tag{2.14}$$

and the normalization properties:

$$D^{(M)} \langle T\, N_\delta [\Theta](o)\tilde{X}(p_1 \dots p_N)\rangle^{prop} \Big|_{p_i=o} = \text{same in zero order} \tag{2.15}$$

for
$$N \leq \delta - M$$

Note that the BPHZ subtraction only acts on the loop contribution to the 1 - irr. vertices and not on their trivial part; this explains the right hand side of (2.15).

By comparing the minimal subtracted R^d in (2.8) with an over-subtracted R^{δ}, Zimmermann[7] obtained the important "Zimmermann identity":

$$N_{\delta}[\Theta] = N[\Theta] + \sum_i r_i \, N[\Theta_i] \qquad (2.16)$$
$$\delta > \dim.\Theta$$

Here the sum goes over all Θ_i's (minimal subtraction!) which on the one hand have the same transformation properties as Θ and in addition fulfill:

$$\dim \Theta < \dim \Theta_i \leq \delta \qquad (2.17)$$

Of course all relations are understood inside Greens functions i.e. for the $<T\ N\ [\Theta]X>$.

By writing (2.16) for the vertex functions and using the normalization properties (2.15) one can easily write explicit formulae for the r_i's in terms of derivations of $< T\ N[O]\tilde{X} >^{prop}$ at zero.

We will have no use for these formulae. An important special case of (2.16) is

$$N_4[A^2]=N[A^2]+r_1\ N[\partial_\mu A\partial^\mu A]+r_2\ N[A\partial^2 A]+r_3\ N[A^4] \qquad (2.18)$$

Remark: Anisotropically subtracted N.P. also fulfill Zimmermann identities[9].

The still missing algebraic relations are the equations of motions[7]

$$\partial^2 <T\ A(x)\ X> = (a-m^2)<T\ A(x)\ X> + \frac{c-\lambda}{3!} <T\ N_3[A^3](x)\ X>$$

$$- i\sum_K \delta(x-x_K) < T X_{\hat{K}} > \qquad (2.19)$$

$X_{\hat{K}}$ = X with $A(x_K)$ deleted

and inside N.P.'s[13]

$$(1+b) < T\ N[A\partial^2 A](x)\ X> \ = \ < T\ N_4[(a-m^2)A^2 + \frac{c-\lambda}{3!} A^4](x)\ X >$$

$$- i \sum_K \delta(x-x_K) < T X > \qquad (2.20)$$

For higher composite fields we have the anisotropic form[14,15]:

$$(1+b)\ < T\ N[A^n\partial^2 A](x)\ X> \ = \ (a-m^2) < T\ N_{n+3}[A^n|A](x)\ X>$$

$$+ \frac{c-\lambda}{3!} < T\ N_{N+3}[A^{n+3}](x)\ X>$$

$$- \delta - \text{fct. terms} \qquad (2.21)$$

Note that apart from the δ-functions and anisotropies these are the Euler-Lagrange equations to L_{eff}(2.5).

The analytically continued euclidean expectation values i.e. the statistical correlation functions obey the same algebraic identity; of course the name "equation of motion" is inappropriate because of the elliptic differential operators. For the Minkowski theory one easily converts the Green function statements into operator statements by using the L S Z asymptotics for the fields in X. The δ-function contact terms in the equation of motion drop out for the L S Z particle-matrixelements of operators. It is interesting to ask what are the corresponding <u>operator</u> properties in the operator formulation of the euclidean theory? We have not investigated this problem.

Gomes and Lowenstein[9] have shown that the minimally subtracted N.P.'s

$$B_{(\mu)_1,\ldots,(\mu)_n} = N[\prod_{i=1}^{n} (D_{(\mu)_i} A(x))] \qquad (2.22)$$

with $D_{(\mu)_i}$ = traceless part of $\partial_{\mu_1} \ldots \partial_{\mu_i}$

(for example: $D_{\mu\nu} = \partial_\mu \partial_\nu - \frac{1}{4} g_{\mu\nu} \partial^\kappa \partial_\kappa$)

form a complete basis of local fields. This means that every N, even the anisotropic ones, can be written as a superposition of B's. An interesting consequence can immediately be drawn: the equation of motion and Zi-identities exhaust the relations between N.P.'s, or to put it in another form, the equations of motions and Zi-identities define a maximal ideal in the formal algebra of quantized composite fields.

All our statements have a statistical mechanics counterpart with only a slight difference in language, for example composite fields are called (composite) fluctuation variables.

3. DERIVATION OF PRE-SCALING EQUATIONS

According to our philosophy developed in the previous section, the fundamental properties of (composite) fields are the algebraic relations between N.P.'s. We are going to use some of these properties for investigating the change of renormalized Greens functions under infinitesimal changes of parameters. We start from the renormalized action principle:

notation $$W = i \int N_4 \, [L_{eff}] \, d^4x \tag{3.1}$$

$$= (1+b)\Delta_2 + (a-m^2)\Delta_1 + (c-g)\Delta_3$$

$$\Delta_1 = \frac{i}{2} \int N_4 \, [A^2] \, d^4x \tag{3.2a}$$

$$\Delta_2 = \frac{i}{2} \int N_4 \, [\partial_\mu A \partial^\mu A] \, d^4x \tag{3.2b}$$

$$\Delta_3 = \frac{i}{4!} \int N_4 \, [A^4] \, d^4x \tag{3.2c}$$

Renormalized action principle: $(s = m^2, \lambda, \mu^2)$

$$\frac{\partial}{\partial s} < T \, X > \; = \; < T \, \frac{\partial W}{\partial s} \, X > \tag{3.3}$$

$$= \{ \frac{\partial(1+b)}{\partial s} \Delta_2 + \frac{\partial(a-m^2)}{\partial s} \Delta_1 + \frac{\partial(c-g)}{\partial s} \Delta_3 \} <T \, X>$$

with

$$\Delta_i < T \, X > \; = \; < T \, \Delta_i \, X >$$

The validity of (3.3) follows directly by graphical

methods[12]. For example for $s = m^2$ the dependence of $\langle T X\rangle$ on m^2 either goes via the m^2 in the Feynman denominator or the m^2 in the counterterm vertices a, b and c. (3.3) just formalizes this statement. As our algebraic input for the pre-scaling equation we now use the Zi-identity

$$\Delta_o = \frac{i}{2} \int N_2[A^2]d^4x = \Delta_1 - r\,\Delta_2 - s\Delta_3\ , \qquad (3.4)$$

which is the integrated form of (2.18) and the integrated equation of motion (2.20)

$$-2(1+b)\ \Delta_2\langle T\ X\rangle = \{2(a-m^2)\ \Delta_1+4(c-\lambda)\ \Delta_3+N1\}\ \langle T\ X\rangle \qquad (3.5)$$

(Remark: for the vertex functions $\Gamma^{(N)} = \langle\, T\ X\,\rangle^{prop}$ we have a change in sign $N \rightarrow -N$)

Because of the occurance of N1 this identity is sometimes referred to[12] as the "counting identity". It is clear that between the 5 operations $\frac{\partial}{\partial s}$ $(s = m,\mu,\lambda)$, N and Δ_o written in terms of the linear independent Δ_i's $i = 1,2,3$[12] there are two relations. The conventional form in the Callan-Symanzik and the renormalization group equation:

$$\text{C.S.}\quad \{2\mu^2 \frac{\partial}{\partial\mu^2} + 2m^2 \frac{\partial}{\partial m^2} + \beta\frac{\partial}{\partial\lambda} + N\gamma\}\langle T\ X\rangle = m^2 2\alpha\Delta_o\ \langle T\ X\rangle \qquad (3.6a)$$

$$\text{R.G.}\quad \{2\mu^2 \frac{\partial}{\partial\mu^2} + \sigma\frac{\partial}{\partial\lambda} + N\tau\}\ \langle\, T\ X\,\rangle = 0 \qquad (3.6b)$$

Equation (3.6a) says that an infinitesimal change of the mass can not be compensated by an infinitesimal change of coupling constant, wave function "re"-normalization and normalization spot, but one needs in addition a "soft" mass insertion Δ_o. On the other hand, according to (3.6b) a μ-change can be compensated by a coupling - and wave function renormalization - change. In the sequel we follow Lowenstein's[12] derivation of (3.6a,b). By writing all operations as a linear superposition of Δ_i's one obtains the equations which determine the α, β, γ, σ, τ, in terms of a, b and c (the use of physical dimension see formula before (2.9a) simplifies the equations):

$$\begin{aligned} 2\alpha m^2 &= 2(a-m^2) + \beta\frac{\partial a}{\partial\lambda} - 2(a-m^2)\gamma \\ -2\alpha m^2 r &= \beta\frac{\partial b}{\partial\lambda} - 2(1+b)\gamma \\ -2\alpha m^2 s &= \beta(\frac{\partial c}{\partial\lambda} - 1) - 4(c-\lambda)\gamma \end{aligned} \tag{3.7a}$$

$$\begin{aligned} 0 &= \mu^2\frac{\partial a}{\partial\mu^2} + \sigma\frac{\partial a}{\partial\lambda} - 2(a-m^2)\tau \\ 0 &= \mu^2\frac{\partial b}{\partial\mu^2} + \sigma\frac{\partial b}{\partial\lambda} - 2(1+b)\tau \\ 0 &= \mu^2\frac{\partial c}{\partial\mu^2} + \sigma(\frac{\partial c}{\partial\lambda} - 1) - 4(c-\lambda)\tau \end{aligned} \tag{3.7b}$$

The a, b, c, r and s start in higher order, so that the determinant of (3.7a) in zero order is nonvanishing. In (3.7b) the determinant of the last two equations for σ

and τ is nonvanishing in zero order, the first equation is no additional restriction, but follows from the last two and the independence of the mass normalization property $\Gamma^{(2)}\Big|_{p^2=m^2} = 0$ on μ and λ. Sometimes it is also convenient to write the coefficients of the differential-equations (3.6a,b) directly in terms of Green functions. For example we obtain directly from the μ normalization property of the two- and four-point function the representation:

$$- 2\tau\, i(\mu^2+m^2) = 2\mu^2 \frac{\partial}{\partial\mu^2} \Gamma^{(2)}\Big|_{p^2=-\mu^2} \tag{3.8a}$$

$$- i\,\sigma + i\,4\,\tau\,\lambda = - 2\mu^2 \frac{\partial}{\partial\mu^2} \Gamma^{(4)}\Big|_{s.p.} \tag{3.8b}$$

For the case of trilinear couplings or couplings involving fields with different masses, the pre-scaling equations do not just consist of the C.S. and R.G. equations. A simple example of three pre-scaling equations will be discussed in section 6.

We now turn to the prescaling equations for composite fields. Consider first $N[A^2]$. A straightforward computation shows that the integrated field equation:

$$(1+b)\langle TN[A\partial^2A](x)\ N[A^2](y)X\rangle = \langle TN_4[(a-m^2)A^2 +$$

$$+ \frac{c-\lambda}{3!}A^4](x)\ N[A^2](y)X\rangle - i\sum_k \delta(x-x_k)\langle TN[A^2](y)X\rangle$$

leads to a counting identity for $\langle TN[A^2]X\rangle$ which is the same as (3.5) if we replace $N \to N+2$. The changed Zimmermann identity is:

$$(\Delta_o-\Delta_1+r\Delta_2+s\Delta_3)\ \langle TN[A^2]X\rangle = \hat{u}\ \langle TN[A^2]X\rangle \tag{3.9}$$

where $\hat{u}$ correction term comes from subgraphs for which the minimal and oversubtracted mass vertex and the new $N[A^2]$vertex are sitting in the same renormalization part:

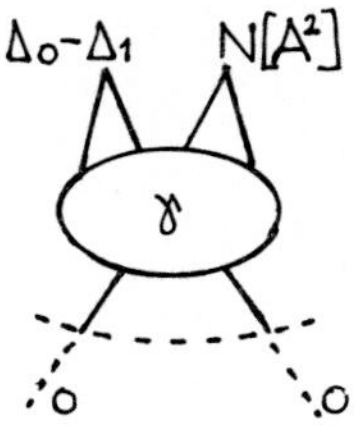

Such subgraphs must have two undifferentiated lines since for all other cases Δ_o and Δ_1 lead both to negative $\delta(\gamma)$.

The $\hat{u}$ is easily determined from the normalization property of the proper parts of (3.9)

$$\hat{u} = \tfrac{1}{2}\Delta_o\ \langle TN[A^2](o)\tilde{A}^2(o)\rangle^{\text{prop}} \tag{3.10}$$

The evaluation of $\hat{u}$ amounts to the computation of "cat"-graphs:

$$\hat{u} = \sum$$ 

The changed C.S. and R.G. equations are obtained by $N \to N+2, \Delta_o \to (\Delta_o-\hat{u})$ hence $(m^2\ \alpha\ \hat{u} = \bar{u})$

$$\left(2\mu^2\frac{\partial}{\partial\mu^2} + 2m^2\frac{\partial}{\partial m^2} + \beta\frac{\partial}{\partial\lambda} + (N+2)\gamma + \bar{u}\right) \langle TN[A^2](x)\ X\rangle$$

$$= 2m^2\alpha\Delta_o \langle TN[A^2](x)\ X\rangle \qquad (3.11a)$$

$$\left(2\mu^2\frac{\partial}{\partial\mu^2} + \sigma\frac{\partial}{\partial\lambda} + (N+2)\tau\right) \langle TN[A^2](x)\ X\rangle = 0 \qquad (3.11b)$$

The normalization of $N[A^2]$ in this procedure is given by the BPHZ subtraction at zero momentum i.e.

$$\langle TN[A^2](o)\tilde{A}(p_1)\tilde{A}(p_2)\rangle^{prop}\Big|_{p_1=p_2=0} = \text{trivial zeroth order contribution}$$

This normalization is quite arbitrary and in fact not appropriate to a discussion of the zero-mass limit. We therefore introduce a μ-normalized normal-product $\hat{N}[A^2]$ by

$$\hat{N}[A^2] = (1-t)N[A^2] \qquad (3.12)$$

t is determined by the normalization condition of $\hat{N}$, i.e.,

$$\langle T\hat{N}[A^2](o)\tilde{A}(p_1)\tilde{A}(p_2)\rangle^{prop}\Big|_{s.p.} = \text{trivial zeroth order contribution}$$

The prescaling equations (3.11) can then be rewritten for $\hat{N}[A^2]$. Defining functions v and w by the equations:

$$\{2\mu^2 \frac{\partial}{\partial\mu^2} + 2m^2 \frac{\partial}{\partial m^2} + \beta\frac{\partial}{\partial\lambda} + \bar{u}\} \; (1-t) \; = \; v(1-t)$$

and

$$\{2\mu^2 \frac{\partial}{\partial\mu^2} + \sigma\frac{\partial}{\partial\lambda}\} \quad (1-t) \; = \; w(1-t)$$

we have the prescaling equations:

$$\{2\mu^2 \frac{\partial}{\partial\mu^2} + 2m^2 \frac{\partial}{\partial m^2} + \beta \frac{\partial}{\partial\lambda} + (N+2)\gamma + v\} \quad \langle T\hat{N}[A^2](x) \; X\rangle = \\ = 2m^2\alpha\Delta_o \; \langle T\hat{N}[A^2](x) \; X\rangle \tag{3.13a}$$

and

$$\{2\mu^2 \frac{\partial}{\partial\mu^2} + \sigma\frac{\partial}{\partial\lambda} + (N+2)\tau + w\} \quad \langle T\hat{N}[A^2](x) \; X\rangle \; = 0 \tag{3.13b}$$

Here v and w may be determined by the normalization condition for $\hat{N}$. It will be considered explicitly in section 5. Note that the differential equations for the corresponding Γ's have -N instead of N. The derivation of prescaling equations may be extended to arbitrary high composite fields. This is most conveniently done for the basis elements (2.22). Instead of u one obtains a matrix since the Zi-identities mix in all other fields with the same canonical dimension. The diagonalization is done with $\hat{N}$ normalized at the μ-symmetry point. We will not use the higher "composite fluctuation variables" in our discussions.

4. THE SHORT AND LONG RANGE LIMIT OF THE THEORY

Starting from the massive theory discussed before, our aim here is to construct a scale invariant limit theory which should be relevant for a description of statistical mechanics at the critical point: $\xi = m^{-1} \to \infty$. Hence we have to investigate the zero mass asymptotic theory first:

$$\lim_{\kappa \to \infty} \Gamma^{(N)}(p_1,..,p_N;\ \kappa^{-1}m,\mu,\lambda).$$

In order to obtain information on the behaviour of the vertex functions in this limit, we may use their homogeneity property

$$\Gamma^{(N)}(\kappa p_1,..,\kappa p_N; m,\mu,\lambda) = \kappa^{4-N}\Gamma^{(N)}(p_1,..,p_N;\ \kappa^{-1}m,\kappa^{-1}\mu,\lambda). \tag{4.1}$$

Then, apart from rescaling of the normalization mass μ and the homogeneity factor, the zero-mass limit is related to the large-momentum limit of the original vertex functions (in perturbation theory):

$$\lim_{\kappa \to \infty} \Gamma^{(N)}(\kappa p_1,..,\kappa p_N;\ m,\mu,\lambda)$$

In this case Weinberg's powercounting theorem[16] tells us that at nonexceptional momenta, $\Gamma^{(N)}$ behaves for large κ as

$$\Gamma^{(N)}(\kappa p_1,..,\kappa p_N;\ m,\mu,\lambda) \simeq \kappa^{4-N}(\ln \kappa)^x \tag{4.2}$$

in every order of renormalized perturbation theory. The power x of the logarithm need not be evaluated for our purposes. This also remains true for

$$\Gamma^{(N)}(\kappa p_1,\ldots,\kappa p_N,m,\kappa\ \mu,\lambda)$$

where, apart from the occurrence of new logarithms, the power behaviour given by dimensionality is the same. Hence

$$\Gamma^{(N)}(\kappa p_1,\ldots,\kappa p_N;\ m,\kappa\ \mu\ ,\lambda)\ \simeq\ \kappa^{4-N}(\ell n\ \kappa)^{x'};$$

or, using (4.1), $\Gamma^{(N)}$ behaves in the zero-mass limit for fixed normalization mass μ as

$$\Gamma^{(N)}(p_1,\ldots,p_N;\ \kappa^{-1}m,\mu,\lambda)\ \simeq\ (\ell\ n\ \kappa)^{x'} \qquad (4.3)$$

Concidering now the difference between the Callan-Symanzik equation and the renormalization group equation

$$\{2m^2\frac{\partial}{\partial m^2} + (\beta-\sigma)\frac{\partial}{\partial\lambda} - (\gamma-\tau)N\}\ \Gamma^{(N)} = 2m^2\alpha\Delta_o\Gamma^{(N)} \qquad (4.4a)$$

we can apply powercounting to the coefficient functions and $\Delta_o\Gamma^{(N)}$. For <u>fixed μ</u> we have for small m:

$$(\gamma-\tau)\ =\ O\ (m^2(\ell\ n\ \ m^2)^x)$$

$$(\beta-\sigma)\ =\ O\ (m^2(\ell\ n\ m^2)^y)$$

$$\alpha\ =\ O\ (1),$$

whereas $\Delta_o \Gamma^{(N)} \simeq (\ln m^2)^z$ and $\Gamma^{(N)} \simeq (\ln m^2)^{x'}$.
The differential equation then tells us that

$$2m^2 \frac{\partial \Gamma^{(N)}}{\partial m^2} \simeq m^2 (\ln m^2)^u ,$$

to which the only solution is

$$\Gamma^{(N)} \simeq \text{const. as } m \to o . \qquad (4.4b)$$

An analogous consideration holds for the $<T\hat{N}[A^2]X>$ in (3.12). Therefore the logarithms are in fact absent to every order of perturbation theory at non exceptional momenta. This is expected to remain true after summing up the perturbation series. Of course the crucial point for the existence of the zero-mass limit is the use of intermediate renormalization at some fixed mass μ. The above consideration shows that we are able to construct for the A^4-theory a zero-mass limit. From the arguments given above we also learn that the two pre-scaling equations coincide to the order $O(m^2 (\ln m^2)^u)$ for $m \to o$. Following Symanzik[17] a preasymptotic zero-mass theory may then be constructed as a solution of the zero-mass pre-scaling equations:

$$\{2\mu^2 \frac{\partial}{\partial \mu^2} + \sigma \frac{\partial}{\partial \gamma} - N\tau\} \Gamma_{as}^{(N)} (p_1,\ldots,p_N;\mu,\lambda) = o \qquad (4.5)$$

Note that because of intermediate renormalization we are able to construct a pre-asymptotic zero-mass theory without investigations of the parquet-approximation type[18].

Now solving equation (4.5) along the characteristic curves[19], the general solution satisfies the relation:

$$\Gamma_{as}(p_1,\ldots,p_N;\kappa^{-1}\mu,\lambda) = a(\kappa)^{-N/2}\Gamma_{as}(p_1,\ldots,p_N;\mu,\lambda(\kappa))$$

or by homogeneity

$$\Gamma_{as}(\kappa p_1,\ldots,\kappa p_N;\mu,\lambda) = \kappa^{4-N}a(\kappa)^{-N/2}\Gamma_{as}(p_1,\ldots,p_N;\mu,\lambda(\kappa)) \tag{4.6}$$

Here the rescaled coupling constant $\lambda(\kappa)$ is implicitly defined by

$$\ln \kappa^2 = \int_{\lambda}^{\lambda(\kappa)} \sigma^{-1}(\lambda')d\lambda' \tag{4.7}$$

and the scaling factor $a(\kappa)$ is given as

$$a(\kappa) = \exp\{\int_{\lambda}^{\lambda(\kappa)} d\lambda'\, \sigma^{-1}(\lambda')\tau(\lambda')\} \tag{4.8}$$

Looking at equation (4.5), $2\mu^2 \frac{\partial}{\partial\mu^2}$ plays the role of the dilation operator and we see that dilation-invariance is prevented only by having $\sigma \neq 0$. Hence we will find scale invariant theories at the zeros of the function $\sigma(\lambda)$. Assuming σ to have a simple zero at a certain value $\hat{\lambda}$ of the coupling constant we have to distinguish two different cases: If $\lambda(\kappa) \to \hat{\lambda}$ the integral (4.7) diverges and

either $\kappa \to \infty$ if $\sigma'(\hat{\lambda}) < 0$, $\hat{\lambda} \equiv \lambda_\infty$

or $\kappa \to 0$ if $\sigma'(\hat{\lambda}) > 0$, $\hat{\lambda} \equiv \lambda_0$

Hence starting with a preasymptotic theory with coupling constant λ and looking at

$$\Gamma_{as}(\kappa p_1,..,\kappa p_N \ \mu,\lambda)$$

we have that on the r.h.s. of equation (4.6) $\lim_{\kappa\to\infty} \lambda\ (\kappa) = \lambda_\infty$ and $\lim_{\kappa\to o} \lambda\ (\kappa) = \lambda_o$ where $\sigma(\lambda_\infty,_o) = o$; $\sigma'\ (\lambda_\infty,_o) \lessgtr o$. We may therefore define the vertex functions of the scale invariant limit theory

$$\Gamma_{lim}\ (p_1,..,p_N;\mu\hat{\lambda}) = \lim_{\lambda\to\hat{\lambda}}\ \Gamma_{as}(p_1,..,p_N,\mu,\lambda) \qquad (4.9)$$

which describe either the high energy (short distance) behaviour of the preasymptotic zero-mass theory for σ' $(\lambda_\infty) < o$ or the low energy (long distance) behaviour for $\sigma'\ (\lambda_o) > o$. The latter case is the relevant one for the description of statistical mechanics near the critical point.

We then observe that the short distance limit of the original massive theory directly leads to the scale-invariant Gell-Mann-Low limit theory (equal to Γ_{lim} at λ_∞)

$$\lim_{\kappa\to\infty} \Gamma(\kappa p_1,..,\kappa p_N;m,\mu,\lambda) = \kappa^{4-N}a(\kappa)^{-\frac{N}{2}}\ \Gamma_{GML}(p_1,..,p_N,\mu,\lambda_\infty)$$

the mass turning out to be irrelevant. On the other hand in order to get the long distance scaling theory[20] we have to start from the zero-mass preasymptotic theory

$$\lim_{\kappa\to o} \Gamma_{as}(\kappa p_1,..,\kappa p_N;\mu,\lambda)=\kappa^{4-N}a(\kappa)^{-\frac{N}{2}}\Gamma_{lim}(p_1,..,p_N,\mu,\lambda_o) \qquad (4.9b)$$

for the construction of the vertex functions. It is clear that one never can get ride of the mass by looking at the low momenta region of the massive theory. For a discussion of correction terms to the limit theories see e.g. Symanzik[17] and Mack[27]. We point out once more that in general scale invariance is prevented by the presence of mass <u>and</u> by a nonvanishing function $\beta(\lambda)$ i.e. $\lambda \neq \hat{\lambda}$. Hence to reach a scale invariant limit there are two steps involved. In one step one has to eliminate the mass, in the other the symmetry breaking by $\beta(\lambda) \neq o$. The later is always achieved by a scale transformation either to high or to low momenta. Concerning the elimination of mass effects one has to remember the completely different role the mass plays in particle physics on one hand and in statistical physics on the other hand. In particle physics the mass m is a given fixed quantity, as well as λ, and if $m \neq o$ the only possibility to get ride of the mass is to go to large momenta. Hence only the high energy region can be scale invariant, and there both breaking effects are eliminated by the large momentum dilation. If it is true (depending on the value of λ) that the dynamical dimension of composite fields occuring in the Lagrangian of a system grow with their canonical values (e.g $d_{A^2} < d_{A^4}$)[27] the scale invariant asymptote is reached by dilatation via the non scale invariant preasymptotic region, where the mass is already negligible. On the other hand in statistical physics near the critical point the "mass" is no fixed parameter. It may be changed by an appropriate change of the thermodynamical parameters of the system (temperature, magnetic field). If one looks for a scale invariant long-distance theory, the mass has to go to zero as a parameter determined by the experimental situation. Note that this picture is in contradiction to the one pursued in recent papers on this subject[21], where it is assumed that one globally has $\lambda = \lambda_o$ and the mass is eliminated by looking at high momenta. Of course the asymptotic theory one reaches by this procedure coincides with the one constructed above. However the physical interpretation is obscured.

Turning to A^4-theory in D = 4 space-(time) dimensions, the function σ can be evaluated as a perturbation series for small λ and it shows a double zero at $\lambda = o$ with $\sigma''(o) > o$. We assume this to be true, at least in the sense of an asymptotic expansion, also for the summed up preasymptotic theory. Again we see from (4.7) that for each positive λ smaller than the possibly existing second zero of σ at λ_∞, the long-distance scaling limit is

$$\lim_{\kappa\to o} \kappa^{N-4} a(\kappa)^{N/2} \Gamma_{as}(\kappa p_1,..,\kappa p_N,\mu,\lambda) = \Gamma_{lim}(p_1,..,p_N,\mu,\lambda_o=o)$$

Hence according to (4.9) it is a free field theory[17]. This is contrary to statements by different authors, who find logarithmic corrections to a mean field theory, mainly by perturbative methods[22].

Of course perturbation theory in our case would also show up logarithms. These are however spurious results of perturbation theory. The correct answer is given by the long-distance limit of the summed up preasymptotic theory. As a result in 4-dimensional space-(time), the long-distance behaviour of the A^4-theory for $\lambda<\lambda_\infty$ describing statistical mechanics, is always a mean field theory. This statement suggests a Wilson perturbative expansion in the dimension of space-(time) to get a nontrivial statistical mechanics*. Hence we will calculate by a perturbation expansion in $D = 4-\varepsilon$ dimensions the function $\sigma(\lambda)$ such that $\sigma(\lambda,\varepsilon) = 0$ yields a curve $\lambda(\varepsilon)$ with

$$\lambda\,(0) = 0 \tag{4.10}$$

and $\quad \sigma'(\lambda,\varepsilon) > 0 \quad$ at $\quad \lambda = \lambda\,(\varepsilon)$

The positive slope of σ guarantees that the preasymptotic theory at $\lambda = \lambda\,(\varepsilon)$ is the long-distance scaling limit of the original theory. So in the construction of a theory of critical phenomena we always have to start from the zero-mass preasymptotic theory.

There are reasons to hope that perturbation expansions of the coefficient functions of the prescaling equations show better convergence properties than those of vertex functions. Therefore our perturbative approach mainly concerns the functions σ,τ and similar quantities whereas other objects have to be calculated possibly by the solution of the corresponding approximate pre-scaling equations.

5. THE CONSTRUCTION OF THE CRITICAL THEORIES OF THE FERROMAGNETIC TYPE

As pointed out in the first section, there are convincing arguments that the ferromagnetic lattice

---

*In the analytic continuation of the BPHZ-renormalized Feynman integrals from D=4 to D=4-ε ($\varepsilon > 0$) dimensions the absolute convergence is preserved, consequently all algebraic relations of the theory remain true.

models of the Lenz-Ising type are described near the critical point by a scalar field theory with the interaction Lagrangian

$$L_{int} = - \frac{\lambda}{4!} A^4 + HA$$

where A(x) is the field theoretic substitute for the spin matrices σ_i and H is the external magnetic field. At the critical point H = o, so that the critical theory is described by the symmetric A^4-theory with large correlation length $\xi = m^{-1} \to \infty$.

A generalization to Heisenberg type models is immediately achieved by taking an arbitrary number of fields $A_1(x) \ldots A_n(x)$ with an interaction Lagrangian:

$$L_{int} = - \frac{\lambda}{4!} \{ \sum_{i=1}^{n} A_i^2 \}^2 + H A_i$$

Also the limit $n \to \infty$ (continuous Heisenberg model) is of interest. Here we limit our considerations to the symmetric A^4-theory, a generalization to many fields beeing immediate. As argued at the end of the last section, the aim is to calculate the functions σ and τ by perturbation theory in $D = 4-\varepsilon$ dimensions. Then we have to look for a solution of

$$\sigma\,(\varepsilon,\lambda) = o$$

describing a long-range theory. Using the critical values of σ and τ in the pre-scaling equations, one can look for solutions of the correlation functions, which approximate the physical correlation functions near the critical point. From the critical behaviour of correlation functions we are then able to calculate relative critical indices. They are relative, because we do not know from the beginning what temperature $t = T - T_c$ in our model means i.e., the index ν, relating temperature to the correlation length at the critical point

$$t \propto \xi^{-\nu}$$

is not known at first.

Let us now turn to the perturbation expansion[23] of σ and τ. In $D = 4-\varepsilon$ dimensions the vertex functions have the canonical dimensions

$$\dim \Gamma^{(N)} = D - N \cdot d$$

where

$$d = \frac{D - 2}{2}$$

is the canonical dimension of the field A. In order to have the coupling constant dimensionless, the adopted normalization conditions are:

$$\Gamma^{(2)}(p,-p)\Big|_{p^2 = m^2} = o \qquad (5.1)$$

$$\Gamma^{(2)}(p,-p)\Big|_{p^2 = -\mu^2} = - i(\mu^2+m^2) \qquad (5.1)$$

$$\Gamma^{(4)}(p\ ,\dots,p_4)\Big|_{sp} = -\,i\,\lambda\mu^{\varepsilon}$$

The Feynman rules in D dimensions are given by the correspondence:

$$\lambda \to \lambda\ \mu^{\varepsilon}$$

$$\frac{1}{(2\pi)^4}\int d^4k \to \frac{1}{(2\pi)^D}\int d^Dk$$

Furthermore we want σ to be dimensionless hence the old σ is always multiplied by a factor $\mu^{-\varepsilon}$.

The functions σ and τ are then given by (3.8).

$$\tau = +i\,\frac{\mu^2}{\mu^2+m^2}\,\frac{\partial\Gamma^{(2)}}{\partial\mu^2}\Bigg|_{p^2 = -\mu^2} \tag{5.2a}$$

$$\sigma = -i2\mu^2\mu^{-\varepsilon}\,\frac{\partial\Gamma^{(4)}}{\partial\mu^2}\Bigg|_{s.p.} + 4\,\lambda\,\tau \tag{5.2b}$$

In order to calculate all relevant quantities to order ε^2, we have to determine $\lambda(\varepsilon)$, considered as a power series in ε, as a solution of $\sigma(\lambda,\varepsilon) = o$ to order ε^2. In lowest order perturbation theory we see that $\lambda \sim \varepsilon$. We therefore need to evaluate $\Gamma^{(2)}$ to second order in 4 dimensions, whereas $\Gamma^{(4)}$ has to be evaluated to order $\lambda^2\ \varepsilon$ and λ^3 (σ being proportional to λ in 4-ε dimensions).

We first calculate $\Gamma^{(2)}(p,-p)$ for small m:

$$\Gamma^{(2)}(p,-p) = i \quad \{a - m^2 + (1+b)p^2 - \tilde{\Pi}(p)\}$$

with $- i \, \tilde{\Pi} = \sum\limits_{\mathrm{PROP}'}$ 

(PROP′ denoting all one-particle irreducible nontrivial Feynman graphs).
a and b are determined recursively by the normalization conditions for $\Gamma^{(2)}$

$$a = + \frac{1}{\mu^2 + m^2} \{\tilde{\Pi}(-\mu^2)\ m^2 + \tilde{\Pi}\ (m^2)\ \mu^2\}$$

$$b = - \frac{1}{\mu^2 + m^2} \{\tilde{\Pi}(-\mu^2) - \tilde{\Pi}(m^2)\}$$

To second order we have

$$\Gamma^{(2)}(p,-p) = i \quad \{ p^2 - m^2 + a^{(2)} + b^{(2)} p^2 - i$$  $\}$

For m → o the leading term is (see Appendix)

$$\tilde{\Pi}\ (\underline{p}^2) \simeq - \frac{\lambda^2}{12} \frac{1}{(4\pi)^4} \underline{p}^2 \quad \ln \frac{\underline{p}^2}{m^2}\ ,$$ $\underline{p}$ euclidean vector.

Using this for m → o

$$\tau \simeq i \frac{\partial \Gamma^{(2)}}{\partial \mu^2} \Bigg|_{\underline{p}^2 = \mu^2} = \frac{\lambda^2}{12} \quad \frac{1}{(4\pi)^4} \qquad (5.3)$$

Similarly we evaluate σ from the fourpoint function:

$$\Gamma^{(4)} = - i \{ \lambda \ \mu^{\varepsilon} - c + \tilde{\lambda} \}$$

where we defined

$$- i \tilde{\lambda} = \sum_{PROP'}$$ 

The vertex normalization defines recursively

$$c = \tilde{\lambda} \Big|_{s.p.}$$

Up to 3rd order we have

$$\Gamma^{(4)} = -i \quad \{ \lambda\mu^{\varepsilon} - c^{(2)} \quad - \quad c^{(3)} + i [$$ 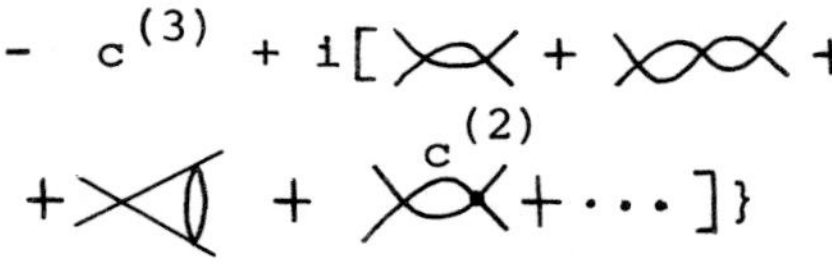 $$]\}$$

with $c^{(2)} = \{$  $\}_{s.p.}$

and $c^{(3)} = \{$  $\}_{s.p.}$

The first graph has to be evaluated to order $\lambda^2\varepsilon$ for small m.

The other integrals being of order λ^3, need only be evaluated in 4 dimensions and for small m. The details of the calculations are given in the appendix.

The infrared divergences of the individual terms drop out as it should be according to (4.4b) and (5.2a,b) and we obtain:

$$\sigma = -i2\mu^2\mu^{-\varepsilon}\left.\frac{\partial\Gamma^4}{\partial\mu^2}\right|_{s.p.} + 4\lambda\tau = -\varepsilon\lambda + \lambda^2\frac{3}{(4\pi)^2}(1+\varepsilon[1+\frac{\gamma}{2} + \frac{1}{2}\ell n\ 3\pi]) - \lambda^3\frac{3}{(4\pi)^4} \qquad (5.4)$$

With $\lambda = \alpha_1\ \varepsilon + \alpha_2\ \varepsilon^2 + O\ (\varepsilon^3)$, the eigenvalue equation $\sigma\ (\lambda,\varepsilon) = o$ has the solutions

$\lambda_\infty = o$ where $\sigma'\ (o) < o$ (short range point)

and

$$\lambda_o = \frac{(4\pi)^2}{3}\varepsilon - \frac{17}{g^2}(4\pi)^2\varepsilon^2 - \frac{(4\pi)^2}{3}\ [1+\gamma/2+1/2\ell n\ 3\pi]\varepsilon^2 + O(\varepsilon^3) \qquad (5.5)$$

where $\sigma'(\lambda_o) > o$ for <u>$\varepsilon > o$;</u>

i.e. in $4-\varepsilon$ dimensions $(\varepsilon>o)\ \lambda_o$ is the desired long-distance point according to (4.10). The function σ is shown in the figure for $4-\varepsilon$ dimensions. For $\varepsilon\to o$ we obtain

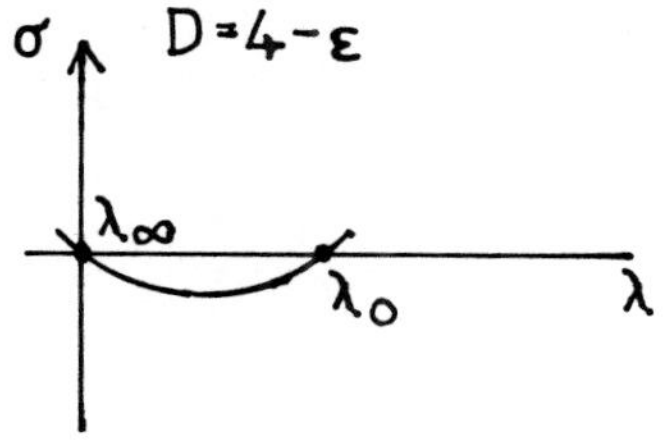

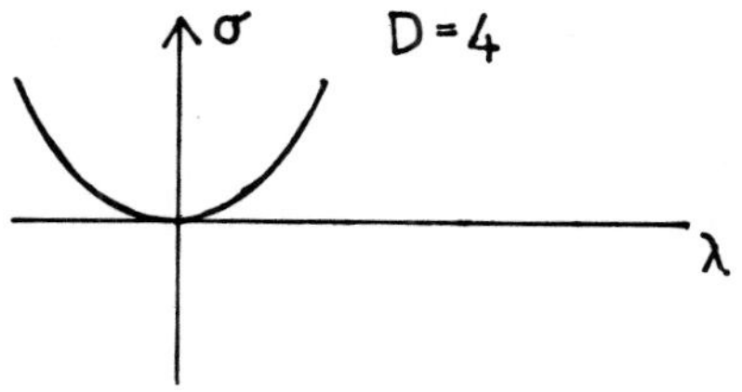

the previously discussed degenerate double-zero at $\lambda = o$, and we see that starting from $\lambda > o$ in the preasymptotic theory we always get the long-distance limit in 4 dimensions. For $\varepsilon > o$ the fields have anomalous dimension

$$\tau(\lambda_o) = \frac{\varepsilon^2}{12.9} + O(\varepsilon^3). \qquad (5.6)$$

We now consider the pre-scaling equation at λ_o:

$$\{2\mu^2 \frac{\partial}{\partial\mu^2} - \tau(\lambda_o)N\} \quad \Gamma_{\text{lim}}^{(N)}(p_1,..,p_N,\mu,\lambda_o) = 0.$$

Hence $\Gamma_{\text{lim}}^{(N)}(p_1,..,p_N;\mu,\lambda) = (\frac{\mu}{\mu_o})^{\tau(\lambda_o)N} \Gamma_{\text{lim}}^{(N)}(p_1,..,p_N,\mu_o,\lambda_o)$,

which shows that the theory only trivially depends on μ through a scaling factor.
Using homogeneity we have

$$\Gamma_{\text{lim}}^{(N)}(\kappa p_1,..,\kappa p_N;\mu,\lambda) = \kappa^{(D-N\cdot d)} \; \Gamma_{\text{lim}}^{(N)}(p_1,..,p_N,\kappa^{-1}\mu,\lambda_o)$$

$$= \kappa^{(D-N\cdot d)} \kappa^{-N\tau(\lambda_o)} \Gamma^{(N)}_{lim}(p_1,\ldots,p_N,\mu,\lambda_o).$$

The dimension of the field is then:

$$\underline{d_A = d + \tau = 1 - \frac{\varepsilon}{2} + \tau} \quad . \tag{5.7}$$

From the field dimension we immediately get the critical exponent η defined by

$$\langle A(x)A(o)\rangle \propto \frac{1}{|x|^{D-2+\eta}} = \frac{1}{|x|^{2d_A}} \quad , \tag{5.8}$$

$$\text{or} \quad \eta = 2\,\tau \ .$$

The field suspectibility χ_A, defined by

$$\chi_A^{-1} = \{\int d^D x \ \langle A(x)A(o)\rangle\}^{-1} = \Gamma^{(2)}(p,-p)\Big|_{p^2=o} = a-m^2, \tag{5.9}$$

then behaves in the limit $m \to o$ as

$$\chi_A^{-1} \simeq m^{2-\eta} \quad .$$

This determines the critical index of the field susceptibility relative to the correlation length m^{-1}.

Within the symmetric theory we are able to determine a second independent critical index, belonging to the energy fluctuation. The leading term of the physical energy in the critical region is the operator of lowest canonical dimension; namely the mass term

$$N[A^2]\ (x) \quad .$$

In section 3 we already derived the pre-scaling equations for this object as well as for the μ-normalized version $\hat{N}[A^2]$, which has to be considered in the preasymptotic limit.

$\hat{N}[A^2]$ is constructed easily out of the original $N[A^2]$ by determining an appropriate counterterm

$$\hat{N}[A^2] = (1-t)\ N[A^2] \quad .$$

Denoting the vertex functions for $\frac{1}{2}N[A^2]$ as $\Gamma_2^{(N)}$, we have the relation

$$\Gamma_2^{(N)} = \sum_{\text{PROP}}$$

with the normalization condition

$$\Gamma_2^{(2)}(p_1,p_2)\Big|_{p_i = 0} = 1 \ . \tag{5.10a}$$

On the other hand the vertex functions $\hat{\Gamma}_2^{(N)}$ corresponding to $\frac{1}{2}\hat{N}[A^2]$ are arranged to be normalized as

$$\hat{\Gamma}_2^{(2)}(p_1,p_2)\Big|_{s.p.} = 1 \tag{5.10b}$$

Hence in perturbation theory we construct $\hat{\Gamma}_2^{(N)}$ by a multiplicative renormalization of $\Gamma_2^{(N)}$

$$\hat{\Gamma}_2^{(N)} = (1-t)\ \Gamma_2^{(N)} , \qquad (5.11a)$$

t is determined by (5.10) and may be evaluated by the recursion

$t = \sum_{\text{PROP}} \{$  $\Big|_{s.p.} - t$  $\Big|_{s.p.} \}$ (5.11b)

In the zero-mass limit, the pre-scaling equations (3.13) coincide. We have

$$\{2\mu^2 \frac{\partial}{\partial\mu^2} + \sigma \frac{\partial}{\partial\lambda} - (N-2)\tau + u\}\ \hat{\Gamma}_2^{(N)} = o, \qquad (5.12)$$

where the coefficient u is determined by the normalization condition to be:

$$- 2\mu^2 \frac{\partial\hat{\Gamma}_2^{(2)}}{\partial\mu^2}\Bigg|_{s.p.} = u \qquad (5.13)$$

We determine u again in 4-ε dimensions by perturbation expansion to second order in λ and ε .

$\hat{\Gamma}_2^{(2)} = \{ 1 - t^{(1)} - t^{(2)} +$  $+$  $+$  $+$

$+$  $c^{(2)} - t^{(1)}$  $\}$ (5.14) with $t^{(1)} = \{$  $\}_{s.p.}$ and $t^{(2)} = \{$  $+$

$+$  $+$  $c^{(2)} - t^{(1)}$  $\}_{s.p.}$

All diagrams have already been calculated for the evaluation of σ. Only the combinatorics and the normalization is modified. We obtain for u the expression

$$u = -2\mu^2 \frac{\partial \hat{\Gamma}_2^{(2)}}{\partial \mu^2}\Bigg|_{s.p.} = + \frac{\lambda}{(4\pi)^2}\left(1+\varepsilon\left(1+\gamma/2+\tfrac{1}{2}\cdot \ln 4\pi\right)\right) \tag{5.15}$$

$$+ \frac{3}{2}\frac{\lambda^2}{(4\pi)^4}\ln\frac{4}{3} + \frac{\lambda^2}{(4\pi)^4}$$

At the critical point $u = u(\lambda_o)$, we have:

$$u = \frac{\varepsilon}{3} + \frac{8}{g^2}\ \varepsilon^2 \tag{5.16}$$

From equation (5.12) we learn, e.g., for N = 2

$$2\mu^2 \frac{\partial \hat{\Gamma}_2^{(2)}}{\partial \mu^2} = -u\ \hat{\Gamma}_2^{(2)}$$

Hence $\hat{\Gamma}_2^{(2)} \propto \mu^{-u}$ and $u + 2\tau$ is the anomalous dimension of $\hat{N}[A^2]$. The dimension of $\hat{N}[A^2]$ is then

$$d_E = 2d + u + 2\tau = 2-\frac{2}{3}\ \varepsilon + \frac{19}{2\cdot 9^2}\ \varepsilon^2 \qquad (5.17)$$

The energy correlation therefore behaves like

$$< E(x)E(o) > \simeq \frac{1}{|x|^{2dE}} \qquad (5.18)$$

and the critical index of the energy susceptibility (specific heat) is

$$\chi_E^{-1} = \{\int d^D x < E(x)E(o) > \}^{-1} \simeq m^{2dE-D} \qquad (5.19)$$

relative to the correlation length.

With the determination of d_A and d_E we computed two independent critical coefficients which are sufficient to determin all others by assuming Kadanoff's scaling laws to be true.

Going beyond the symmetric theory and introducing "temperature and magnetic field-perturbations", we will be able to derive the Kadanoff scaling laws (see section 7).

In order to calculate directly the critical indices γ and α, defined by

$$\begin{aligned} \chi_A^{-1} &\propto |t|^\gamma \\ \chi_E^{-1} &\propto |t|^\alpha \end{aligned} \qquad (5.20)$$

we have to introduce a temperature t in our model. We use the correspondence to the discrete Lenz-Ising model in the critical region to establish heuristically a direct introduction of t in Lagrangian field theory. In the discrete case the physical energy reads:

$$\frac{J}{kT} \sum_{<ij>} \sigma_i \sigma_j = \text{critical value} - t \frac{J}{kT} \sum_{<ij>} \sigma_i \sigma_j$$

with $t = \frac{T-T_c}{T_c}$. Correspondingly the action W in the continuous version may be written as

$$W = W_{critical} - t \, E(x)$$

where E(x) is the continuous analog to the physical energy density:

$$\frac{J}{kT} \sum_{<ij>} \sigma_i \sigma_j \rightarrow \int dx \; E(x) = \frac{1}{2} \int d^3x \; \{\partial_i \sigma(x) \partial^i \sigma(x) - m^2 \sigma^2(x)\}$$

In the critical region the leading contribution from E(x) is the one of lowest dimension i.e. the $N_2[\sigma^2]$ content of E(x). The critical index of E(x) is therefore $d_E = d_{A^2}$ calculated above. On the other hand the critical index ν relating the temperature t to the correlation length m^{-1} is defined by

$$|t| \propto m^{1/\nu} \quad . \tag{5.21}$$

Hence the dimension of W demands for the relation

$$\nu^{-1} + d_E = D \quad . \tag{5.22}$$

Therefore in Lagrangian field theory we may introduce

explicitly a parameter t (called temperature in the following) that approximates a linear measure of the temperature, to the extent that E(x) is approximated by $N_2[A^2]$, in the form

$$L\ (x) = L(x)_{crit.} - t\ N[A^2](x) \qquad (5.23)$$

This will be used to discuss temperature-perturbations to the critical theory in section 7. Here we only give the expressions for γ and α. Using (5.22), (5.9) and (5.19) we have

$$\gamma = \nu\ (D-2d_A) = \frac{D-2d_A}{D-d_E} \qquad (5.24)$$

$$\alpha = \nu\ (D-2d_E) = \frac{D-2d_E}{D-d_E}$$

6. MAGNETIC FIELD-PERTURBATIONS ON THE SYMMETRIC THEORY

6.1. Ferromagnetic model with external field.

In this section we want to study magnetic field-perturbations on the critical theory in order to get other critical indices and an equation of state near the critical point. In the presence of an external magnetic field H, which in our ferromagnetic model gives rise to a magnetization M, the equation of state relates M to H and t:

$$M = f\ (H,t)$$

The critical indices δ and β are then defined by

$$M \Big|_{t=o} \propto H^{1/\delta} \tag{6.1}$$

and

$$M \Big|_{H=o} \propto t^{\beta} \quad \text{for} \quad t < o$$

Explicitly we study A^4-theory with a coupling to an external magnetic field H = const. In the formulation of a cut-off field theory this model has been discussed by Brezin et. al.[24]. The effective Lagrangian has the form

$$L_{eff}(\bar{A}) = L_o(\bar{A}) - \frac{\lambda-c}{4!} A^4 - H\bar{A} + \frac{\bar{a}}{2}\bar{A}^2 + \frac{\bar{b}}{2} \partial\bar{A}\ \partial\bar{A} \tag{6.2}$$

The field $\bar{A}$ has nonvanishing vacuumexpectation value

$$< \bar{A} > = M , \tag{6.3}$$

that has to be identified with the magnetization. To keep the usual form of BPH quantization it is convenient to write $L_{eff}(A)$ in terms of the interpolating fields A for the free fields; hence we substitute

$$\bar{A} \to A = \bar{A} - M ,$$

such that

$$L_{eff}(A) = L_o(A) - \frac{\lambda-c}{4!}A^4 - \frac{(\lambda-c)M}{3!} A^3 + \frac{a}{2} A^2 + \frac{b}{2} \partial A\ \partial A \tag{6.4}$$

$$+((a-m^2)M-\frac{1}{3}(c-\lambda)\ M^3-H)\ A$$

In the construction of the Green-functions by the Gell-Mann-Low formula:

$$G^{(N)}(x_1,..,x_N) = \langle T\ X \rangle$$

$$= FP \langle \Phi_o | T \{ \prod_{i=1}^{N} A_o(x_i)\ e^{i\int :L_{int}^{(o)}:(x)} \} | \Phi_o \rangle_x \ ,$$

we have to specify the degree function $\delta(\gamma)$ for the subtraction procedure. Because the A^3-coupling as well as parts of the A^2 counterterms are induced by $N_4[\bar{A}^4]$ of the original Lagrangian these terms have to be quantized with N_4 also, i.e. with oversubtractions. This procedure guarantees that the original meaning of the physical parameters is not changed. Hence to each Feynman graph γ the assigned subtraction degree is

$$\delta(\gamma) = 4 - N \ .$$

This leads to a vanishing of all tadpol contributions, and, because the field A is introduced to satisfy

$$\langle A \rangle = o$$

the linear term in the Lagrangian vanishes; or

$$H = (a-m^2)\ M-\frac{1}{3}(c-\lambda)\ M^3 \tag{6.5}$$

which represents the equation of state of our model.

We have now to go through the same arguments as in the symmetric case concerning the derivation of pre-scaling equations. The physical parameters are now m, λ, $g = \lambda M$ and the normalization mass μ. a, b, c and $d = cM$ are again finite counterterms to be determined by the normalization conditions. In addition to the previous normalization conditions (2.3), we have a new one defining the normalization of M or g at the symmetry point of the three point function

$$\Gamma^{(3)}(p_1,p_2,p_3,m,\mu,g,\lambda)\Big|_{p_i p_j = -\frac{1}{2}(3\delta_{ij}-1)\mu^2} = -ig\mu^{\varepsilon/2} \qquad (6.6)$$

g being chosen to have dimension 1 in $4-\varepsilon$ dimensions. In addition to the operations Δ_1, Δ_2, Δ_3 we have now a new one:

$$\Delta_4 = \frac{i}{3!} \int dx \, N_4 \, [A^3] \, (x) \qquad (6.7)$$

By the Zimmermann identities we are able to relate the oversubtracted Δ's to the minimally subtracted one's. However in order to keep the discussion simple we will use, instead of

$$\Delta_o = \frac{i}{2} \int dx \, N_2 \, [A^2] \, (x) \, ,$$

the oversubtracted insertion

$$\bar{\Delta}_1 = \frac{i}{2} \int dx \, N_3 \, [A^2] \, (x) \qquad (6.8)$$

which is also a "mild" object compared to Δ_1. Furthermore we have the minimally subtracted

$$\bar{\Delta}_4 = \frac{i}{3!} \int dx \, N_3 \quad [A^3] \, (x) \tag{6.9}$$

The integrated Zimmermann identities read

$$\bar{\Delta}_1 = \Delta_1 + t_2 \, \Delta_2 + t_3 \, \Delta_3$$

$$\bar{\Delta}_4 = \Delta_4 + s_2 \, \Delta_2 + s_3 \, \Delta_3 \, .$$

Together with the differential vertex operations

$$\frac{\partial}{\partial \lambda}, \frac{\partial}{\partial g}, \frac{\partial}{\partial m^2}, \frac{\partial}{\partial \mu^2}$$

and the integrated equation of motion we have 7 operations expressed in terms of the 4 linearly independent $\Delta_1, .., \Delta_4$. Therefore we have 3 linear relations between the 7 operations. The derivation being straight forward, we write down these pre-scaling equations with an appropriate choice of the coefficients:

$$\{2\mu^2 \frac{\partial}{\partial \mu^2} + \sigma \frac{\partial}{\partial \lambda} - \tau \, N + (1+\omega) \, g \frac{\partial}{\partial g}\} \, \Gamma^{(N)} = o \tag{6.10a}$$

$$\{D + \beta \frac{\partial}{\partial \lambda} - \gamma N + \delta \, g \frac{\partial}{\partial g}\} \, \Gamma^{(N)} = 2m^2 \, \alpha \bar{\Delta}_1 \, \Gamma^{(N)} \tag{6.10b}$$

$$\{g \quad \xi \, D + g\eta \frac{\partial}{\partial \lambda} - g \, \zeta N + \phi \frac{\partial}{\partial g}\} \, \Gamma^{(N)} = \bar{\Delta}_4 \, \Gamma^{(N)} \tag{6.10c}$$

D is the dilatation operator $D = 2\mu^2 \frac{\partial}{\partial \mu^2} + 2m^2 \frac{\partial}{\partial m^2} + g \frac{\partial}{\partial g}$

Apart from the extra terms proportional to $g \frac{\partial}{\partial g}$, the first two equations are recognized as the ordinary pre-scaling equation of the symmetric theory. The third equation reads in the limit $g \to o$:

$$\phi \left. \frac{\partial \Gamma^{(N)}}{\partial g} \right|_{g = o} = \bar{\Delta}_4 \, \Gamma^{(N)} \qquad (6.11)$$

just giving

$$< T \, N_3 \, [A^3](o) \, \tilde{A}(p_1) .. \tilde{A}(p_N) > ^{\text{PROP}}$$

in the symmetric theory. Generally it describes how one may compensate a change in the trilinear coupling g in the vertex functions by other operations.

We now turn to the question of the existence of a zero-mass limit for the vertex functions. The reasoning is exactly along the lines of section 4. There is however a main difference in that the zero-mass limit only exists if the trilinear coupling constant g tends to zero linearly in m. This comes from the homogeneity relation

$$\Gamma^{(N)}(\kappa p_1, .., \kappa p_N; m, \mu, g, \lambda) = \kappa^{4-N} \Gamma^{(N)}(p_1, .., p_N; \kappa^{-1} m, \kappa^{-1} \mu, \kappa^{-1} g, \lambda),$$

and from the application of power counting to the l.h.s. for large κ. Only the physically irrelevant mass μ can be rescaled without changing the powers in the behaviour of $\Gamma^{(N)}$. The pre-scaling equation then assures us the existence of

$$\lim_{\kappa \to o} \Gamma^{(N)}(p_1, .., p_N; \kappa^{-1} m, \mu, \kappa^{-1} g, \lambda)$$

The theory defined by this limit is the preasymptotic theory of the nonsymmetric model. This theory may have a reminicent dependence on the trilinear coupling $\bar{g}=\frac{g}{m}$.

In perturbation theory, indeed one can find diagrams, which give surviving contributions arising however only from oversubtraction terms; e.g.,

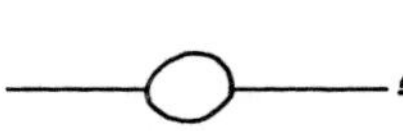
$$\simeq i \frac{\bar{g}^2}{12} \frac{1}{(4\pi)^2} \underline{p}^2 = - i \tilde{\Pi} (-\underline{p}^2).$$

However, looking at $\Gamma^{(2)}$ for m = o, we have

$$\Gamma^{(2)}(p,-p) = i \{ - (1+b) \underline{p}^2 - \tilde{\Pi}(-\underline{p}^2) \}$$

with $b = \frac{\bar{g}^2}{12} \frac{1}{(4\pi)^2}$ as determined by the normalization condition. Hence $\Gamma^{(2)}$ is in fact independent of $\bar{g}$; similarly we have in the four point function constant contributions, because all graphs have $\delta(\gamma) = o$ (one subtraction). An example of such a graph is

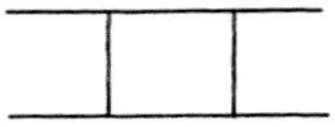
$$\simeq i \frac{1}{(4\pi)^2} \bar{g}^4$$

Of course by the normalization condition, this constant is subtracted out by a corresponding counterterm.

It is just the third pre-scaling equation which relates the Green-functions constructed by oversubtraction to the minimally subtracted 3-vertex insertions, and, because the $\bar{g}$ terms just arise from oversubtractions, the third equation should tell us something about the $\bar{g}$ dependence of the vertex functions.

Indeed the coefficient functions ξ, η, ζ, ϕ, which are finite in the preasymptotic limit, lead to the degeneration of the equation (6.10c)

to
$$\phi \bar{g} \frac{\partial \Gamma^{(N)}}{\partial \bar{g}} \simeq \bar{g} \, m \, \bar{\Delta}_3 \, \Gamma^{(N)} \simeq o \qquad (6.12)$$

in the limit $g = m\bar{g} \to o$. This shows that the vertex functions are independent of $\bar{g}$ in the pre-scaling limit, and hence they coincide with those of the symmetric theory. Thus we are back to the symmetric case.

So what do we learn from this discussion? First we showed that the model leads to the same preasymptotic and hence to the same critical theory. There is no critical theory in $4-\varepsilon$ dimensions having contributions from the trilinear coupling. At first considering the preasymptotic Callan-Symanzik equation

$$\{ D + \beta \frac{\partial}{\partial \lambda} - \gamma N + \delta \bar{g} \frac{\partial}{\partial \bar{g}} \} \Gamma^{(N)} = 0$$

one could think of a scale invariant theory by solving

$$\beta \, (\varepsilon, \lambda, \bar{g}) = o$$

$$\delta \, (\varepsilon, \lambda, \bar{g}) = o$$

for $\lambda(\varepsilon)$ and $\bar{g}(\varepsilon)$.

The above consideration shows that there is no such solution. In fact we have $\delta = -1$ in $D = 4$ dimensions. It is of course reasonable from the point of view of dimensional analysis that the trilinear coupling has to be absent in a scale invariant theory near D = 4.

On the other hand we may now investigate perturbations in m and M by using the new pre-scaling equations

$$\{ 2\mu^2 \frac{\partial}{\partial \mu^2} + \sigma \frac{\partial}{\partial \mu} - \tau N + (1+\omega) g \frac{\partial}{\partial g} \} \Gamma^{(N)} = 0 \qquad (6.13)$$

and taking σ, τ and ω at the critical point.

We first have to calculate ω. By the normalization condition of $\Gamma^{(3)}$ we get, from the renormalization group equation, the relation

$$g(1+\omega) = i 2\mu^2 \left. \frac{\partial \Gamma^{(3)}}{\partial \mu^2} \right|_{s.p.} + 3\, g\, \tau . \qquad (6.14)$$

In perturbation theory we have to go to the order $\lambda\varepsilon$ and λ^2 because g drops out of the equation.

$$\Gamma^{(3)} = - i \{ g\mu^{\varepsilon/2} - d + \tilde{\Sigma} \}$$

(6.15a)

$- i \tilde{\Sigma} = \sum_{\text{PROP}'}$ 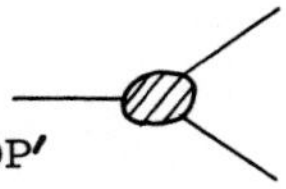

and $\qquad d = \left. \tilde{\Sigma} (p_1, p_2, p_3) \right|_{s.p.}$

To second order we get:

$\Gamma^{(3)} = - i \{ g\mu^{\varepsilon/2} - d^{(1)} - d^{(2)} +$  $+$ 

$+$ 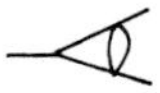 $+$ 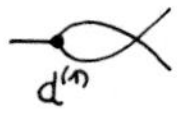 $+$ 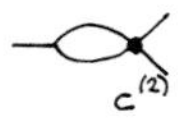 $\}$

(6.15b)

$d^{(1)} = \{$  $\}_{s.p.}$

$d^{(2)} = \{$  $+$  $+$ 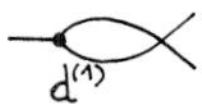 $+$ 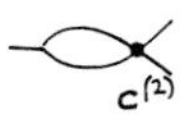 $\}_{s.p.}$

Again apart from the different combinatories for the 3-vertex and a different normalization, we have to evaluate the same terms as for σ before. The result is:

$$(1+\omega) = -\frac{\varepsilon}{2} + \lambda 3 \frac{1}{(4\pi)^2}(1+\varepsilon(1+\gamma/2 + 1/2 \ \ln 4\pi)) \qquad (6.16a)$$

$$- \lambda^2 \frac{g}{2} \frac{1}{(4\pi)^4} \ln \frac{4}{3} - \frac{6\lambda^3}{(4\pi)^4} + \frac{1}{4} \frac{\lambda^2}{(4\pi)^4}$$

at $\lambda_\infty = 0$ we have $\omega = -1 + \frac{\varepsilon}{2}$ (6.16b)

at λ_o we have $\omega = -1 + \frac{\varepsilon}{2} - \tau = - d_A$. (6.16c)

Hence ω is just the negative of the dimension of the field. This relation we will understand later on.

From (6.14) and the fact that $\Gamma^{(2)}_{as}, \Gamma^{(3)}_{as}$ are independent of g it follows that ω does not depend on oversubtractions. This becomes clear, when we realize that $(1+\omega)$ is the anomalous dimension of $\hat{N}[A^3]$. On the other hand $\hat{N}[A^3]$ is the contribution of lowest dimension to $N_4[A^3]$ (see section 7) and hence the leading one in the asymptotic large distance region.

6.2. The equation of state in the critical region.

We now turn to the consideration of the equation of state:

$$H = (a-m^2)\ M - \frac{1}{3}(c-\lambda)\ M^3 \tag{6.17}$$

For the functions $a-m^2 \equiv \tilde{a}$ and $c-\lambda \equiv \tilde{c}$ we have the prescaling equations

$$D\ \tilde{a} + \beta\frac{\partial\tilde{a}}{\partial\lambda} - 2\gamma\tilde{a} + \delta\ M\ \frac{\partial\tilde{a}}{\partial M} = \text{inh.}$$

$$2\mu^2\ \frac{\partial\tilde{a}}{\partial\mu^2} + \sigma\frac{\partial\tilde{a}}{\partial\lambda} - 2\tau\tilde{a} + (1+\omega)\ M\ \frac{\partial\tilde{a}}{\partial M} = 0 \tag{6.18}$$

$$D\ \tilde{c} + \beta\frac{\partial\tilde{c}}{\partial\lambda} - 4\gamma\tilde{c} + \delta\ M\ \frac{\partial\tilde{c}}{\partial M} = \text{inh.}$$

$$2\mu^2\ \frac{\partial\tilde{c}}{\partial\mu} + \sigma\frac{\partial\tilde{c}}{\partial\lambda} - 4\tau\tilde{c} + (1+\omega)\ M\frac{\partial\tilde{c}}{\partial M} = 0\ .$$

We now look for a solution of these equations where the leading critical behaviour is given by terms starting with the coefficient functions at the critical point, whereas $\tilde{a}$ and $\tilde{c}$ are perturbed by small mass m and small magnetization M. The equations for $\tilde{a}$ and $\tilde{c}$ near the critical point are given by (6.18) with $\beta = \sigma = 0$, $\gamma = \tau = \tau\ (\lambda_0)$ and $\delta = \omega = \omega\ (\lambda_0)$.

We split $\tilde{a}$ and $\tilde{c}$ into a part being independent of M and an M-dependent part.

$$\tilde{a} = \tilde{a}_o + \tilde{a}_1, \quad \tilde{c} = \tilde{c}_o + \tilde{c}_1$$

$\tilde{a}_o$ and $\tilde{c}_o$ are solutions of the symmetric equations (M=o). Hence

$$\tilde{a}_o \propto \mu^{2\tau} m^{2-2\tau}$$

$$\tilde{c}_o \propto \mu^{4\tau} m^{\varepsilon-4\tau}$$

From this we may calculate the first correction to the inhomogeneity of the C.S.-equation and solve (6.18) for $\tilde{a}_1$ and $\tilde{c}_1$. In this way we get selfconsistent solutions of the pre-scaling equations near the critical point.

$$\tilde{a} = c_1 m^{2-2\tau} \mu^{2\tau} + c_2(\mu^2)(\mu^{1-\varepsilon/2})^{2(1-\tau)\omega^{-1}} M^{-2(1-\tau)\omega^{-1}} \tag{6.19}$$

$$\tilde{c} = 3b_1 \mu^{4\tau} m^{\varepsilon-4\tau} + 3b_2\mu^{\varepsilon}(\mu^{1-\varepsilon/2})^{(\varepsilon-4\tau)\omega^{-1}} M^{(4\tau-\varepsilon)\omega^{-1}} \tag{6.20}$$

Hence the leading terms to the equation of state are

$$H = c_1\mu^{2\tau}m^{2-2\tau}M + c_2 \mu^2(\mu^{1-\varepsilon/2})^{2(1-\tau)\omega^{-1}} M^{1-2(1-\tau)\omega^{-1}}$$
$$-b_1\mu^{4\tau} m^{\varepsilon-4\tau}M^3 - b_2 \mu^{\varepsilon}(\mu^{1-\varepsilon/2})^{(\varepsilon-4\tau)\omega^{-1}} M^{3-(\varepsilon-4\tau)\omega^{-1}} \tag{6.21}$$

At the coexistence curve H = o, we find from (6.21), using (6.16c), the relation

$$M \propto m^{dA}$$

and m^{-1} is proportional to the correlation length. By (6.1) and (5.21) we obtain

$$\begin{aligned} \beta &= \nu d_A \\ \delta &= 1-2(1-\tau)\omega^{-1} \end{aligned} \tag{6.22}$$

We may then check by comparison with (5.24) that Kadanoff's scaling sum rules[25)]

$$\beta(\delta+1) = \gamma+2\beta = 2-\alpha \tag{6.23}$$

are satisfied.

The physically irrelevant parameter μ can be eliminated by going over to the new quantities

$$\bar{H}=H\mu^{-3+\varepsilon/2}, \quad \bar{m}=m\mu^{-1}, \quad \bar{t}=t\mu^{-2}, \quad \bar{M}=M\mu^{-1+\varepsilon/2}$$

of canonical dimension o. Near the critical point (6.21) then takes the form

$$\bar{H}=\bar{c}_1\bar{t}^{\gamma}\bar{M} - \bar{b}_1\bar{t}^{\,(\gamma-2\beta)}\bar{M}^3 + (\bar{c}_2-\bar{b}_2)\bar{M}^{\delta}$$

the bared coefficients being multiples of the unbared ones in (6.21).

The coefficients c_1, c_2, b_1 and b_2 are evaluated by comparison with low order perturbation expansion of $\tilde{a}$ and $\tilde{c}$. The main contributions in the ε-expansion are:

$$c_1=-1; \; c_2= \frac{(4\pi)^2}{2.9}\left(4+\frac{\pi}{\sqrt{3}}\right)\varepsilon^2; \; b_1=-\frac{(4\pi)^2}{3}\varepsilon; \; b_2=o$$

In the calculation of these coefficients we observe that they depend on the renormalization subtraction procedure (oversubtractions of M-vertices). Hence the physical interpretations of the parameters m, λ, M of our model becomes relevant. So fare we interpreted them freely as the corresponding physical quantities; this however is justified only concerning dimensional analysis (critical indizes). The ambiquity is controled within one model by the algebraic identities, this however does not help for the specific physical interpretation.

The only thing we know is that the mass is the inverse correlation length. In general our parameters λ, t, M, H are related in a complicated manner to the corresponding physical quantities. These problems are better understood in the lattice formulations and the closely related cut-off field theories[1] and need further investigations in renormalized quantum theory.

6.3. The perturbed scale-invariant theory as an approximation to the physical theory in the critical region.

We have to ask the question in what sense mass- and magnetic field perturbations at the critical theory ($\lambda=\lambda o$) describe the behaviour of the physical theory ($\lambda\neq\lambda o$) under such perturbations. In statistical mechanics criticality is expected in a region

$$\xi^{-1} << \kappa \; |\underline{p}| << \Lambda \quad \text{resp.} \quad \Lambda^{-1} << \kappa^{-1} \; |\underline{x}| << \xi \tag{6.22}$$

where Λ is the lattice cut-off, ξ the correlation length and κ the dilation parameter. In the construction of the scale-invariant long-distance theory we first let $\xi^{-1}\to o$ and then $\kappa\to o$ in compatibility with (6.22). In the κ-ξ^{-1}-plane the vertex-functions are not continuous at $\kappa=\xi^{-1}=o$ and the limits cannot be interchanged. Particularly one cannot go to $\kappa=o$ and look at $\xi\to\infty$ in order to see critical behaviour.

If we look e.g. at the susceptibility expressed in the form

$$"\chi_A^{-1}(\xi) = \Gamma^{(2)}(p;\xi^{-1})\big|_{p=o} = a-\xi^{-2}" \tag{6.23}$$

we do not find the scaling behaviour

$$\chi_A(\xi) \propto \xi^{2-\eta} \quad (\xi\to\infty) \tag{6.24}$$

at least if $\lambda\neq\lambda o$. In order to see how (6.24) is obtained from the massive theory with $\lambda\neq\lambda o$ we look at (6.23) in configuration space.

$$\chi_A(\xi) = \int d\underline{x} \quad G^{(2)}(\underline{x}, \xi^{-1}) \; .$$

For large ξ and $|\underline{x}|$ we may write

$$G^{(2)}(\underline{x},\xi^{-1}) \simeq f(\underline{x}, |\underline{x}|/\xi)\, e^{-|\underline{x}|/\xi} \quad .$$

By (5.8) we have

$$G^{(2)}(\underline{x},o) \simeq f(\underline{x},o) \propto |\underline{x}|^{-2d_A}, \quad |\underline{x}| \to \infty$$

whereas for fixed ξ

$$G^{(2)}(\underline{x},\xi^{-1}) \propto \frac{e^{-|\underline{x}|/\xi}}{|\underline{x}|}, \quad |\underline{x}| \to \infty$$

the scaling behaviour will never be seen. However we can find to any given ε a small ratio $|\underline{x}|\cdot\xi^{-1} = \delta_\varepsilon$ and an $|\underline{x}_o|$ such that

$$\left|\frac{f(\underline{x},\delta)-f(\underline{x},o)}{f(\underline{x},o)}\right| \lesssim \left|\frac{G^{(2)}(\underline{x},\xi^{-1})-G^{(2)}(\underline{x},o)}{G^{(2)}(\underline{x},o)}\right| < \varepsilon \quad \text{for } |\underline{x}|>|\underline{x}_o|$$

whenever $f(\underline{x},\delta)$ is continuous in δ for large $|\underline{x}|$. The susceptibility is then well approximated by

$$\chi_A^\delta(\xi) = \int_{\Lambda^{-1}}^{\xi\delta^{-1}} d\underline{x}\; G^{(2)}(\underline{x},\xi^{-1})$$

and to any ε there is a ξ_o such that

$$\left|\frac{\chi_A^\delta - \chi_A^o}{\chi_A^o}\right| < \varepsilon \quad \text{for} \quad \xi > \xi_o$$

Translated to momentum space we see that in the physical theory (6.24) is obtained from (6.23) not for fixed $p = o$ but for $\xi\to\infty$ with

$$\xi^{-1}/|\underline{p}| = \delta$$

small. On the other hand in the perturbed scale-invariant-theory (6.24) is directly obtained from

$$\tilde{\chi}_A^{-1}(\xi) = \tilde{\Gamma}^{(2)}_{\lim}(p,\xi^{-1}) \Big|_{p=o} = \tilde{a}\text{-}\xi^{-2}$$

in the limit $\xi\to\infty$. Here the tilded quantities are the perturbed solutions of the dilation invariant scaling equations (see section 7). So we are able to state in which sense our perturbed quantities $\tilde{F}$ of the scale-invariant theory approximate the corresponding quantities F of the massive theory with $\lambda\neq\lambda_o$. Namely, to any ε we can give a δ and a $\underline{p}_o$ such that

$$\left| \frac{F(\{\underline{p}_i\},\ \xi^{-1},\lambda) - \tilde{F}(\{\underline{p}_i\},\ \xi^{-1},\ \lambda_o)}{\tilde{F}(\{\underline{p}_i\},\ \xi^{-1},\ \lambda_o)} \right| < \varepsilon \qquad (6.25)$$

for $\xi^{-1}/|\underline{p}_o| = \delta$ and $|\underline{p}_i| < |\underline{p}_o|$.

We want to emphasize that there is no reason that the critical theory is identical to the globally dilatation invariant theory with $\lambda=\lambda_o$. Concrete computations in the two dimensional Ising model indicate that the theories agree only in the long-range limit[30].

In the next section we will discuss the perturbated scale-invariant theory in more detail.

7. TEMPERATURE- AND MAGNETIZATION-PERTURBATIONS ON THE SCALE INVARIANT THEORY AND KADANOFF's SCALING LAWS.

In this section we discuss perturbations on the scale-invariant long-range theory in a similar way to that proposed by Wegener[31] in a phenomenological context. In our field theoretical frame-work we choose the "temperature" t and the "magnetization" $g = \lambda M$ as perturbation parameters, the magnetization being chosen instead of the magnetic field for technical simplicity. We start from the symmetric A^4-Lagrangian L_s and add the relevant perturbation terms with appropriate BPH counterterms:

$$L = L_s + \frac{t'-t}{2} N[A^2](x) + \frac{g'-g}{3} N[A^3](x) + hA(x). \qquad (7.1)$$

The corresponding vertex functions

$$\Gamma^{(N)}(\{p_i\};\ \lambda, m, \mu, t, g)$$

are defined by the Gell-Mann-Low formula and the normalization conditions:

$$\Gamma^{(1)} = 0 \qquad (7.2a)$$

$$\Gamma^{(2)}\Big|_{p^2=\xi^{-2}} = 0 \qquad (7.2b)$$

$$\Gamma^{(2)}\Big|_{p^2=-\mu^2} = -i(\mu^2+\xi^{-2}) \qquad (7.2c)$$

$$\Gamma^{(3)}\Big|_{sp} = -ig \tag{7.2d}$$

$$\Gamma^{(4)}\Big|_{sp} = -i\lambda\mu^{\varepsilon} \tag{7.2e}$$

Here the correlation length ξ is an unknown function of the independent parameters of the theory. A further condition will be necessary to fix t relative to m.

The prescaling equations are derived by the methods presented in section 3 to have the form:

$$\begin{aligned} &\{D+\beta\frac{\partial}{\partial\lambda} - N\gamma + pt\frac{\partial}{\partial t} + \phi g\frac{\partial}{\partial g}\}\Gamma^{(N)} = 2m^2\alpha\Delta_o\Gamma^{(N)} \\ &\{2\mu^2\frac{\partial}{\partial\mu^2} + \sigma\frac{\partial}{\partial\lambda} - N\tau + (q+1)t\frac{\partial}{\partial t} + (\omega+1)g\frac{\partial}{\partial g}\}\Gamma^{(N)} = o \end{aligned} \tag{7.3}$$

The usual power counting arguments in the construction of a preasymptotic theory only give direct information on the limit $m\to o$ together with $t\propto m^2$ and $g\propto m$ leading to the preasymptotic theory of L_s. This is not what we are interested here, we rather would like to think of (7.1) for small t and g in the limit

$$L_s \to L_{lim} ,$$

where L_{lim} symbolically denotes the long-range scale-invariant Lagrangian corresponding to L_s. Or starting from the massive theory, (7.1) gives rise to a perturbation expansion

$$\Gamma^{(N)} = \sum_{n,m=o}^{\infty} \frac{t^n g^m}{n!m!} \Delta_o^n \bar{\Delta}_4^m \Gamma_s^{(N)} , \qquad (7.4)$$

with Δ_o and $\bar{\Delta}_4$ the vertex operations introduced in sections 3 and 6, where the "composite vertex functions" $\Delta_o^n \bar{\Delta}_4^m \Gamma_s^{(N)}$ are determined fully by L_s. We then may investigate wether (7.4) makes sense in the scale-invariant limit.

Unfortunately there are no operations $\Delta_{o\ as}$ and $\bar{\Delta}_{4\ as}$ giving sense to (4.7) in the preasymptotic zero mass limit. This has been shown by Symanzik[17] in the case of Δ_o. In fact $\Delta_o \Gamma_s^{(2)}$ asymptotically satisfies the inhomogeneous CS equation

$$\{2m^2 \frac{\partial}{\partial m^2} +\beta(\lambda)\frac{\partial}{\partial \lambda} +v(\lambda)\} \Delta_o \Gamma_{s\ \underline{as}}^{(2)} = \chi(\lambda)\Gamma_{s\ \underline{as}}^{(4)}(p,-p,o,o,m^2,\lambda), \qquad (7.5)$$

with appropriate coefficients $v(\lambda)$ and $\chi(\lambda)$.

$\Gamma_{\underline{as}}^{(A)}(p,-p,o,o)$ is asymptotically of the same order as $\Delta_o \Gamma_{\underline{as}}^{(2)}$ and hence $\Delta_o \Gamma_{\underline{as}}^{(2)}$ is not a solution of a homogeneous equation and therefore can not have a dimension in the scaling-limit.

Furthermore there is no multiplicative renormalization diagonalizing the operation Δ_o. Thus there is no expansion (7.4) for the critical theory.

The difficulties encountered are coming from the fact that the $\Delta_o^n \bar{\Delta}_4^m \Gamma^{(N)}$ are vertex functions of corresponding composite-field quantities at exceptional momenta.

For non exceptional momenta the composite Green-functions

$$<T\ N[A^2](y_1)..N[A^2](y_n)N[A^3](z_1)..N[A^3](z_m)X^{(N)}>_s \tag{7.6a}$$

can be shown to have a scale-invariant limit, when they are normalized at μ. We will discuss this in the following.

The vertex functions corresponding to (7.6a) are denoted by

$$\hat{\Gamma}^{(N)}_{n,m}\ (p_1,\dots,p_N,p_{N+1}..p_{N+n},p_{N+n+1},\dots,p_{N+n+m},\lambda,\mu,m) \tag{7.6b}$$

or graphically

$$\hat{\Gamma}^{(N)}_{n,m} = \sum_{\text{PROP}}$$ 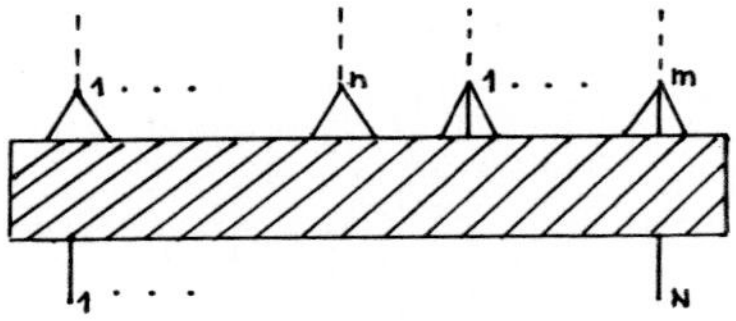

The relation to $\Delta^n_o\ \bar{\Delta}^m_4\ \Gamma^{(N)}$ is given by

$$(-i)^{n+m}\Gamma^{(N)}_{n,m}\ (p_1,..,p_N;o,\dots,o,\lambda,\mu,m)=\Delta^n_o\bar{\Delta}^m_4\Gamma^{(N)}\ (p_1,..,p_N;\lambda,\mu,m)$$

For m≠o (7.4) can therefore be used to derive the prescaling equations for $\hat{\Gamma}^{(N)}_{n,m}$ from (7.3) and appropriate normalization conditions. We choose

$$\hat{\Gamma}^{(N)}_{o,o} \equiv \Gamma^{(N)}_s$$

to fulfil the normalization conditions (2.3). This determines β,γ,σ,τ and α to be the symmetric one's. p,ϕ,q and ω are then determined by two further normalization conditions, which for convenience are chosen to be

$$\hat{\Gamma}^{(2)}_{1;o}\Big|_{sp} = 1$$

and (7.7)

$$\hat{\Gamma}^{(3)}_{o;1}\Big|_{sp} = 1$$

From (2.3) and (7.7) the "normalizations" (7.2) are then computed with the aid of (7.4).

For $N>1$ and $N+2n+m\geq 4$ the $\hat{\Gamma}^{(N)}_{n;m}$ are then uniquely defined and have a preasymptotic limit $\hat{\Gamma}^{(N)}_{as}$ for non exceptional momenta. The preasymptotic $\hat{\Gamma}_{as}$'s satisfy the differential equation

$$\{2\mu^2 \frac{\partial}{\partial\mu^2} + \sigma\frac{\partial}{\partial\lambda} - N\tau + n\tau_2 + m\tau_3\}\ \hat{\Gamma}^{(N)}_{n,m\ as} = o \qquad (7.8)$$

where

$$\tau_2 = -2\mu^2 \frac{\partial\hat{\Gamma}^{(2)}_{1;o}}{\partial\mu^2}\Bigg|_{sp} + 2\tau = (q+1) = u \qquad (7.9)$$

and

$$\tau_3 = -2\mu^2 \frac{\partial\hat{\Gamma}^{(3)}_{o;1}}{\partial\mu^2}\Bigg|_{sp} + 3\tau = (\omega+1)$$

are the anomalous dimensions of $\hat{N}[A^2]$ and $\hat{N}[A^3]$ calculated in (5.15) and (6.16). The $\hat{N}$'s are computed from the N's, subtracted according to BPHZ at zero momentum, by determining recursively t' and g' from (7.7) and the relations

$$t\hat{N}[A^2] = (t-t')\, N[A^2]; \quad g\hat{N}[A^3] = (g-g')\, N[A^3].$$

So far we have not yet defined $\hat{\Gamma}^{(N)}_{n,m}$ for the cases

$$(N,n,m) = (o,2,o),\ (o,o,2),\ (o,o,4)$$
$$(1,o,3),\ (1,1,1) \text{ and } (2,o,2)\ ,$$

where additional renormalization parts occur. If we demand (7.8) to hold with (7.9) for all (N,n,m), we can show all $\hat{\Gamma}$'s to be defined uniquely by this requirement. We only illustrate this for $\hat{\Gamma}^{(o)}_{2;o}$. In order to get a finite preasymptotic limit, we have to impose a μ-normalization condition

$$\hat{\Gamma}^{(o)}_{2;o}\Big|_{sp} = Z$$

and $\hat{\Gamma}^{(o)}_{2;o}$ is supposed to satisfy

$$\{D+\beta\frac{\partial}{\partial\lambda} +2(p+1)\}\ \hat{\Gamma}^{(o)}_{2;o} = 2m^2\ \alpha\Delta_o\ \hat{\Gamma}^{(o)}_{2;o}$$

$$\{2\mu^2\ \frac{\partial}{\partial\mu^2} + \sigma\frac{\partial}{\partial\lambda} + 2(q+1)\}\ \hat{\Gamma}^{(o)}_{2;o} = o\ .$$

Therefore we have

$$(p-q)\ Z = 2m^2\alpha\Delta_o\hat{\Gamma}^{(o)}_{2;o}\Big|_{sp}$$

and assuming Z to have a zero-mass limit, we conclude that $\hat{\Gamma}^{(o)}_{2;o}$ has a preasymptotic limit for nonexceptional momenta. Z is then given recursively by

$$Z = - \frac{\mu^2}{q+1} \frac{\partial \hat{\Gamma}^{(o)}_{2;o}}{\partial \mu^2} \Bigg|_{sp}$$

Similarly finite μ-normalizations can be determined for the remaining cases to save (7.8).

We are now able to construct the scale-invariant long-range-limit $\hat{\Gamma}^{(N)}_{n;m\ Lim}$ starting from the preasymptotic theory, as discussed in section 4. (7.8) then holds with σ, τ, τ_2 and τ_3 at λ_o:

$$\{2\mu^2 \frac{\partial}{\partial \mu^2} - N\tau + n\tau_2 + m\tau_3\} \hat{\Gamma}^{(N)}_{n;m\ Lim} = o \qquad (7.10)$$

Hence using homogeneity (dim $\Gamma^{(N)}_{n;m}$ = D-Nd+2nd+3md-(n+m)D) the composite-field vertex-functions obey the scaling relations

$$\hat{\Gamma}^{(N)}_{n;m\ Lim}(\{\kappa p_i\}, \mu, \lambda_o) =$$

$$= \kappa^{D-Nd_A+n(d_A 2-D)+m(d_A 3-D)} \hat{\Gamma}^{(N)}_{n;m}(_{Lim}\{p_i\}, \mu, \lambda_o)$$

or

$$t^n g^m \hat{\Gamma}^{(N)}_{n;m\ Lim}(\{p_i\}, \mu, \lambda_o) = (t\kappa^{D-d_A 2})^n \ (g\kappa^{D-d_A 3})^m \ \kappa^{Nd_A-D} \cdot$$

$$\cdot \hat{\Gamma}^{(N)}_{n;m\ Lim}(\{\kappa p_i\}, \mu, \lambda) \qquad (7.11a)$$

The physically irrelevant parameter μ can be eliminated by introducing quantities of canonical dimension zero and we have the Kadanoff type relations

$$\bar{t}^n \bar{g}^m \hat{\Gamma}^{(N)}_{n;m\ \mathrm{Lim}}(\{\bar{p}_i\},\lambda_o) = (\bar{t}\kappa^{D-d_A 2})^n (\bar{g}\kappa^{D-d_A 3})^m \kappa^{Nd_A - D} \cdot$$

$$\cdot\ \hat{\hat{\Gamma}}^{(N)}_{n;m\ \mathrm{Lim}}(\{\kappa\bar{p}_i\},\lambda_o) \qquad (7.11b)$$

i.e. the r.h.s. is invariant under the substitution

$$\bar{t} \to \bar{t}\ \kappa^{D-d_A 2}$$

$$\bar{g} \to \bar{g}\ \kappa^{D-d_A 3} \quad (\text{or } \bar{M}=\bar{g}/\lambda_o \to \bar{M}\ \kappa^{D-d_A 3}) \qquad (7.12)$$

$$\hat{\hat{\Gamma}}^{(N)} \to \hat{\hat{\Gamma}}^{(N)} \kappa^{Nd_A - D}$$

The Kadanoff coefficients are

$$x = D-d_A = d_A 3 \quad \text{and} \quad y = D-d_A 2 = \nu^{-1}$$

Particularly for N=o,1 it is impossible to construct an extension of $\hat{\Gamma}^{(N)}_{n;m\ \mathrm{Lim}}$ to the exceptional case necessary to give (7.4) sense near the critical point. Otherwise a Kadanoff scaling law[25]

$$\tilde{\Gamma}^{(N)}_{\text{"Lim"}}(\{p_i\},\mu,\lambda_o,t,g) = \kappa^{-D+Nd_A} \tilde{\Gamma}^{(N)}_{\text{"Lim"}}(\{\kappa p_i\},\mu,\lambda_o t\kappa^{y}, g\kappa^{D-x})$$

$$(7.13)$$

would directly follow from (7.11) assuming asymptotic

convergence. So (7.13) has to be shown in a different way.

It follows from our discussion that the limit $m\to o$ can be studied only after resumming (7.4). This amounts to an incorporation of $N[A^2]$ into L_o and it is $N[A^2]$ that replaces the mass term $N_4[A^2]$ for $m=o$. We are then confronted with the problem to quantize the theory with a $N[A^2]$ "mass-term" rather than an $N_4[A^2]$, coming out inavoidably by the usual BPHZ quantization procedure. We will consider this problem in detail in a later publication. Here we only want to give a procedure for the construction of a theory fulfilling (7.13) globally. For the summed-up theory (7.1) the limit $m\to o$ exists and we know from our discussion in section 6.1, that for t and g going to zero simultaneously in an appropriate way, the pre-scaling equations approach

$$\{2\mu^2 \frac{\partial}{\partial\mu^2} + \sigma\frac{\partial}{\partial\lambda} - N\tau + \tau_2\, t\frac{\partial}{\partial\tau} + \tau_3\, g\frac{\partial}{\partial g}\}\,\Gamma^{(N)} = o\ , \qquad (7.14)$$

with coefficients being independent of t and g. (The normalization conditions can be chosen in accordance with (7.7) such that the former calculations for the coefficients can be used). This suggests to construct vertex-functions $\Gamma^{(N)}_{Kad.}$ as perturbation solutions of the scale invariant limit i.e. with τ,τ_2 and τ_3 at λ_o, $\Gamma^{(N)}_{Kad.}$ satisfies

$$\{2\mu^2 \frac{\partial}{\partial\mu^2} - N\tau + \tau_2\, t\frac{\partial}{\partial t} + \tau_3\, g\frac{\partial}{\partial g}\}\ \Gamma^{(N)}_{Kad.} = o$$

with the general solution

$$\Gamma^{(N)}_{Kad.}(\{\kappa p_i\},\mu^2,\lambda_o,t\kappa^{D-d_A 2},g\kappa^{D-d_A 3}) =$$

$$= \kappa^{D-Nd_A}\,\Gamma^{(N)}_{Kad.}(\{p_i\},\mu^2,\lambda_o,t,g) \quad .$$

Eliminating μ we obtain the Kadanoff relation

$$\bar{\Gamma}^{(N)}_{Kad.}(\{\kappa\bar{p}_i\},\lambda_o,\bar{t}\kappa^y,\bar{g}\kappa^{D-x}) = \kappa^{D-N(D-x)}\,\bar{\Gamma}_{Kad.}(\{\bar{p}_i\},\lambda_o,\bar{t},\bar{g}) \tag{7.15}$$

The Kadanoff vertex functions $\Gamma^{(N)}_{Kad.}$ and the corresponding Green functions satisfy Kadanoff's scaling laws globally. The $\Gamma^{(N)}_{Kad.}$ are supposed to approximate the physical vertex functions $\Gamma^{(N)}$ in a region $\xi^{-1} \ll |\underline{p}_i| \ll \Lambda$ in the sense of (6.25).

8. OTHER MODELS AND OUTLOOK

For a A^4 coupling the function σ has for $\lambda = 0$ a double zero. If one considers only a positive λ as physically acceptable, then this zero determines the long distance behaviour in 4 dimensions. If on the other hand, following Symanzik[26], negative couplings are permitted, the free field becomes a short distance asymptote:

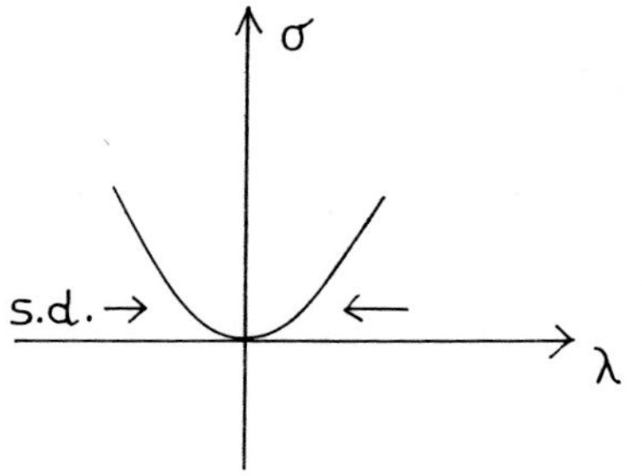

In 4 ∓ ε dimensions the double zero splits in the following way:

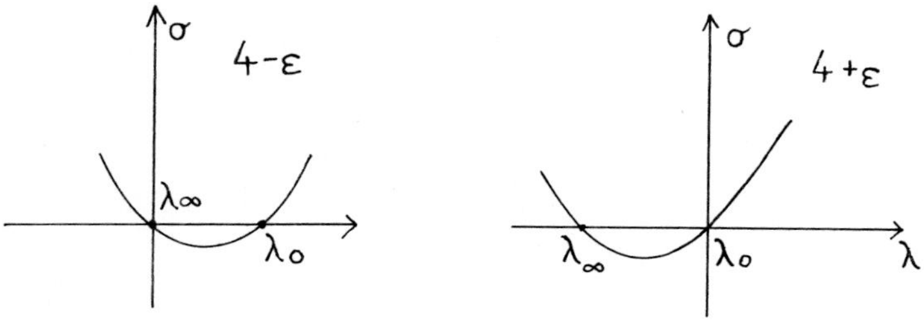

Such figures may be drawn for <u>any</u> renormalizable coupling (in d dimensions) continued to d ± ε dimension. Since the number of different renormalizable couplings with the exception of scalar fields in two dimensions is finite, one can in principle discuss all cases. Unfortunately this has only been done for some special cases:

1. A^3 coupling in d = 6 dimensions[27]

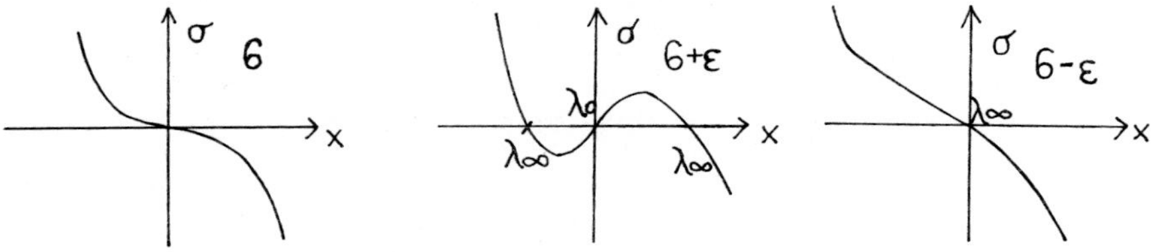

2. A^6 coupling in $3 \pm \varepsilon$ dimensions[28]

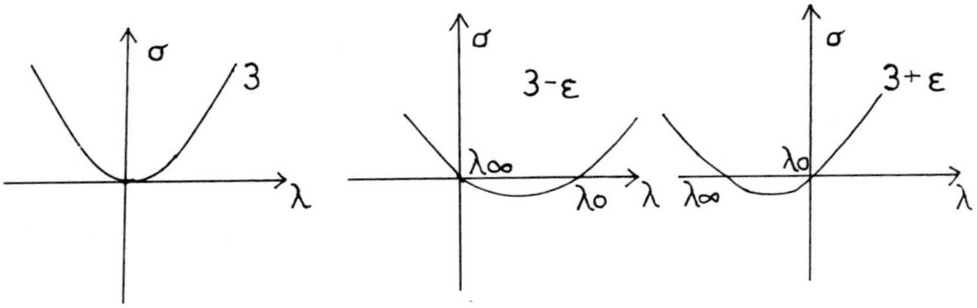

3. Thirring model[29]

In this case the two dimensional version has $\sigma = 0$ and there is no $\lambda(\varepsilon)$ with $\lambda(0) = 0$ besides the usual $\varepsilon = 0$ and $\lambda = 0$ curve on which σ vanishes.

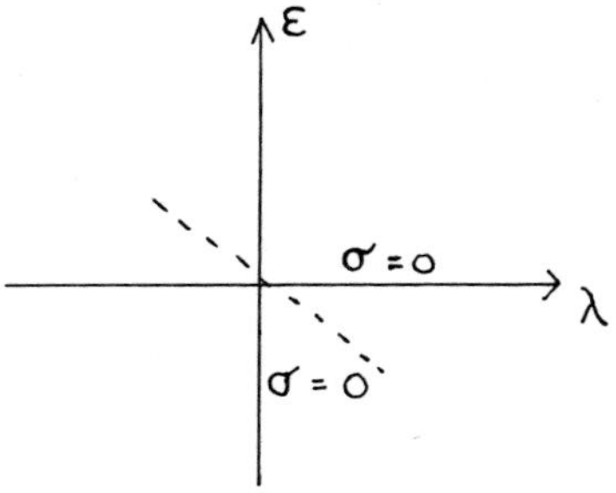

The broken line indicates for comparison the non-trivial curve on which σ vanishes in the analytically continued $4 - \varepsilon$ dimensional A^4 coupling. On the ε-axis ($\lambda = 0$) σ vanishes for every model. The second curve $\sigma = 0$ degenerates in the Thirring model to the λ-axis. The question of whether there can be a curve $\lambda(\varepsilon)$ which emerges not from the origin but from a value $\lambda = \lambda_0 \neq 0$ is an open question.

It is plausible that if one also considers several

renormalizable couplings at the same time (i.e. for theories involving several fields) one may get situations in which the l.d. respectively s.d. limit point turns into a higher dimensional manifold. In the l.d. case such models could be used for studying tri-critical behaviour.

APPENDIX

The Evaluation of Integrals

In the calculation of τ, σ, ω and u we have to evaluate several Feynman integrals. This may be done most conveniently in the euclidean region using Bogoliubov's parametrization with a regulator r in intermediate steps:

$$\frac{1}{\underline{p}^2 + m^2} \longrightarrow \int_r^\infty d\alpha \quad e^{-\alpha(\underline{p}^2+m^2)} \quad .$$

The Taylorsubtractions are carried out at zero momentum according to Bogoliubov's R-operation. In our parametrization the momentum integrations are all of the Gauss form i.e.

$$\int d^D \ell e^{-a(\underline{\ell}^2+2b\underline{p}\cdot\underline{\ell})} = \int \prod_{i=1}^{D} d\ell_i e^{-a(\ell_i^2+2bp_i\ell_i)} = (\sqrt{\frac{\pi}{a}})^D e^{ab^2\underline{p}^2}$$

with an immediate analytic continuation from integer D to arbitrary values of D. In the achieved form of the

D-dimensional Feynman-integral we may turn to Feynman parameters ξ_ℓ by a successive application of transformations:

$$(\alpha_i, \alpha_k) \to (\xi_\ell, \alpha_\ell), \text{ with } (\alpha_i, \alpha_k) = (\xi_\ell \cdot \alpha_\ell, (1-\xi_\ell)\alpha_\ell)$$

hence
$$\int_0^\infty d\alpha_i \quad d\alpha_k \to \int_0^\infty d\alpha_\ell \; \alpha_\ell \int_0^1 d\xi_\ell$$

where ℓ labels the pair (i,k). In order to get simple expressions one has to care of the symmetry of the graph in the above transformation. For the last α-integration we use:

$$\int_0^\infty d\alpha \; \alpha^a \; e^{-\alpha x} = \Gamma\,(a+1)\; x^{-(a+1)} \quad .$$

In the following we give the D-dimensional Feynman integrals contributing in low order to τ and σ, as well as their expansion for small $\varepsilon = 4 - D$ and m to the extent we need them (in m^2 up to terms $O(m^2 \; \ell n \; m^2)$):

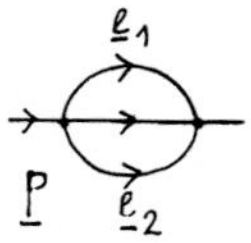

$$= \frac{1}{(2\pi)^{2D}} \int d^D\underline{\ell}_1 \quad d^D\underline{\ell}_2 \; \{ \; \frac{1}{\underline{\ell}_1^2+m^2} \; \frac{1}{\underline{\ell}_2^2+m^2}$$

$$\frac{1}{(\underline{p}-\underline{\ell}_1-\underline{\ell}_2)^2+m^2} - \begin{matrix}\text{subtr.}\\ \text{terms}\end{matrix} \} = \frac{\Gamma(3-D)\,(m^2)^{D-3}}{(4\pi)^D} \int_0^1 d\alpha \; d\beta \; \beta \cdot$$

$$[\alpha(1-\alpha)\ \beta^2+\beta(1-\beta)]^{D/2}\qquad\{[(1+\frac{\alpha(1-\alpha)\beta(1-\beta)}{\alpha(1-\alpha)\beta+(1-\beta)}\frac{p^2}{m^2})^{D-3}\quad -1]\quad +$$

$$+(3-D)\frac{\alpha(1-\alpha)\beta(1-\beta)}{\alpha(1-\alpha)\beta+(1-\beta)}\frac{p^2}{m^2}\ \}\simeq\ +\frac{1}{2}\frac{1}{(4\pi)^4}p^2\ \ln\frac{p^2}{m^2}\ ,(\varepsilon=0;m\to 0).$$

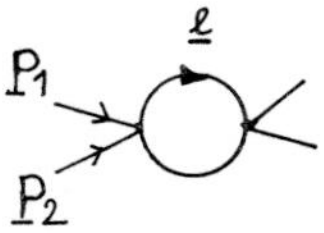

$$=\frac{1}{(2\pi)^D}\int d^D\ell\quad\{\frac{1}{\ell^2+m^2}\frac{1}{(p-\ell)^2+m^2}-\text{subtr.term}\}$$

$p=p_1+p_2$

$$=\frac{\Gamma(2-D/2)\,(m^2)^{D/2-2}}{(4\pi)^{D/2}}\int_0^1 d\alpha\ \{(1+\alpha(1-\alpha)\frac{p^2}{m^2})^{D/2-2}\quad -1\}$$

$$\simeq\frac{m^{-\varepsilon}}{(4\pi)^2}\{[1+\varepsilon(\gamma/2+\frac{1}{2}\ln\ 4\pi)]\quad(\ln\frac{p^2}{m^2}-2)$$

$$-\frac{\varepsilon}{4}I+\varepsilon\ \ln\frac{p^2}{m^2}-\frac{\varepsilon}{4}(\ln\frac{p^2}{m^2})^2\}\quad +O(\varepsilon^2),(m\to 0)$$

with γ the Euler constant and $I=\int_0^1 d\alpha\ \ (\ln\alpha(1-\alpha))^2$

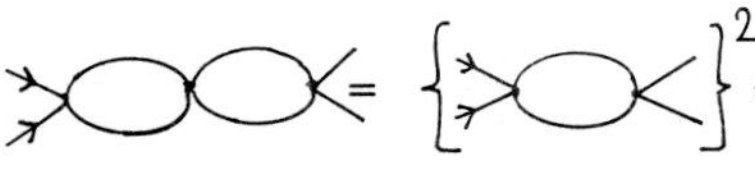

$$\simeq\ \frac{1}{(4\pi)^4}\ (\ln\frac{p^2}{m^2}-2)^2,\quad(\varepsilon=0;m\to 0)$$

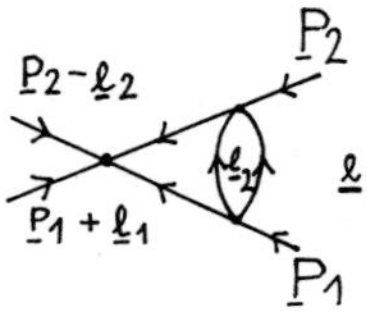

$$\underline{\ell}_1-\underline{\ell}_2 = \frac{1}{(2\pi)^{2D}} \int d^D\underline{\ell}_1 \, d^D\underline{\ell}_2 \, \{\frac{1}{(\underline{\ell}_1-\underline{\ell}_2)^2+m^2}$$

$$\frac{1}{\underline{\ell}_2^{\,2}+m^2} \, \frac{1}{(\underline{p}_1+\underline{\ell}_1)^2+m^2} \, \frac{1}{(\underline{p}_2-\underline{\ell}_2)^2+m^2} \; - \text{subtr.terms}\}$$

$$= \frac{\Gamma(4-D)\,(m^2)^{D-4}}{(4\pi)^D} \int_0^1 d\,\beta_1 \, d\beta_2 \, d\gamma\,(1-\gamma)\,\gamma^{1-D/2} \quad .$$

$$. \{((1-\gamma)+\gamma\beta_2(1-\beta_2))^{-D/2} [\, (1+\frac{\underline{p}_1^2}{m^2}(1-\gamma)\,\beta_1+\frac{\underline{p}_2^2}{m^2}(1-\gamma)\,(1-\beta_1) - \frac{\underline{q}^2}{m^2} \, .$$

$$. \frac{(1-\gamma)^2}{(1-\gamma)+\gamma\beta_2(1-\beta_2)})^{D-4} -1 \,]$$

$$- (1-\gamma)^{-D/2} \quad [(1+\frac{\underline{p}_1^2}{m^2}(1-\gamma)\,\beta_1+\frac{\underline{p}_2^2}{m^2}(1-\gamma)\,(1-\beta_1)-\frac{\underline{q}^2}{m^2}(1-\gamma))^{D-4} -1]\}$$

with $\quad \underline{q} = (\beta_1 \, \underline{p}_1 - (1-\beta_1)\, \underline{p}_2)$.

Even for $\varepsilon = o$ and $m \to o$ it would be tedious to carry out the integrations in this case. As follows from section 4 however, we need only evaluate the derivative $\frac{\partial}{\partial\mu^2}$ at the symmetry point of the above expression. The result is:

$2\mu^2 \frac{\partial}{\partial \mu^2} \{$ 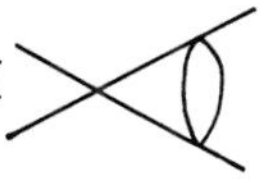 $\}_{sp} \simeq \frac{2}{(4\pi)^4}\{\ell n\frac{4\mu^2}{3m^2} - 3\}$, $(\varepsilon = o;\ m \to o)$.

In order to get the ordinary Feynmanintegrals in Minkowski space, the euclidean values have to be multiplied by

$$(-i)^a \ (i)^b$$

with a the number of internal lines
and b the number of loop integrations.
Furthermore $\underline{p} \cdot \underline{q}$ has to be substituted by - p.q.

Completed by the usual combinatorial factors and coupling constants we may evaluate from the given expressions τ, σ, ω and u for m = o to appropriate order in λ and ε.

We first evaluate τ. We have for $\varepsilon = o$ and $m \to o$:

$$\Gamma^{(2)}(p,-p) = i\ \{p^2 - m^2 + a^{(2)} + b^{(2)} p^2\} \quad +$$

$$= i\ \{-\underline{p}^2 - m^2 + \frac{\lambda^2}{12(4\pi)^4}(-m^2\ \ell n\frac{\mu^2}{m^2} + \underline{p}^2\ \ell n\ \frac{p^2}{\mu^2})\}$$

$$\Gamma^{(2)}_{as}(p,-p) = i\ \ \{-\underline{p}^2 + \frac{\lambda^2}{12(4\pi)^4}\ \underline{p}^2\ \ell n\ \frac{p^2}{\mu^2}\ \}$$

Hence we have (5.3):

$$\tau = i\ \frac{\partial \Gamma^{(2)}_{as}}{\partial \mu^2}\Bigg|_{p^2=\mu^2} = \frac{\lambda^2}{12(4\pi)^4} \quad .$$

The function σ has to be evaluated to order λ^3 and $\lambda^2\varepsilon$. We then have for $m \to o$:

$\Gamma^{(4)} = -i\{\lambda\mu^{\varepsilon} - c^{(2)} - c^{(3)} + i[$  $+$  $+$ 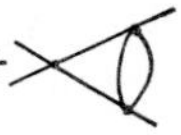

$+$  $_{c^{(2)}}$ $+$ crossed terms $]\}$

with

$c^{(2)} = \{$  $+$ crossed terms$\}_{sp} = \frac{\lambda^2}{(4\pi)^2}\frac{3}{2}\{[1+\varepsilon(\gamma/2+\frac{1}{2}\ell n 4\pi)](\ell n\frac{4\mu^2}{3m^2} - 2)$

$$- \frac{\varepsilon}{4} I + \varepsilon \ell n\frac{4\mu^2}{3m^2} - \frac{\varepsilon}{4}(\ell n\frac{4\mu^2}{3m^2})^2\} \quad (1+\frac{\varepsilon}{2}\ell n\frac{\mu^2}{m^2})\mu^{\varepsilon}$$

and

$c^{(3)} = \{$  $+$ 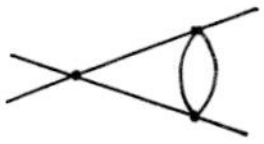 $+$  $_{c^{(2)}}$ $+$ crossed terms$\}_{sp,\varepsilon=o}$

The contribution of $\Gamma^{(4)}$ to σ is given by (5.26):

$$-i\mu^{-\varepsilon}2\mu^2\frac{\partial\Gamma^{(4)}}{\partial\mu^2}\Bigg|_{sp} = -\{\varepsilon\lambda - \mu^{-\varepsilon}2\mu^2\frac{\partial c^{(2)}}{\partial\mu^2} - 2\mu^2\frac{\partial c^{(3)}}{\partial\mu^2}$$

$+ \mu^{-\varepsilon}2\mu^2[\frac{\partial}{\partial\mu^2}$  $]_{sp} + 2\mu^2[\frac{\partial}{\partial\mu^2}$ 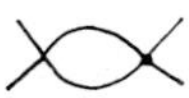 $_{c^{(2)}}]_{sp}\}$

Or using our expressions for the integrals we have:

$$-\mu^{-\varepsilon} 2\mu^2 \; [\frac{\partial c^{(2)}}{\partial \mu^2} - \{ \frac{\partial}{\partial \mu^2}$$  $$\}_{sp} \;] = \frac{\lambda^2 3}{(4\pi)^2} [1+\varepsilon(1+\gamma/2+\frac{1}{2} \; \ln \; 3\pi)]$$

and

$$-2\mu^2 [\; \frac{\partial c^{(3)}}{\partial \mu^2} - \{\frac{\partial}{\partial \mu^2}$$  $$_{c^{(2)}} \; \}_{sp} \;] =$$

$$= - \frac{6\lambda^3}{(4\pi)^4} \; (\ln \frac{4\mu^2}{3m^2} - 2-1)$$ 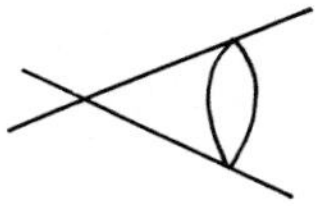

$$- \frac{3\lambda^3}{(4\pi)^4} \; (\ln \frac{4\mu^2}{3m^2} - 2)$$ 

$$+ \frac{3.6\lambda^3}{(4\pi)^4} \; (\ln\frac{4\mu^2}{3m^2} - 2) \qquad 2\mu^2 \frac{\partial}{\partial \mu^2}\{$$  $$_{c^{(2)}} \; \}_{sp}$$

$$- \frac{9\lambda^3}{(4\pi)^4} \; (\ln \frac{4\mu^2}{3m^2} - 2) \qquad 2\mu^2 \; \{\frac{\partial}{\partial \mu^2}$$  $$_{c^{(2)}} \; \}_{sp}$$

$$= - \frac{6\lambda^3}{(4\pi)^4}$$

Here we explicitly observe the cancellation of infrared singular terms. Thus we have for m = o:

$$\sigma = -i\mu^{-\varepsilon} \; 2\mu^2 \; \frac{\partial \Gamma_{as}{}^{(4)}}{\partial \mu^2} \Bigg|_{sp} + 4\lambda\tau = -\varepsilon\lambda + \lambda^2 \; \frac{3}{(4\pi)^2} (1+\varepsilon[1+\gamma/2+\frac{1}{2}\ln 3\pi])$$

$$-\lambda^3 \frac{6}{(4\pi)^4} + \lambda^3 \frac{3}{(4\pi)^4} \quad .$$

Analogously we evaluate ω and u. They are determined by the same integrals as σ, as seen from (5.14) and (6.15).

ACKNOWLEDGEMENTS

We have greatly profited from a visit of F. Wegner. We thank R. Seiler and W. Theis for many discussions concerning asymptotic properties of fields. Without the never ending patience of Mrs.I.Spielvogel, the typing of the manuscript would not have been completed.

REFERENCES

1. K.G. Wilson and J.Kogut, Institute for Advanced Study Lecture Notes C O O 2220-2 1972.
2. K. Osterwalder and R. Schrader, Havard University preprint 1972 and literature quoted therein.
3. K. G. Wilson, Phys. Rev. B4, 3174 (1971).
4. For those of our mathematically minded collegues for which the word "identical" is too strong we refer to 2).
5. K. Wilson, Phys. Rev. D3, 1818 (1971).
6. R. Brandt, Ann. Phys. 52, 122 (1969).
7. W. Zimmermann, Brandeis Lectures, Cambridge 1970.
8. J. H. Lowenstein and B. Schroer, NYU report 18172,

to be published in the comment section of Phys. Rev.D.

9. M. Gomes and J. Lowenstein, Nucl. Phys. B, Sept. 1972.

10. See for example reference 7.

11. Y. M. Lam, University of Pittsburgh preprint, Sept.1972.

12. J. H. Lowenstein, University of Maryland Lecture Notes 1972.
J. H. Lowenstein, Commun. Math. Phys. 24, (1971).

13. J. H. Lowenstein, Phys. Rev. D4, 2281 (1971).

14. Y. M. Lam, Phys. Rev. D6, 2145 (1972).

15. M. Gomes and J. H. Lowenstein, "Linear Relations among Normal Product Fields", University of Pittsburgh preprint 1972.

16. S. Weinberg, Phys. Rev. 118, 838 (1960).

17. K. Symanzik, Springer Tracts in Modern Phys., Vol. 57, (1971) and Commun. Math. Phys. 23, 49 (1971).

18. J. T. Diatlov, V.V. Sudakov, K.A. Ter-Martirosian JETP 5, 631 (1957).

19. The following discussion is taken from Symanzik ref.17 with the slight, but for our purpose important modification, that our μ normalization permits the direct identification with a μ-normalized zero-mass theory.

20. B. Schroer, "A Theory of Critical Phenomena based on the Normal Product Formalism", FU Berlin preprint 1972.
See distinction between short- and long-distance scaling limits according to the slope of σ has been first discussed by G. Mack, Kaiserslautern Lectures, Lecture Notes in Physics Vol. 17, Springer Verlag.

21. E. Brezin, I.C. Guillon and J. Zinn-Justin, "Wilson Theory of Critical Phenomena and Callan-Symanzik Equations in 4-ε Dimensions", Saclay preprint. P.K. Mitter, "Callan-Symanzik Equations and ε-Expansions". University of Maryland Technical Report No. 73-020.
The scale invariant correlation functions for statistical mechanics cannot be constructed by dilatation processes starting from the massive theory. The computational part of the determination of the critical indices is correct in these papers because the authors pick among the many scale invariant limits the one with anomalous dimension which moves continously into zero for $\varepsilon \to 0$. The values of their critical coupling constant in 2nd order in ε is different from our $\lambda_o(\varepsilon)$ and cannot be used to construct the correlation function at the critical temperature.

22. A.I. Larkin and D.E. Khmel'nitzky, JETP 29, 1123 (1969).
These authors obtain for logarithmic factors in the leading term. If perturbation theory is only done in the coefficients of the differential equations, one does not obtain a logarithm. Our statement is relevant for the critical correlation functions only if our idealized picture, which led to the elimination of the cut-off (lattice distance), is correct in 4 dimensions. Compare K. Symanzik, "On theories with Massless Particles" DESY preprint.

23. In the following we present the computation of critical indices directly for the m = 0 preasymptotic theory. This computation is simpler than the

massive one in ref. 21.

24. K.G. Wilson, Phys. Rev. Letters 28, 548 (1972).
E. Brezin, D.J. Wallace and K.G. Wilson, Phys. Rev. Letters 29, 591 (1972).

25. L.P. Kadanoff et al., Rev. Mod. Phys. 39, 395 (1967).

26. K. Symanzik, DESY preprint 1973.

27. G. Mack, Kaiserslautern lectures "Lecture Notes in Physics" Vol. 17, Springer Verlag, chapter 7, p.314.

28. Private communication from J. Geicke, S. Meyer and R. Seiler.

29. J. Geicke and S. Meyer, FU Berlin preprint to be published in Phys. Letters.

30. T.T. Wu, Phys. Rev. 149, 380 (1966).

31. F. Wegener, Phys. Rev. B5, 4529 (1972).

Acta Physica Austriaca, Suppl.XI, 475–492 (1973)

ON PHASE TRANSITIONS IN OPEN SYSTEMS FAR FROM THERMAL EQUILIBRIUM*

BY

K. Hepp
Department of Physics, ETH, CH-8049 Zürich

1. INDRODUCTION: CONVENTIONAL WISDOM

In these lectures I shall outline some rigorous results, obtained in collaboration with E.H. Lieb [H4], concerning a class of non-trivial quantum mechanical many-body systems with irreversible behavior far away from thermal equilibrium.

According to the conventional wisdom in the statistical mechanics of irreversible processes (see e.g. [D1], [G3], [L1]) there are two levels in which a simplified description of an N-body system is expected in the thermodynamic limit, where $N \to \infty$, $V \to \infty$, N/V = const: There should exist a complete set of "extensive" (or "gross" or "macroscopic") observables $A^1_{(N)}, \ldots A^\ell_{(N)}$,

$$A^k_{(N)} = \sum_{n=1}^{N} A^k_n , \tag{1}$$

*Lecture given at XII. Internationale Universitätswochen für Kernphysik, Schladming, February 5 - 17, 1973.

formed out of "local" observables A_n^k, for which in a suitable sense the limit

$$N^{-1} A^k_{(N)} \equiv \alpha^k_{(N)} \to \alpha^k \tag{2}$$

exists and where the time evolution of the limiting "intensive" observables $\underline{\alpha} = (\alpha^1, \ldots, \alpha^\ell)$ is described by a system of first order differential equations

$$\dot{\underline{\alpha}} = \underline{F} (\underline{\alpha}) \ . \tag{3}$$

In most realistic systems the number of independent intensive observables is not finite. For a viscous one-component isotropic fluid the usual phenomenological theory [D1] requires the knowledge of the density $\rho(\underline{x},t)$, velocity $\underline{v}(\underline{x},t)$ and temperature $T(\underline{x},t)$ at every point $\underline{x}$ for a fixed time t. Then the future is uniquely determined by the five conservation laws of mass, energy and momentum under some phenomenological assumptions, using the thermodynamical equations of state for the energy $u = u(\rho,T)$ and the pressure $p = p(\rho,T)$. In these lectures we restrict ourselves to systems like the laser, where a macrostate is characterized by a finite number of intensive observables.

In an autonomous system, (3) should be autonomous. Furthermore one expects (3) to be dissipative and to have strong stability properties for a range of values of the parameters of the system, such as damping constants and pump parameters, which characterize hidden microscopic aspects of the system. One expects a "thermal branch" [G1] with an asymptotically stable equilibrium state $\underline{\alpha}_o$, satisfying $\underline{F}(\underline{\alpha}_o) = 0$ and $\underline{\alpha}(\underline{\gamma},t) - \underline{\alpha}_o \to 0$ for $t \to \infty$ and for every initial condition $\underline{\gamma}$. At a critical value η_c of some parameter η, $\underline{\alpha}_o$ could become unstable and a periodic solution $\underline{\beta}_o(t)$ could bifurcate from $\underline{\alpha}_o$ [H5], such that for almost all $\underline{\gamma}$, $\underline{\alpha}(\underline{\gamma},t) - \underline{\beta}_o(t+t(\underline{\gamma})) \to 0$ for $t \to \infty$ and some $t(\underline{\gamma})$. The physical interpretation of this bifurcation phenomenon in the macroscopic description of the system has been given in terms of a phase transition to a self-oscillatory non-thermal branch far away from thermal equilibrium [G1]. More complicated bifurcations might give a qualitatively correct description of phenomena such as turbulence [R2].

Einstein [E1] and Onsager [O1] have recognized that the fluctuations of the intensive observables around the limiting vales

$$a^k_{(N)}(t) = \frac{1}{\sqrt{N}} \sum_{n=1}^{N} (A^k_n(t) - \alpha^k(t)) \tag{4}$$

should "normally" be well-behaved in the limit $N \to \infty$. The expected equations of motion for the $a^k(t) = \lim a^k_{(N)}(t)$ are linear and inhomogeneous:

$$\dot{\underline{a}}(t) = X(t)\, \underline{a}(t) + \underline{f}(t) \,, \tag{5}$$

and the connection between (3) and (5) has been formulated [O1] as the hypothesis of "regression of fluctuations": (5) should be a stochastic process with "white noise" $\underline{f}(t)$ independent of the initial condition $\underline{a} = \underline{a}(t = 0)$. The average $\langle \underline{a}(t) \rangle$ over the reservoir coordinates should satisfy the same equations as $\underline{\alpha}(t)$ linearized around $\underline{\alpha}(\underline{\gamma},t)$.

Even in quantum mechanics, the limiting intensive observables should behave classically. On the $\sqrt{N}$-level, the difference between quantum and classical mechanics should become important. In quantum mechanics, the fluctuation observables should have bose character, if $[A_m^i, A_n^j] = 0$ for $m \neq n$, since then

$$[a^i_{(N)}, a^j_{(N)}] = N^{-1} \sum_{n=1}^{N} [A_n^i, A_n^j] \tag{6}$$

becomes an intensive observable. Similarly, one expects that the "fluctuation forces" $\underline{f}(t)$ should behave as boson fields, according to some "law of large numbers".

2. EXAMPLE: THE LASER

The laser is a highly non-trivial example to which the above qualitative ideas should apply [D3], [G2], [S1]. To a fairly reasonable approximation this quantum mechanical system can be described by the Dicke-Haken-Lax model [D2], [H1], [L2]. Here the radiation field in the cavity, which interacts with matter, is restricted to a fixed finite number of boson modes. For simplicity, we only consider one radiation oscillator of frequency ν with creation and annihilation operators $a^{\#}$. There are N active molecules of M levels, with only uninteresting translational degrees of freedom. The relevant electron states have fermion creation and annihilation operators $a_n^{m\#}$, where $1 \leq m \leq M$, $1 \leq n \leq N$. For simplicity, let $M = 2$ and introduce the quasi-spin operators

$$S_n^+ = a_n^{2*} a_n^1 = (S_n^-)^* , \qquad S_{(N)}^+ = \sum_{n=1}^{N} S_n^+ ,$$

$$S_n^3 = \frac{1}{2}(a_n^{2*} a_n^2 - a_n^{1*} a_n^1) , \qquad S_{(N)}^3 = \sum_{n=1}^{N} S_n^3 . \tag{7}$$

If ε is the energy difference between the upper and lower level, then the Hamiltonian $H_{(N)}^S$ describing the closed system of N two-level molecules interacting with one spatially independent mode of the radiation field becomes

$$H_{(N)}^S = \nu \; a^* a + \varepsilon \; S_{(N)}^3 + N^{-1/2}(\lambda S_{(N)}^+ a + \mu \; S_{(N)}^+ a^* + \text{h.c.}) . \tag{8}$$

Here we have assumed ε, $\nu > 0$ and that the volume V of the cavity to be proportional to N, so that $H_{(N)}^S$ should describe the closed system for all N with N-independent coupling constants λ and μ.

We introduce as intensive observables

$$\alpha_{(N)}^{\#} = N^{-1/2} \; a^{\#} ,$$

$$\tau_{(N)}^i = N^{-1} \; S_{(N)}^i \; , \quad 1 \le i \le 3 . \tag{9}$$

The Heisenberg equations of motion for (9) propagated under (8) are

$$\dot{\alpha}_{(N)} = i[H_{(N)}^S , \alpha_{(N)}] = - i\nu\alpha_{(N)} - i\lambda \; \tau_{(N)}^- - i\mu\tau_{(N)}^+ ,$$

$$\dot{\tau}^{-}_{(N)} = -i\varepsilon\tau^{-}_{(N)} + 2i\lambda\tau^{3}_{(N)}\alpha_{(N)} + 2i\mu\tau^{3}_{(N)}\alpha^{*}_{(N)} , \tag{10}$$

$$\dot{\tau}^{3}_{(N)} = -i\lambda\tau^{+}_{(N)}\alpha_{(N)} - i\mu\tau^{+}_{(N)}\alpha^{*}_{(N)} + \text{h. c.}.$$

We observe that the five operators (9) are sufficient for defining a closed set of non-linear operator differential equations. In certain states ω of the closed system, e.g. in the ground and Gibbs states for all temperatures and all values of the coupling constants, one can see [H2], [H3] that the expectation values of products of intensive observables, $\omega(\beta^{i}_{(N)}(s)\ldots\beta^{j}_{(N)}(t))$, $\beta^{k}_{(N)} \varepsilon \{ \alpha^{\#}_{(N)}, \underline{\tau}_{(N)} \}$, converge for $N \to \infty$ and can be computed using the solutions of (1o) with $\alpha^{\#}_{(N)}, \tau^{i}_{(N)}$ replaced by c-number quantities $\alpha^{\#}$, τ^{i}.

The real laser is a "small" system S, the cavity with Hamiltonian $H^{S}_{(N)}$ in a Hilbert space H_{S}, coupled by $H^{R}_{(N)}$ to a reservoir R, a quantum system of a large number of degrees of freedom, into which the photons can dissipate and which should try to pump the molecules into an inversion value η of τ^{3}. The semiclassical theory [H1] is expected to correctly describe the form of the equations of motion for the intensive observables in the limit $N \to \infty$, if they are propagated under $H_{(N)} = H^{S}_{(N)} + H^{R}_{(N)}$. The conventional procedure is to add a linear dissipative term to the Hamiltonian part on the right hand side of (1o) and nothing more for $N \to \infty$:

$$\dot{\alpha} = -(i\gamma + \kappa)\alpha - i\lambda^{*}\tau^{-} - i\mu\tau^{+} ,$$

$$\dot{\tau}^{-} = -(i\varepsilon + \gamma)\tau^{-} + 2i\lambda\tau^{3}\alpha + 2i\mu\tau^{3}\alpha^{*}, \tag{11}$$

$$\dot{\tau}^3 = - \delta(\tau^3 - \eta) - (i\lambda\tau^+\alpha + i\mu\tau^+\alpha^* + h.\,c.)\ .$$

Here the damping constants satisfy:

$$\kappa > 0, \quad \gamma > 0, \quad \delta > 0\ , \tag{12}$$

and the pump parameter η:

$$- 1/2 \leq \eta \leq 1/2\ . \tag{13}$$

If $\lambda = \mu = 0$, the reservoir would drive the intensive observables from any physical initial state $\alpha \varepsilon C$, $|\underline{\tau}| \leq 1/2$ into the stationary state of (11), the "thermal branch"

$$\alpha = \tau^- = 0\ , \qquad \tau^3 = \eta\ . \tag{14}$$

If $(\lambda,\mu) \neq (0,0)$, the solution (14) no longer needs to be stable. If one makes the rotating wave approximation, $\mu = 0$, then there exists a non-trivial harmonic periodic solution of (11)

$$\alpha(t) = \alpha\, e^{-i\omega t}\ , \qquad \tau^-(t) = \tau^-\, e^{-i\omega t}\ , \tag{15}$$

with time independent α, $\tau^- \ \varepsilon\ C$, $\tau^3 \ \varepsilon\ R$, if $\eta_c < \eta \leq 1/2$, where

$$\tau^3 = \eta_c = \frac{\kappa\gamma}{2|\lambda|^2}\left(1 + \frac{(\varepsilon-\gamma)^2}{(\kappa+\gamma)^2}\right)\ ,$$

$$\omega = (\kappa\varepsilon + \gamma\nu)/(\kappa + \gamma) ,$$

$$|\alpha|^2 = \delta(\eta - \eta_c)/\kappa , \tag{16}$$

$$\tau^- = (\omega - \gamma + i\kappa)/\lambda^* .$$

At the laser threshold $\eta = \eta_c$, this solution bifurcates from (11). The linearized equations of (11) around (14) show that, for $\eta < \eta_c$, (14) is stable. By the Hopf bifurcation theorem [H5], (16) is the unique stable attractor in the neighborhood of η_c, and we can show that this self-oscillatory behavior of the laser is asymptotically stable for almost all initial conditions and all values of $\eta_c < \eta \leq 1/2$, if $\mu = 0$ and $\kappa = \gamma = \delta/2$. For small values of $|\mu|$, a periodic anharmonic laser oscillation with $|\alpha|^2 \sim \eta - \eta_c$ persists. The formulae (16) for η_c, ω and the linear dependence of the mean photon number per atom $|\alpha|^2$ on $\eta - \eta_c$ are in quite good agreement with the results of experiments on finite lasers.

The central question is whether there exist quantum mechanical models for reservoirs and a mathematically rigorous limiting procedure which in the limit $N \to \infty$ leads to the phenomenological equations (11) for the intensive observables, and, furthermore, whether the fluctuations are well-behaved for $N \to \infty$ and for all times if they are at $t = 0$, as well as whether they satisfy Onsager's regression hypothesis. This is not an entirely trivial problem, as the historical development of irreversible statistical mechanics shows. In order to have $O(N)$ effects on the "small" system S instantaneously, the reservoir R has to be rather singular. Metaphorically, the laser is a lady riding on a tiger and her safe homecoming depends on whether the errors can be kept at $o(\sqrt{N})$ in the limit $N \to \infty$.

It turns out [H4] that the Dicke-Haken-Lax model of the laser with suitable reservoirs becomes exactly soluble in the limit $N \to \infty$, with $o(\sqrt{N})$ errors, if one introduces a cut-off in the photon numbers at ζN, where ζ is finite but arbitrarily large. The limiting solutions are very insensitive with regard to changes in ζ for large ζ. In such a model, which is no more unrealistic for describing a real laser than (8), all operators are bounded and we can use the mathematically elegant theory of C^*-algebras for discussing the limit $N \to \infty$. This will be outlined in the following section.

3. MEAN FIELD MODELS WITH LINEAR DISSIPATION

We shall consider a large number, N, of identical subsystems. Each subsystem $\{R_n, S_n\}$, $n \in Z$, consists of a system part S_n and its private reservoir R_n. S_n is a fermion system of M degrees of freedom with creation and annihilation operators $a_n^{m\#}$, $1 \leq m \leq M$. R_n consists of 2M fermion fields $A_{nw}^{m\#}$, $B_{nw}^{m\#}$, where $w \in R$ and with the conventional anticommutation relations

$$\{a_n^m, a_n^{m*}\} = 1 ,$$
$$\{A_{nw}^m, A_{nw'}^{m*}\} = \{B_{nw}^m, B_{nw'}^{m*}\} = \delta(w-w'), \tag{17}$$

and vanishing anticommutators between all other first order fermion operators.

The coupling between elements from $\{S_n\}$ should be non-linear, but of the mean-field type, i.e. without distinction of the localization of the different S_n. Clearly, such mean field couplings should simplify for $N \to \infty$. The coupling between the system $S = \{S_n\}$ and the reservoir $R = \{R_n\}$ should be a linear interaction between each a_n^m and the corresponding fields A_{nw}^m, B_{nw}^m. If one assumes equal coupling strengths for all n, then in the limit $N \to \infty$ the law of large numbers should lead to a strong simplification for this dissipative mechanism. The total coupling scheme is depicted below:

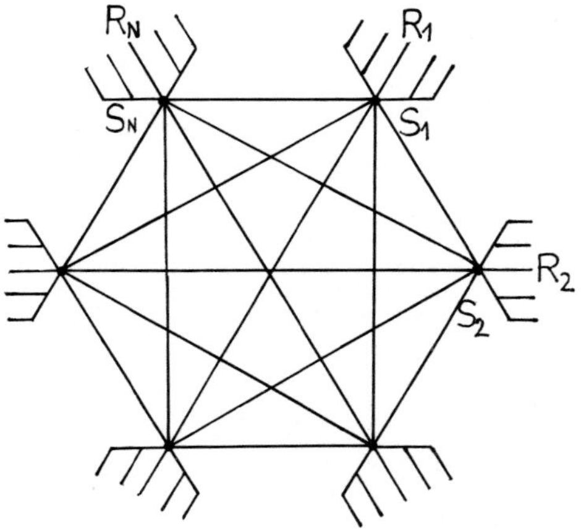

Fig. 1

Structure of a system $S = \{S_1, \ldots S_N\}$ with mean field interaction and independent reservoirs $R = \{R_1, \ldots R_N\}$.

Let A_n be the C^*-algebra generated by all smeared-out polynomials in these operators which conserve the total number of fermions. Then $\cup\ A_n$ generates a quasilocal C^*-algebra A. A_n contains the Lie algebra L_n with basis

$$S_n^{k\ell} = a_n^{k*} a_n^{\ell} \quad , \qquad 1 \le k,\ell \le M \ . \tag{18}$$

For all n, $L_n \simeq$ Lie U(M). In our models, a complete set of macroscopic observables is given by

$$S_{(N)}^{k\ell} = N\ \tau_{(N)}^{k\ell} = \sum_{n=1}^{N} S_n^{k\ell} \ . \tag{19}$$

The extensive observables $S_{(N)}^{k\ell}$ again form a representation of Lie U(M), while the $\tau_{(N)}^{k\ell}$ have $O(N^{-1})$ commutators. It is known [K1], [R1] that the $\tau_{(N)}^{k\ell}$ become c-numbers if one takes the limit $N \to \infty$ in any translation invariant state ω on A: for all monomials $\tau_{(N)}^{ik},\dots,\tau_{(N)}^{pq}$

$$\lim_{N\to\infty} \omega(\tau_{(N)}^{ik} \dots \tau_{(N)}^{pq}) = \int \mu_\omega(d\underline{\alpha})\ \alpha^{ik} \dots \alpha^{pq} \ , \tag{20}$$

where μ_ω is a probability measure on the "phase space" R^{M^2} with the coordinates $\alpha^{ii} = \alpha^{ii*}$, $\alpha^{ij} = \alpha^{ji*}$, $|\alpha^{ij}| \le 1$ for $1 \le i,j \le M$. If ω is an extremal translation invariant state, then μ_ω is concentrated in one point $\underline{\alpha} \in R^{M^2}$, and for all monomials

$$\lim_{N\to\infty} \omega(\tau_{(N)}^{ik} \dots \tau_{(N)}^{pq}) = \alpha^{ik} \dots \alpha^{pq} \ . \tag{21}$$

We call a state ω on A (which is not necessarily translation invariant) "classical" with respect to $\{\tau_{(N)}^{ik}\}$ if (20) holds for all monomials, and "pure classical" if (21) holds. A stronger requirement for ω is to have normal fluctuations around some $\underline{\alpha} \in R^{M^2}$, i.e. if for

all monomials in the fluctuation observables
$s^{ik}_{(N)} = \sqrt{N}(\tau^{ik}_{(N)} - \alpha^{ik})$

$$\lim_{N\to\infty} \omega(s^{ik}_{(N)} \dots s^{pq}_{(N)}) = (\Omega_s, s^{ik} \dots s^{pq} \Omega_s) \tag{22}$$

holds. The identification of the right hand side as "vacuum" expectation values of possibly unbounded operators s^{ik} is possible by the GNS construction, and the s^{ik} are generalized bosons:

$$[s^{ik}, s^{pq}] = \delta^{kp} \alpha^{iq} - \delta^{iq} \alpha^{pk} , \tag{23}$$

where δ^{kp} is the Kronecker symbol.

In this kinematical framework we consider mean field interactions of the type

$$H^S_{(N)} = N \beta (\underline{\tau}_{(N)}), \tag{24}$$

where β is a self-adjoint polynomial in the $\underline{\tau}_{(N)}$ with coefficients independent of N. By the Lie structure of the $\{S^{ik}_{(N)}\}$, the equations of motion for the $\underline{\tau}_{(N)}$, propagated under $H^S_{(N)}$, close for ever N.

The dissipation is introduced by

$$H^R_{(N)} = \sum_{m=1}^{M} \sum_{n=1}^{N} H^m_n , \tag{25}$$

$$H^m_n = \int dw (A^{m*}_{nw} A^m_{nw} + B^{m*}_{nw} B^m_{nw}) w +$$

$$+ \int dw \, \{ (g_A^{m*} \, A_{nw}^{m*} + g_B^{m*} \, B_{nw}^{m*}) \, a_n^m + h.c.\} \quad . \tag{26}$$

$H_{(N)}^R$ introduces a 1-parameter group of strongly continuous automorphisms of A, and $H_{(N)} = H_{(N)}^R + H_{(N)}^S$ too, since $H_{(N)}^S \in A$. The reason for introducing two fermion fields as reservoirs is for reaching all possible pump values η^k (see (29)) for τ^{kk} by suitable choices of the coupling constants g_A^k and g_B^k and by taking initial states ω on A of the type

$$\omega = \omega_s \otimes \omega_R \, , \qquad \omega_R = (\Omega_R, \bullet \, \Omega_R) \, ,$$

$$A_{nw}^m \, \Omega_R = B_{nw}^{m*} \, \Omega_R = 0 \quad . \tag{27}$$

The Fock state Ω_R is empty of A-quanta and full of B-quanta, and we take ω_s to be a state on the C^*-algebra generated by all $a_n^{m\#}$.

The system of equations satisfied by the $\underline{\tau}_{(N)}$ propagated under $H_{(N)}$ is no longer closed:

$$\begin{aligned} \dot{\tau}_{(N)}^{k\ell}(t) &= \delta^{k\ell} \, \eta^k - \gamma^{k\ell} \, \tau_{(N)}^{k\ell}(t) \\ &\quad + i \, [H_{(N)}^S(t), \tau_{(N)}^{k\ell}(t)] + \phi_{(N)}^{k\ell}(t) \, , \end{aligned} \tag{28}$$

where

$$\begin{aligned} \eta^k &= 2\pi \, |g_B^k|^2 \, , \\ \gamma^{k\ell} &= \pi (|g_A^k|^2 + |g_B^k|^2 + |g_A^\ell|^2 + |g_B^\ell|^2) \, , \end{aligned} \tag{29}$$

and where the $\phi^{k\ell}_{(N)}$ are complicated expressions involving the $H_{(N)}$-evolved $a^{m\#}_{n(N)}(t)$. However, these "fluctuation forces" have the form

$$\phi^{k\ell}_{(N)}(t) = \phi^{k\ell}_{(N)+}(t) + \phi^{k\ell}_{(N)-}(t) ,$$

$$\phi^{k\ell}_{(N)-}(t)\ \Omega_R = 0 = (\phi^{k\ell}_{(N)+}(t))^*\ \Omega_R . \tag{30}$$

The following two theorems can be proved [H4] by an iteration scheme which converges uniformly in N:

Theorem A: Let $H_{(N)} = H^S_{(N)} + H^R_{(N)}$ be of the form (24), (25) and let ω satisfy (27), ω_s being a classical state with respect to $\underline{\tau}_{(N)}$ with probability measure μ_ω. Then for all monomials $\tau^{ik}_{(N)}(r),\dots,\tau^{pq}_{(N)}(t)$

$$\lim_{N\to\infty} \omega(\tau^{ik}_{(N)}(r)\dots\tau^{pq}_{(N)}(t)) = \int \mu_\omega(d\underline{\alpha})\,\tau^{ik}(\underline{\alpha},r)\dots\tau^{pq}(\underline{\alpha},t), \tag{31}$$

and the convergence is uniform on compact sets in $(r,..,t)$ $\underline{\tau}(\underline{\alpha},t)$ is the solution of the c-number equation

$$\dot{\tau}^{ik} = \delta^{ik}\,\eta^k - \gamma^{ik}\,\tau^{ik} + \{\beta,\ \tau^{ik}\} , \tag{32}$$

where the Poisson bracket between the polynomial β (see (24)) and τ^{ik} is evaluated using

$$\{\tau^{ik},\ \tau^{pq}\} = i(\delta^{kp}\,\tau^{iq} - \delta^{iq}\,\tau^{pq}) . \tag{33}$$

<u>Theorem B:</u> Under the assumptions of Theorem A let $\underline{\tau}_{(N)}$ have normal fluctuations in ω_s around $\underline{\alpha}$. Then for all monomials in $s^{ik}_{(N)}(t) = \sqrt{N}(\tau^{ik}_{(N)}{}^{s}(t) - \tau^{ik}(\underline{\alpha},t))$

$$\lim_{N\to\infty} \omega(s^{ik}_{(N)}(r)\ldots s^{pq}_{(N)}(t)) = (\Omega, s^{ik}(r)\ldots s^{pq}(t)\Omega) \qquad (34)$$

uniformly on compacts in $(r,\ldots,t)$. The $\underline{s}(t)$ are solutions of the linearized equation of (32) around $\underline{\tau}(\underline{\alpha},t)$ with inhomogeneous term $\underline{f}(t)$ and initial condition $\underline{s}(0) = \underline{s}$. $\underline{s}$ operates on the GNS Hilbert space H_S with vacuum Ω_S, as defined by (22). The $\underline{f}(t)$ are generalized boson fields and they are defined on a Hilbert space H_R with cyclic vacuum Ω_R by

$$f^{k\ell}(t) = f^{k\ell}_{+}(t) + f^{k\ell}_{-}(t) ,$$

$$f^{k\ell}_{-}(t)\,\Omega_R = 0 = (f^{k\ell}_{+}(t))^{*}\,\Omega_R, \qquad (35)$$

$$[f^{k\ell}_{-}(s), f^{pq}_{-}(t)] = [f^{k\ell}_{+}(s), f^{pq}_{+}(t)] = 0 ,$$

$$[f^{k\ell}_{-}(s), f^{pq}_{+}(t)] = 2\pi\delta(s-t)\ \{\delta^{kq}\,\delta^{\ell p}\,|g^{k}_{B}|^2$$

$$- \delta^{kq}\,\tau^{p\ell}(\underline{\alpha},t)\,|g^{k}_{B}|^2 + \delta^{\ell p}\tau^{kq}(\underline{\alpha},t)\ |g^{\ell}_{A}|^2\} .$$

The total Hilbert space H for $\underline{s}(t)$ is $H_S \otimes H_R$ with $\Omega = \Omega_S \otimes \Omega_R$.

We see that in our class of models a clear separation of classical and quantum effects is possible. In $O(N)$ one obtains the phenomenological classical dissipative equations which have qualitative features as described in section 1 and 2. In $O(\sqrt{N})$ around the classical macroscopic observables one sees quantum fluctuations, which are bosons evolving in time by linear inhomogeneous equations driven by Gaussian and Markovian fluctuation forces, the quantum analog to the Ornstein-Uhlenbeck process [U1].

The Dicke-Haken-Lax model of the laser is a quantum mechanical model for a clock or a steam engine. For $\eta < \eta_c$ the system is to $O(N)$ friction dominated and comes to rest from every initial state. For $\eta > \eta_c$, the laser is self-oscillatory and the crossing of the laser threshold leads to a reservoir induced phase transition of an open system far from thermal equilibrium.

ACKNOWLEDGEMENT

In conclusion I would like to thank E. H. Lieb for many enlightening discussions and for his kind permission to present unpublished results in these lectures. Of course, all misrepresentations of the ultimate truth are entirely my responsibility.

REFERENCES

D1. S.R. de Groot, P. Mazur, "Non-equilibrium Thermodynamics", North-Holland Publ. Co., Amsterdam (1962).

D2. R.H. Dicke, Phys. Rev. 93, 99 (1954).

D3. V. Dohm, Solid State Comm. 11, 1273 (1972).

E1. A. Einstein, Ann. d. Physik 33, 1275 (1910).

G1. P. Glanssdorff, I. Prigogine, "Thermodynamic Theory of Structure, Stability and Fluctuations", Wiley, London (1971).

G2. R. Graham, H. Haken, Z. Physik 237, 30 (1970).

G3. M. S. Green, J. Chem. Phys. 20, 1281 (1952).

H1. H. Haken, Handbuch der Physik Vol. XXV/2C, Springer-Berlin (1970).

H2. K. Hepp, E.H. Lieb, Annals of Phys. 76, 360 (1973).

H3. K. Hepp, E.H. Lieb, "Statistical mechanics of matter coupled to a quantized radiation field", preprint, Bures (1973).

H4. K. Hepp, E.H. Lieb, to appear in Helv. Phys. Acta.

H5. E. Hopf, Ber. d. math.-phys. Kl. d. Sächs. Akad. d. Wiss., Leipzig 94, 3 (1942).

K1. D. Kastler, D.W. Robinson, Comm. Math. Phys. 3, 151 (1966).

L1. O.E. Lanford III, in "Statistical Mechanics and Mathematical Problems" (Ed. A. Lenard), Springer, Berlin (1973).

L2. M. Lax, in "Phase Transitions and Superfluidity", Brandeis Lectures 1965 (M. Chrêtien et al., ed.), Gordon and Breach, NY (1968).

O1. L. Onsager, Phys. Rev. 37, 405; 38, 2265 (1931).

R1. D. Ruelle, Comm. Math. Phys. 3, 133 (1966).

R2. D. Ruelle, F. Takens, Comm. Math. Phys. 20, 167 (1971).

S1. M. O. Scully, V. Degiorgi, Phys. Rev. A2, 1170 (1970).

U1. G. E. Uhlenbeck, L.S. Ornstein, Phys. Rev. 36, 823 (1930).

Acta Physica Austriaca, Suppl.XI, 493–525 (1973)

EXACT RESULTS ON THE STRUCTURE OF MATTER*

BY

W. THIRRING
Institut für Theoretische Physik, Universität Wien

1. INTRODUCTION AND RESULTS

i) Position of the Problem

A system of N non-relativistic particles interacting with Coulomb and Newton potentials is governed by the Hamiltonian

$$H_N = \sum_{i=1}^{N} \frac{p_i}{2m_i} + \sum_{\substack{i,j=1\\ i>j}}^{N} (e_i e_j - \kappa m_i m_j)|x_i - x_j|^{-1} = K + V. \tag{1.1}$$

Treated quantummechanically it should give a good description ($\sim$ 1% accuracy) of atoms, molecules, macroscopic and cosmic bodies: The non-relativistic treatment of electricity and gravitation should be a useful approximation in most circumstances. Strong and weak forces are usually hidden in nuclei which on this level of description are treated as elementary particles. We shall simply call matter systems described by (1.1) and ask what can be rigorously deduced from this Hamiltonian.

In quantum mechanics where x_i and p_j are unbounded operators we first have to define the domain $\mathcal{D}$ on which

*Lecture given at XII. Internationale Universitätswochen für Kernphysik, Schladming, February 5 - 17, 1973.

H_N is self-adjoint. For K this is known to consist of functions whose first and second derivatives are square integrable. On some occasions we shall consider our system enclosed in a box Ω. Due to the singularity of $1/|x|$ V is also an unbounded operator. However the usual intuitive argument with the uncertainty relation suggests that in quantum mechanics this singularity is harmless. This argument can be phrased mathematically [1] in the form $\forall\ b > 0\ \exists\ a > 0$ such that

$$||V\psi|| < a||\psi|| + b||K\psi|| \qquad \forall\ \psi \in D . \qquad (1.2)$$

As a consequence one can deduce readily that K+V is self-adjoint on the same domain and as K bounded from below. Thus the problem is mathematically well defined:

$e^{iH_N t}$ is unitary and the spectrum of H_N contained in an interval $[E_0,\infty)$. With respect to the latter Combes and Balslev [2] proved the following results which have been partly anticipated as physically reasonable:

1) The singular continuous part (e.g. the one concentrated on a set of (Lebesgue) measure zero) is absent.

2) $(H-z)^{-1}$ can be continued through the real axis and may have poles in the second sheet ("resonances") but only for $\mathrm{Re}\ z < 0$.

3) There are no discrete points of the spectrum for $\mathrm{Re}\ z > 0$ (but for $N \geq 3$ there are some embedded in the continuum for $\mathrm{Re}\ z < 0$. They correspond to states whose decay is forbidden by selection rules).

I called the poles in the second sheet resonances although the usual scattering theory does not apply:

$e^{-iKt} e^{iHt}$ does not [3] strongly converge for $t \to \pm\infty$.

In the following I shall concentrate on the most elementary properties of H_N, namely the position of E_o and later some properties of level densities as they are reflected in the thermodynamic functions.

ii) Atoms

For N = 2 we have the Hamiltonian of the hydrogen atom and everybody knows how to solve that. But for more complicated atoms no exact solution is available and one might ask whether one can at least give rigorous bounds within which E_o has to be. For Helium there are indeed excellent bounds available, only about 1‰ apart [4]. For more complicated atoms there are very good upper limits as result of Hartree-Fock calculations. However, I am not aware of any useful lower bounds. Of course, there are always the trivial lower bounds one obtains just neglecting the Coulomb repulsion, but they are about 40 % too low.

iii) Macroscopic Bodies ($N \sim 10^{24}$)

The first empirical fact to account for is the saturation character of the chemical forces, e.g. one wants an inequality of the type

$$- A \leq \frac{E_o}{N} \leq - B \quad . \tag{1.3}$$

An upper limit within some percent of the empirical value is easily available by taking atomic (or molecular) trial functions. The lower bounds are notoriously hard to get and only Dyson and Lenard [5] were able to prove (1.3), however with $A \sim 10^{14}$! It may be a matter of dispute whether we can say that we understand a system when we have reached 50 % or 10 % or 1 % accuracy, but a factor 10^{14} clearly means we have not solved the problem.

The difficulty of proving (1.3) is connected with the fact that it is generally not true. As we shall see shortly it is spoiled by gravitation, thus we need $\kappa = 0$. Furthermore, the system has to be electrically neutral (if in a box with volume $\sim N$), i.e.

$$\sum_{j=1}^{N} e_j = 0$$

(or small in some sense). Even then one finds for bosons [6]

$$- A\, N^{5/3} \leq E_o \leq - B\, N^{7/5} \tag{1.4}$$

To get (1.3) one needs the additional assumption that the particles of one sign of charge (say all negative ones, the electrons) obey the exclusion principle. The factor 10^{14} is just due to the complexity of the analysis. If on every page you give away by an estimate a factor 2 or 3, after 40 pages you have lost 10^{14}.

Regarding the thermodynamic behaviour of matter it has been proved by Lebowitz and Lieb [7] that the relevant quantities exist and have the desired convexity properties. Their analysis is based on the result of Dyson and Lenard and thus holds iff:

1) $\kappa = 0$
2) $\sum_j e_j = 0$ (or small)
3) All negative (or positive) particles are fermions.

Starting with a finite system in a box Ω one defines a corresponding Hamiltonian $H_N(\Omega)$. It has a purely discrete spectrum and its eigenvalues (in ascending order) are denoted $\varepsilon_i(\Omega)$. The thermodynamic quantities in the canonical and microcanonical ensemble are

$$e^{-\beta F_N(\beta,\Omega)} = \mathrm{Tr}\, e^{-\beta H_N(\Omega)} \tag{1.5}$$

$$E_N(S,\Omega) = e^{-S} \sum_{i=1}^{e^S} \varepsilon_i(\Omega). \tag{1.6}$$

Lebowitz and Lieb could show the existence of

$$\lim_{N\to\infty} N^{-1} F_N(\beta,\frac{N}{\rho}) = f(\beta,\rho) \tag{1.7}$$

$$\lim_{N\to\infty} N^{-1} E_N(Ns,\frac{N}{\rho}) = \eta(s,\rho). \tag{1.8}$$

In fact, lower bounds for the sequences (1.7) and (1.8) can be derived from (1.3) and all the considerable effort goes into showing that these sequences are monotonically decreasing. Furthermore they show that $\eta(s,\rho)$ is convex increasing in s and ρ. This implies that temperature, specific heat, pressure and compressibility are positive. These fundamental properties of matter are by no means automatic and one can easily find examples where one or the other of these quantities become negative.

If we do not neglect gravitation ($\kappa > 0$) the picture changes completely. For fermions (if $\sum e_i = 0$) one can show [8]

$$- N^{7/3} A \leq E_o \leq - N^{7/3} B \tag{1.9}$$

for N sufficiently large ($N > 10^{54}$) so that gravitation dominates electricity. Thus the N behaviour is the same as for atoms. It comes about in the same way, namely the radius of the system goes as $N^{-1/3}$ and there are $\sim N^2$ contributions to V. For bosons where the zeropoint pressure is smaller $R \sim N^{-1}$ and $E \sim N^3$. The fact that the system contracts to a state of high density can be used to compute the thermodynamic quantities exactly [9]. I shall demonstrate in the last section that for fermions $N^{-7/3}E_N(Ns, N^{-1}\Omega)$ and $N^{-7/3}F_N(N^{4/3}\beta, N^{-1}\Omega)$ tend (for $N \to \infty$) to what one gets from solving the (temperature dependent) Thomas-Fermi equation. A computer solution [10] of this non-linear integral equation shows indeed the behaviour of f and η one expects from a star. In particular there is a region where η is not convex in s. This implies a negative heat capacity and reflects the interesting feature of gravitating systems to become hotter when they give off energy.

Since a computer only spits out numbers and does not give explanations I shall try to give some insight into this remarkable happening by a rough argument. Let us guess the free energy of fermions in a box Ω with gravitational interactions*. The well known expression of the thermodynamic quantities for free fermions involve the somewhat awkward Fermi function, but it is sufficient for our argument to interpolate between the high temperature limit where the system behaves like a classical ideal gas and the zero temperature limit where the zero point motion dominates. Therefore, we will construct the free energy per particle by adding to the zero point energy the free energy of the ideal Fermi gas and the potential energy corresponding to a

*From now on we shall use natural units ($k=m=\hbar=c=1$).

uniform density in the volume V. It turns out that this procedure is numerically an acceptable approximation for the exact free energy of free fermions and the approximation of uniform density is not so bad because the system collapses altogether once the density becomes too inhomogeneous. In this way we get for the free energy per particle

$$\frac{F}{N} = \left(\frac{N}{V}\right)^{2/3} - T \ln \frac{V}{N} T^{3/2} - \frac{\alpha_G N}{V^{1/3}} \quad . \tag{1.10}$$

From this we can deduce by differentiation with the help of the standard thermodynamical relations the following values for entropy per particle, energy per particle and pressure:

$$\frac{S}{N} = \ln \frac{V}{N} T^{3/2} + \frac{3}{2} = \frac{3}{2} \ln\left(\frac{V}{N}\right)^{2/3} \frac{2}{3}\left(E + \frac{\alpha_G N}{V^{1/3}} - \left(\frac{N}{V}\right)^{2/3}\right) + \frac{3}{2} \tag{1.11}$$

$$\frac{E}{N} = \left(\frac{N}{V}\right)^{2/3} + \frac{3}{2} T - \frac{\alpha_G N}{V^{1/3}} \tag{1.12}$$

$$P = \left(\frac{N}{V}\right)^{5/3} \frac{2}{3} + \frac{NT}{V} - \frac{\alpha_G N^2}{3V^{4/3}} = \frac{E_K + \frac{1}{2} E_p}{3V/2} = \frac{E - \frac{1}{2} E_p}{3V/2} \tag{1.13}$$

The pressure has been rewritten in various ways to show that our approximation conforms with the virial theorem. Note that the pressure has three contributions. The thermal pressure $\sim V^{-1}$, the zero point pressure $\sim V^{-5/3}$, and the negative contribution due to the gravitational interaction $\sim V^{-4/3}$. Because of their different V-dependence the thermal pressure will dominate for large V and for small V the zero point pressure will win, but for sufficiently large N there will be a region in between where the pressure becomes negative. This is certainly physically impossible because the system is not glued to the walls but what will happen is that the system contracts by itself and thereby reaches a state of lower free energy.

Let us first consider the microcanonical case where the energy is the free variable. From Eq. (1.13) we see that for

$$E > -\frac{\alpha_G N}{2V^{1/3}} \tag{1.14}$$

the pressure is positive whereas for

$$E < -\frac{\alpha_G N}{2V^{1/3}} \tag{1.15}$$

it becomes formally negative. But this means that the system will contract to a volume V_o such that

$$E = -\frac{\alpha_G N}{2V_o^{1/3}} \quad . \tag{1.16}$$

This volume has now also to be inserted in the other thermodynamic relations and the relation between energy and temperature

$$\frac{3}{2}\,T = \frac{E}{N} + \frac{\alpha_G N^2}{V^{1/3}} - \left(\frac{N}{V}\right)^{2/3} \tag{1.17}$$

becomes by substituting V_o for V

$$\frac{3}{2}\,T = -\,\frac{E}{N} - \frac{4E^2}{\alpha_G^2 N^{4/3}} \quad . \tag{1.18}$$

Thus, the temperature is a linear function of energy for high energies and for low energies a parabolic function as indicated in Fig. 1.

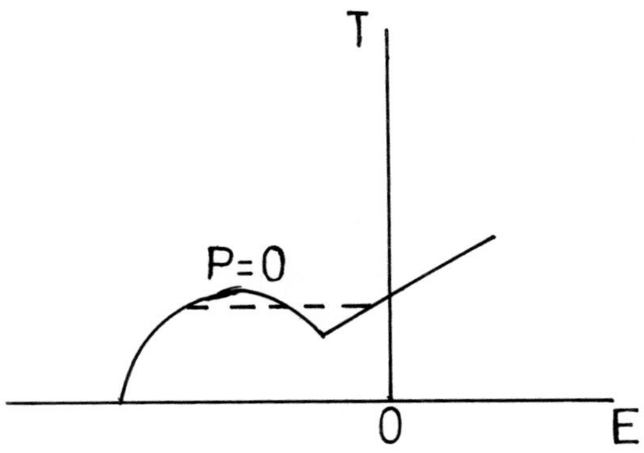

Fig. 1

One sees that actually the region with pressure zero begins with negative specific heat (T increasing with decreasing E), but at low energies the dominating zero point pressure makes the behaviour again normal.

Let us now consider the situation if we keep the temperature fixed and change the radius. The condition P = 0 now gives two solutions for the value of the volume. Since the pressure is proportional to the derivative of the free energy with respect to the volume, we see that the free energy is not monotonic in V. This implies that at a certain volume we can gain free energy by collapsing the system and thus the non-monotonic part of the curve will be bridged by a straight line. This corresponds to a phase transition where at constant temperature a finite amount of energy is released. This situation is illustrated in Fig. 2 and in Fig. 1 we have also drawn the corresponding pointed line which bridges the region of negative specific heat.

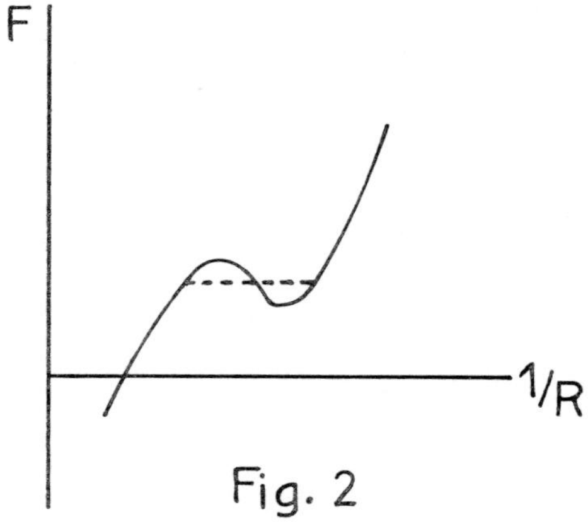

Fig. 2

In the following two sections I shall give some more details on the results for those systems where one can obtain more explicit results and not only abstract existence theorems. Although I will not be able to give all ε's of the proofs I hope to convey the ideas behind the methods used. I shall only quote exact bounds and consider results of uncontrolled approximations as irrelevant.

2. THE METHOD OF MINIMAL CONCAVITY

i) General Inequalities for Coulomb Systems

It is known that E_o is a concave function of the coupling constant (e.g. $e^2 = \alpha$). More generally the average of the first e^S energy levels (= energy at fixed entropy) possesses this property. This is a simple consequence of the generalized Minimax 11 principle:

$$\frac{1}{n}\sum_{1}^{n} E_n = \frac{1}{n}\inf_{H_n} \mathrm{Tr}_{H_n} H \ , \ n \in Z \tag{2.1}$$

where H_n is an n-dimensional subspace and $\mathrm{Tr}_{H_n} H$ is the trace of H restricted to this subspace. H and any trace of it is linear in α and the infimum of linear functions is concave. We shall denote the quantity defined in (2.1) by $E(\alpha)$, and concavity tells us for instance

$$\alpha (V)_\alpha < E(\alpha) - E(0) < \alpha (V)_0 \tag{2.2}$$

where $(V)_\alpha = \frac{\partial E}{\partial \alpha}$ is the microcanonical average of the potential evaluated for $H = H_0 + \alpha V$. In general α cannot be varied in reality but is a God-given constant. However, for Coulomb systems the α-dependence is linked with the volume dependence for dimensional reasons. Since the only lengths in the game are the Bohr-radius α^{-1} (if $\hbar = m = 1$) and the size of the system L we have

$$E(\alpha) = \alpha^2 f(\alpha L) \tag{2.3}$$

where

$$f(x) \sim \begin{cases} 1/x^2 & \text{for} \quad x \to 0 \\ \text{const} & \text{for} \quad x \to \infty \end{cases} .$$

The convexity inequality (2.2) can then be expressed in terms of the pressure $P = -\frac{1}{3L^2}\frac{\partial E}{\partial L}$:

$$P \geq \frac{E(\alpha,L) + E(0,L)}{3L^3} . \tag{2.4}$$

Similarly since $E(\alpha)$ is concave

$$\frac{\partial^2 E}{\partial \alpha^2} \leq 0 \tag{2.5}$$

which gives for the compressibility

$$\kappa^{-1} = -\frac{L}{3}\frac{\partial P}{\partial L} \leq 2P - \frac{2E}{9L^3} \quad ; \tag{2.6}$$

as mentioned above the quantities in (2.4) and (2.6) are understood at fixed entropy. For free fermions they become in the limit $\alpha \to 0$ (or $L \to 0$) the equalities

$$P = \frac{2E}{3L^3} \quad , \qquad \kappa^{-1} = \frac{10E}{9L^3} \quad . \tag{2.7}$$

For normal densities (2.4) and (2.6) are easily satisfied, which just means that then one is rather far from the region where first order perturbation theory is any good. In the next section we shall therefore try to strengthen (2.5).

It should be remarked that these inequalities hold for any Coulomb system irrespective of its size. They also apply to one diatomic molecule in the Born-Oppenheimer approximation where the extremal length is now the distance between the two atoms. The equilibrium position r_o is defined by $f'(\alpha r_o) = 0$ and there (2.5) becomes

$$\alpha^2 r_o^2 f''(\alpha r_o) \leq -2f(\alpha r_o) \quad . \tag{2.8}$$

Let M be the reduced mass of the molecule. (2.8) then can be rewritten as an upper limit of the vibrational frequency ω:

$$\omega \leq \sqrt{\frac{4|E|}{M\, r_o^2}} \quad . \tag{2.9}$$

Similarly (2.6) can be expressed as a general upper limit of the velocity of sound in (non-relativistic) Coulomb systems.

ii) Bogoliubov-type Inequalities for Unbounded Operators. There is a general inequality [13] for $\partial^2 E/\partial\alpha^2$ which is just what we are after:

$$\frac{\partial^2 E}{\partial\alpha^2} \leq - \frac{(\alpha|[D,V]|\alpha)^2}{(\alpha|[D^\dagger[H,D]]|\alpha)} \quad . \tag{2.10}$$

Here $|\alpha)$ is the eigenvector belonging to the eigenvalue $E(\alpha)$ and, surprisingly enough, D is an arbitrary operator. This relation is usually deduced by methods legitimate in a finite dimensional space. In Hilbert space one can easily deduce absurd results from (2.10) if V, H and D are unbounded operators. Paradoxes are well-known [14] which one can obtain with formal manipulations of the canonical commutation relations by ignoring domain questions. Before using (2.10) further we shall prove (2.10) such that our application [12] is covered.

For arbitrary potentials (2.10) is probably useless since the calculation of the expectation values with the disturbed eigenvectors $|\alpha)$ will be impossible. For the Coulomb potential we may, however, use for D the generator of the dilatation group. There the commutators can be expressed by H and V and correspondingly the expectation values by E and $\partial E/\partial\alpha$. The corresponding differential inequality will be integrated in the next section.

For a rigorous deduction of (2.10) we follow the general strategy that it is better to use the group elements rather than the (unbounded) generators. The group in question is ($\beta \in R$)

$$x \to e^{\beta} \cdot x \quad , \qquad p \to e^{-\beta} \cdot p \quad . \tag{2.11}$$

It is now essential whether we work in infinite space and our Hilbert space are the (properly antisymmetrized) functions of $L^2(R^{3N})$ or whether we work in a finite volume $-L \leq x_i \leq L$. In the first case (2.11) is represented by a group of unitary transformations whereas in the second case it exists only for $\beta \leq 0$ and then gives a semi-group of partial isometries. The generator is in both cases a hermitian operator of the form xp+px. However, the domains are different: in the first case it is self-adjoint whereas in the second case it is maximal symmetric. This shows immediately that formal manipulations with the canonical commutation rules are no good because they would not make any distinction between these possibilities. However, we know that they should lead to the virial theorem with external virial for the finite volume and without it in infinite space. Thus one pays for ignoring domain questions by loosing something not of the size ε but of the size of the effect one wants to calculate.

We shall only consider here the case $L = \infty$ which is appropriate for atoms, for instance. We shall derive a generalization of (2.10) for the sum of the first n levels. Consider a Hamiltonian depending linearly on two parameters

$$H(\alpha,\beta) = K(1+2\beta) + V_p(1+\beta) + V_e(1+\alpha+\beta) \tag{2.12}$$

with

$$V_p = -Z \sum_{i=1}^{N} |x_i|^{-1} , \qquad V_e = \sum_{i>j} |x_i - x_j|^{-1} . \tag{2.13}$$

According to the previous argument the matrix of the second derivatives of

$$E(\alpha,\beta) = \frac{1}{n} \inf_{H_n} \mathrm{Tr}_{H_n} H(\alpha,\beta) \tag{2.14}$$

must be non positive*. Using that $\inf_{H_n} \mathrm{Tr}_{H_n}$ is invariant under unitary transformations and that the scale transformation (2.11) produces $T \to e^{-2\beta} T$, $V_{e,p} \to e^{-\beta} V_{e,p}$ we have ($V = V_e + V_p$)

$$E(\alpha,\beta) = \frac{1}{n} \inf_{H_n} \mathrm{Tr}_{H_n} (T(1+2\beta)e^{-2\beta} + V_p(1+\beta)e^{-\beta} +$$

$$+ V_e(1+\alpha+\beta)e^{-\beta})$$

$$= \frac{1}{n} \inf_{H_n} \mathrm{Tr}_{H_n} \left(H - \frac{\beta^2}{2}(4T+V) + \alpha(1-\beta)V_e + O_3\right) . \tag{2.15}$$

O_3 are terms of higher than second power in α, β. By the usual rules of perturbation theory, which are applicable because $V_{p,e}$ are bounded with respect to K, and by the

*The second derivatives may not exist for a finite number of points. In any case they will exist as negative distributions which will be good enough for our purpose.

virial theorem we have

$$\left.\frac{\partial^2 E}{\partial \beta^2}\right|_o = - <4T + V> = 2E \tag{2.16}$$

$$\left.\frac{\partial^2 E}{\partial \beta \partial \alpha}\right|_o = - <V_e> = - \frac{\partial E}{\partial \alpha}$$

where <> is the average of the expectation values of the first n levels. (In case of level crossing take n sufficiently large to cover the degenerate levels). Thus we have finally

$$\frac{\partial^2 E}{\partial \alpha^2} \leq (\frac{\partial^2 E}{\partial \alpha \partial \beta})^2 / \frac{\partial^2 E}{\partial \beta^2} = (\frac{\partial E}{\partial \alpha})^2 / 2E \tag{2.17}$$

as could be deduced by using formally D = xp+px in (2.10). We have shifted in our derivation the point $\alpha = 0$ but obviously (2.17) holds $\forall \ \alpha$.

iii) Integration of (2.17)

We put

$$E(\alpha) = E(0) \ e^{\int_0^\alpha d\alpha' \rho(\alpha')} \quad ; \tag{2.18}$$

since $V_e > 0$ and $E(0) < 0$ we have $\rho(0) = \frac{E'(0)}{E(0)} < 0$.
(2.17) becomes

$$\rho' \geq - \rho^2/2 \ . \tag{2.19}$$

If we take a solution $\bar{\rho}$ of

$$\bar{\rho}' = - \bar{\rho}^2/2 \tag{2.20}$$

with $\bar{\rho}(0) = \rho(0)$ we have

$$\bar{\rho}(\alpha) \leq \rho(\alpha)$$

or

$$E(\alpha) \leq E(0)\, e^{\int_0^\alpha d\alpha' \bar{\rho}(\alpha')} \quad . \tag{2.21}$$

(2.20) can easily be integrated and we obtain a parabolic upper bound

$$E(\alpha) \leq E(0)\left(1 + \frac{\alpha}{2}\,\frac{E'(0)}{E(0)}\right)^2 \qquad \forall \alpha > 0 \quad .$$

Similarly one deduces $\forall \alpha_o > 0$

$$E(\alpha) \leq E(0)\left[1 + \frac{\alpha}{\alpha_o}\left(\sqrt{\frac{E(\alpha_o)}{E(0)}} - 1\right)\right]^2 \qquad \forall\, 0 \leq \alpha \leq \alpha_o \quad . \tag{2.22}$$

Thus a lower bound for $\alpha = \alpha_o$ can be used to give a lower bound for all α inbetween and this is better than the trivial linear bound

$$E(\alpha) \geq \frac{\alpha}{\alpha_o}\,(E(\alpha) - E(0)) \tag{2.23}$$

which follows from concavity.

As illustration let us just quote the simple example of N = 2, n = 1. The problem can be rescaled such that

$$HZ^{-2} \rightarrow H = \frac{p_1^2+p_2^2}{2} - r_1^{-1} - r_2^{-2} + \frac{1}{Z}\, r_{12} \quad .$$

This means that changing Z with N fixed we can actually

vary α, although not continuously. There $E'(0) = 5/8Z$ and for $E(\alpha) < E_1(0) = -5/8$ one has the linear bound $E(\alpha) \geq E(0) + \alpha(0|V_e^{-1}|0)^{-1} = -1 + \alpha \frac{16}{35Z}$. Taking $\alpha_o = \frac{105}{128} Z$ such that $E(\alpha_o) = E_1(0)$ we have from (2.22) and (2.21)

$$-[1 - \frac{0.2553}{Z}]^2 = -[1 - \frac{128}{105Z}(1 - \sqrt{\tfrac{5}{8}})]^2 \leq E \leq - [1 - \frac{5}{16Z}]^2$$

$$= -[1 - \frac{0.315}{Z}]^2 \quad . \qquad (2.24)$$

In Fig. 3 I have given the parabolic and the linear bounds together with some experimental numbers. The concavity aspects are obvious.

Of course, using elaborate trial functions these results can be improved [4]. However, the present method also works in more complicated situations [12] where useful trial functions are not available. Furthermore our inequalities also give an upper bound for $(\alpha|V_e|\alpha) = \alpha\frac{\partial E}{\partial \alpha}$, in our example (N = Z = 2)

$$|(\alpha|V_e|\alpha)/(\alpha|V_p|\alpha)| \leq \frac{1}{6.4} \quad .$$

In the Thomas-Fermi model the latter ratio is 1/7. In any case we see that for atomic systems we talk about differences between the rigorous bounds of the % order whereas for macroscopic systems the lower one is off by 10^{14}.

Fig. 3

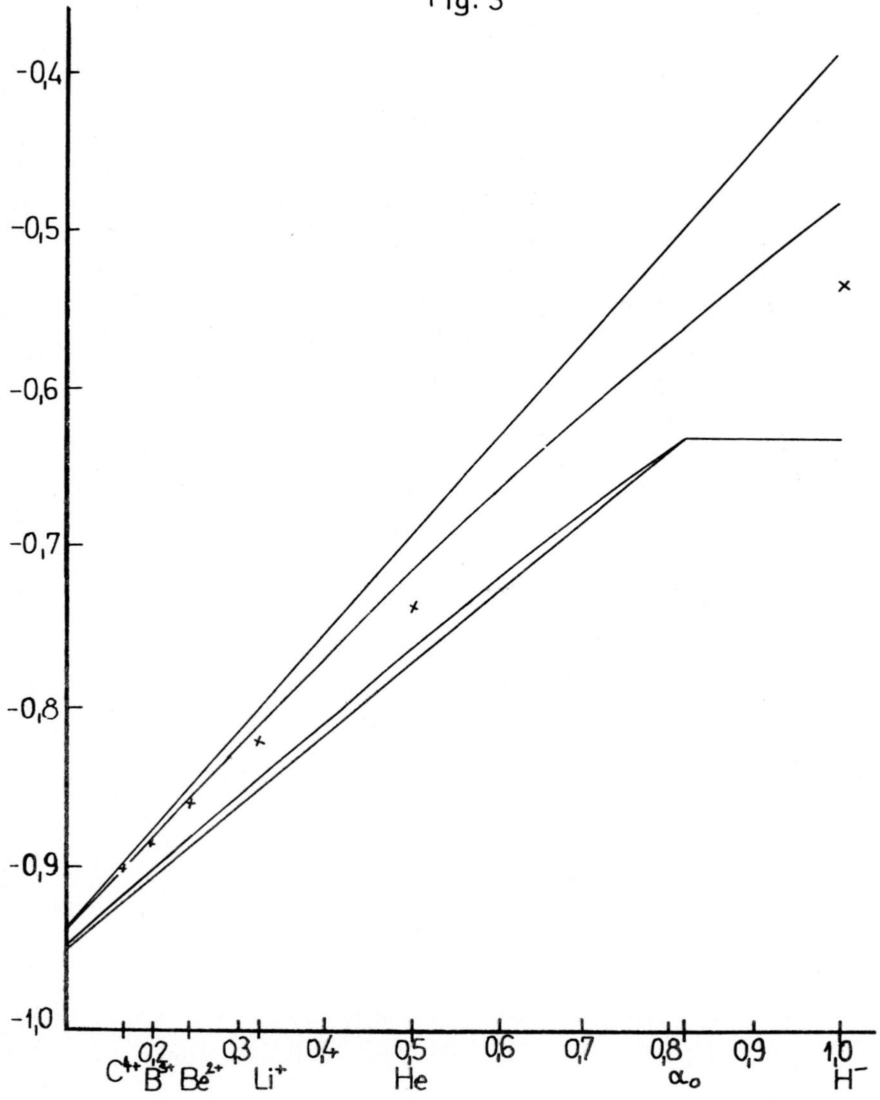

3. COSMIC BODIES

i) Finally we shall consider systems dominated by gravitation ((1.1) with $N \sim 10^{57}$). They have some similarity with atoms insofar that their radius shrinks $\sim N^{-1/3}$ and therefore their energy goes like $N^{7/3}$. By this compression atoms are squashed and in the resulting high-density plasma the motion of the particles is mainly determined by the mean gravitational field. Thus one expects the Thomas-Fermi equation to become exact in the limit $N \to \infty$. This can indeed be rigorously demonstrated by a rather lengthy analysis. I can give here only the main points and have to refer you for details to [9]. The strategy is the following: You want to calculate the thermodynamic functions and for this you have first to get rid of the singularity in the potential since this will lead to local inhomogeneities. Furthermore for our method a finite self-energy is essential. In fact one shows that approximating the potential by a step function things change little. Little means small compared with the leading term which is $\sim N^{7/3}$. Secondly one separates by infinite walls the boxes within which the potential is constant. Then one has separate free systems where the constant potentials give a contribution proportional to the occupation number of the boxes. The partition functions are now sums over the occupation numbers and one has to demonstrate that the former are almost given by their value for a self-consistently determined set of occupation numbers. These self-consistency equations become, in the limit of infinitely many boxes, the Thomas-Fermi-equations.

The tactics consists mainly in keeping track of

the various ε's and watching the order in which the different limits are to be taken.

ii) Regularizing the Potential

As basis for our further analysis we have to establish the lower bound (1.11). This can be done [8] by the simple device of rewriting ($\hbar = m_i = 1$)

$$H \geq \sum_{i=1}^{N} \frac{p_i^2}{2} - \alpha \sum_{i \neq j} |x_i - x_j|^{-1} \quad , \quad \alpha = (\kappa + e^2)\frac{1}{2} \tag{3.1}$$

in the form

$$H \geq \sum_{i=1}^{N} \sum_{j \neq i} \left(\frac{p_j^2}{2N} - \alpha |x_i - x_j|^{-1}\right) = \sum_{i=1}^{N} h_i \quad . \tag{3.2}$$

The

$$h_i = \sum_{j \neq i} \left(\frac{p_j^2}{2N} - \alpha |x_i - x_j|^{-1}\right) \tag{3.3}$$

represent N-1 non-interacting particles (with mass N) moving in the Coulomb field of the particle i (which has no kinetic energy). h_i can be bounded from below with the reasoning we have used for atoms. Filling the Balmer levels $-\frac{1}{2} N\alpha^2 \frac{1}{n^2}$ according to the exclusive principle up to

$$n_o \sim N^{1/3} \text{ we find } h_i \geq -\gamma\, N^{4/3}\, \alpha^2$$

and thus

$$H \geq -\gamma\, N^{7/3}\, \alpha^2 \quad . \tag{3.4}$$

The numerical factor γ is of the order 1.

Remark: H is, of course, invariant under permutations of the particles, but h_i is not. Thus the exclusion principle has not been fully taken into account in treating h_i. Nevertheless the bound (3.4) holds a fortiori for all totally antisymmetric functions. With the same method one can show that the addition of a narrow Yukawa potential $- e^{-\mu r}/r$ to $1/r$ changes little. Suppose one uses the latter in (3.1) to (3.3) instead of $1/r$. The argument goes through in the same way except that we now can use $n_o \sim N_b^{1/3}$ where N_b is the number of bound states which for a central potential is known to be bounded by [15]

$$N_b \leq 2\int_o^\infty dr \; 2mr|V(r)|[\sup_r r^2 2m|V(r)|]^{1/2} \quad . \qquad (3.5)$$

We shall not derive this formula but only remark that for the $\alpha e^{-\mu r}/r$ the corresponding bound $\sim (\frac{2m\alpha}{\mu})^{3/2}$ is about the "classical" value

$$\int \frac{d^3p}{(2\pi)^3} d^3x \; \Theta(2m|V(r)| - p^2) \quad .$$

Using this formula we have to remember that m in (3.3) is N and that our system will shrink $\sim N^{-1/3}$. Correspondingly we put $\mu = sN^{1/3}$ (s^{-1} a fixed fraction of the total size of the system) and obtain

$$N_b \leq \gamma_1 N \left(\frac{\alpha}{s}\right)^{3/2}$$

and thus

$$H \geq - \gamma_2 N^{7/3} \alpha^2 \left(\frac{\alpha}{s}\right)^{1/2} \quad , \qquad (3.6)$$

again $\gamma_{1,2} = O(1)$. Thus for s sufficiently large the influence of a Yukawa potential becomes arbitrarily small on the $N^{7/3}$-scale. This can be rephrased by saying that even if $\alpha \sim s^{1/5}$ its influence is bounded $\forall$ s:

$$0 \leq \sum_{i>j} \frac{e^{-s|x_i-x_j|N^{+1/3}}}{|x_i-x_j|} \leq s^{-1/5} (K + \gamma_2 N^{7/3}) \tag{3.7}$$

or

$$K - \sum_{i>j} |x_i-x_j|^{-1} \leq K - \sum_{i>j} \frac{1-e^{-s|x_i-x_j|N^{1/3}}}{|x_i-x_j|} \leq$$

$$K(1 + s^{-1/5}) - \sum_{i>j} |x_i-x_j|^{-1} + s^{-1/5} \gamma_2 N^{7/3} . \tag{3.8}$$

Thus replacing the 1/r by the continuous function $\frac{1-e^{-\mu r}}{r}$ has no larger effect than adding a small constant and changing the mass slightly.

Since the thermodynamic functions F (1.5) and E (1.6) are monotonic in H the corresponding inequalities also hold for them.

The next step is approximating

$$\frac{1 - e^{-\mu|x_i-x_j|}}{|x_i-x_j|}$$

by a piecewise constant function. This can be done such that the operators change in norm arbitrarily little so that there is no question that F and E also survive this operation with negligible change.

iii) Putting Up Walls

Next we have to separate the various regions by infinitely high δ-function barriers so that the subsystems communicate only via the potential. These walls are mathematically represented by the condition that the wave function has to vanish at the boundaries. To see in this formulation that the walls are something positive one has to express the Hamiltonian as a quadratic form rather than as an operator. (Although they are the same formal expression they differ in domain). There the domain is restricted by the introduction of walls and thus the eigenvalues increase since they can be expressed as infima taken over the domain. The harder part is to show that the energies do not increase too much. For this one constructs [16] for each state of the system without walls a wave-function for the system with walls such that the expectation value of the Hamiltonian is only slightly increased. This can be done since the number of walls is kept fixed in the limit of $N \to \infty$. Of course, one cannot brutally cut the wave functions at the 3h walls since this would cost too much kinetic energy. To do this in a gentle fashion the wave-function is multiplied by a function f which goes smoothly to zero at the walls. Since this would decrease the norm of the state one increases the size of the system with walls to $L' = L + 2b(h-1)$. The region within a distance b from the walls where the wave-function goes to zero is used twice in the system with walls by folding it back. This means we define a mapping $L'^3 \to L^3$ which in one dimension looks like Fig. 4 and is given in formulae by

Fig. 4. The mapping R from L' to L for h = 3.

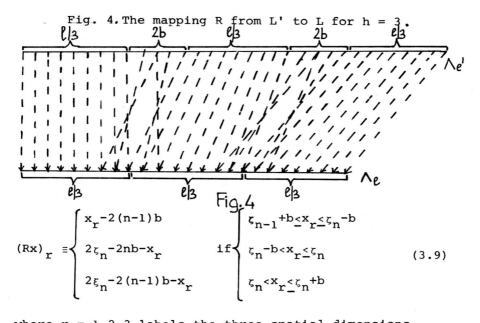

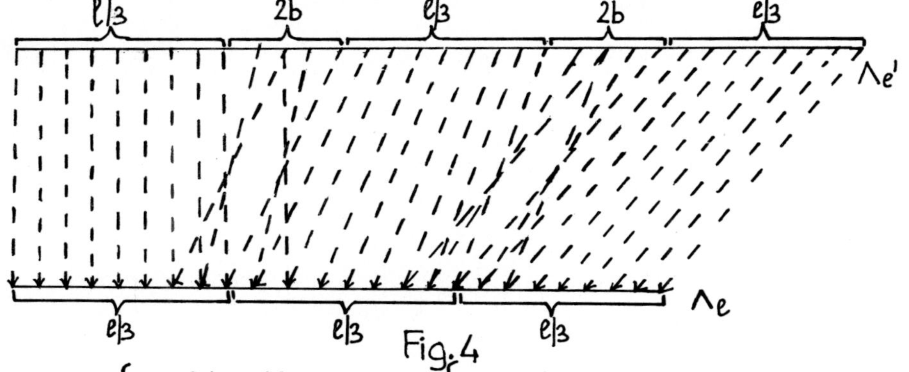

$$(Rx)_r \equiv \begin{cases} x_r-2(n-1)b & \\ 2\zeta_n-2nb-x_r & \text{if } \\ 2\xi_n-2(n-1)b-x_r & \end{cases} \begin{cases} \zeta_{n-1}+b\le x_r\le\zeta_n-b \\ \zeta_n-b<x_r\le\zeta_n \\ \zeta_n<x_r\le\zeta_n+b \end{cases} \tag{3.9}$$

where r = 1,2,3 labels the three spatial dimensions, n = 0,1,...,h the walls (including those of the original cube) in a given spatial dimension, and $\zeta_n = n\frac{L}{h}(2n-1)b$ the location of such a wall.

Consider next a function $f: [0,L+2(h-1)b] \to R$ with the following properties: f is once continuously differentiable, $f^2(x) \le 1$; for n = 1,2,...,h-1: $f(x) \neq 1$ for $x \in [\zeta_n-2b,\zeta_n+2b]$ only; $f(\zeta_n) = 0$; $\xi \to f(\zeta_n+\xi)$ is even for $|\xi| \le 2b$; $f^2(\zeta_n+\xi) + f^2(\zeta_n+2b-\xi) = 1$ for $0 \le \xi \le b$; $(f'(x))^2 \le 1/b^2$. One can then show by an elementary calculation that $I: H_{N,L^3} \to H_{N,L'^3}$ defined by

$$(I\Phi)(\vec{x}_{11},\ldots,\vec{x}_{2N}) \equiv \prod_{\alpha i}\prod_{r=1}^{3} f(x_{\alpha i,r})\,\Phi(R\vec{x}_{11},\ldots,R\vec{x}_{2N}) \tag{3.10}$$

is an isometry:

$$(I\Phi, I\psi) = (\Phi, \psi) \tag{3.11}$$

One can show that the conditions imposed on f ensure that the range of I is in the appropriate domain and that the expectation value of the energy is only changed by a small amount. From the standard Minimax principles for the thermodynamic quantities it follows that they also are shifted only slightly and one can give exact bounds on the change.

iv) The Result

We have now arrived at a situation where particles move in separated cubes within which the potential is constant and depends only on the occupation numbers. Since the kinetic energy is explicitly known the thermodynamic quantities can be written as sums over all possible occupation numbers of well-known expressions. By the usual manipulations of statistical mechanics one then shows that in the limit $N \to \infty$ they are given by the leading term. The latter is obtained if the occupation numbers satisfy a certain self-consistency relation. This goes over into the Thomas-Fermi equation if one retraces the various steps, i.e. goes to the limit of infinitely many boxes so that the potential becomes continuous again and then takes away the $e^{-\mu r}/r$ so that one is left with the Coulomb potential. I have to refer you to [9], [17] for a demonstration that the result survives all these manipulations. The outcome of all this can be stated as follows:

For given $\beta > 0$, $L > 0$ there are solutions of the Thomas-Fermi equation

$$\rho_\alpha(\vec{x}) = \int \frac{d^3q}{(2\pi)^3} \frac{1}{1 + e^{\beta[\frac{q^2}{2M_\alpha} + W_\alpha(\vec{x}) - \mu_\alpha]}} \tag{3.12}$$

with

$$W_\alpha(\vec{x}) = \sum_\beta \int d^3x' \frac{e_\alpha e_\beta - \kappa M_\alpha M_\beta}{|\vec{x} - \vec{x}'|} \rho_\beta(\vec{x}') \quad , \tag{3.13}$$

which are normalized

$$\int_{L^3} d^3x \quad \rho_\alpha(\vec{x}) = 1. \tag{3.14}$$

One then calculates

$$u = \sum_\alpha \int_{L^3} d^3x \quad \rho_\alpha(\vec{x}) \; W_\alpha(\vec{x}) \quad , \tag{3.15}$$

$$f = -u + \sum_\alpha \{\mu_\alpha - \frac{1}{\beta}\int_{L^3} d^3x \int \frac{d^3q}{(2\pi)^3} \log(1+e^{-\beta[\frac{q^2}{2M_\alpha}+W_\alpha(\vec{x})-\mu_\alpha]})\} \tag{3.16}$$

$$\eta = u + \sum_\alpha \int_{L^3} d^3x \int \frac{d^3q}{(2\pi)^3} \frac{q^2}{2M_\alpha} \frac{1}{1+e^{\beta[\frac{q^2}{2M_\alpha} + W_\alpha(\vec{x})-\mu_\alpha]}} \quad , \tag{3.17}$$

and

$$s = \beta(\eta - f) \quad . \tag{3.18}$$

If (3.12) to (3.14) has more than one solution that which

minimizes f has to be chosen, and the minimal f coincides with $f(\beta,L)$. From $\beta \to s$, $\beta \to \eta$ the graph $\eta(s,L)$ is constructed by choosing the smaller one of two possible η's. The microcanonical and the canonical thermodynamical functions are Legendre-transforms of each other if the Legendre-transform is possible, i.e. for those values of the entropy s where

$$\frac{\partial^2 \eta(s,L)}{\partial s^2} > 0 \quad .$$

To solve these equations analytically is impossible but I can show you the results of a computer solution for the special case $e = 0$ and only one species of Fermions: For amusement we used for the mass of particles the mass of neutrons and the number N of particles corresponding to a star to illustrate that for this situation these phenomena happen at energies of several MeV per particle and of sizes for the system of several km. In Fig. 5 we plot the entropy as a function of the energy and show that for a radius of 100 km we have indeed a region where the function becomes convex.

For the smaller radius this phenomenon is quenched by the zero point energy. The energy as the function of the derivative of this curve, the inverse temperature, is shown in Fig. 6. Thus, we have a region where we have three energy values corresponding to one temperature, and at a certain temperature the system in the canonical ensemble will jump from one branch of the curve to the other. The exact analysis verifies the usual statement that this happens when the free energy of the lower branch becomes lower than the free energy of the upper branch.

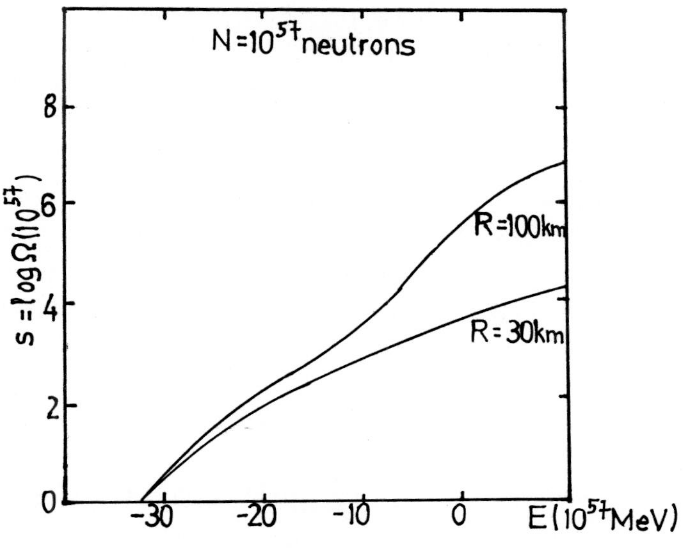
N=10^{57}neutrons
s = log Ω(10^{57})
8
6
4
2
0
R=100km
R= 30km
-30
-20
-10
0
E(10^{57}MeV)

Fig. 5

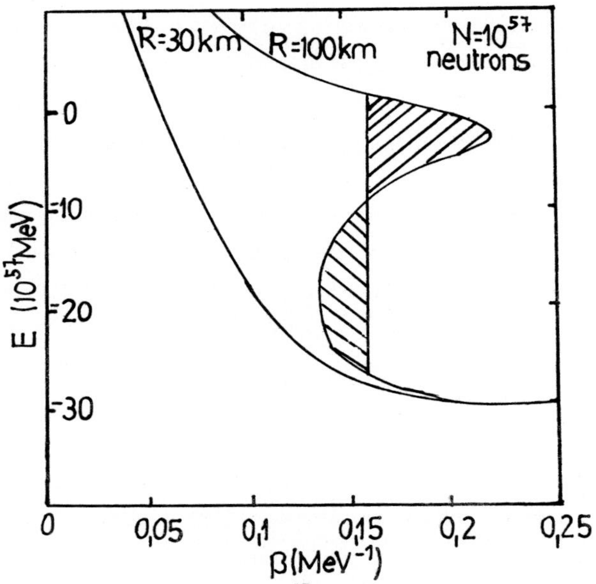
R=30km
R=100km
N=10^{57}
neutrons
0
-10
-20
-30
E (10^{57}MeV)
0
0,05
0,1
0,15
0,2
0,25
β(MeV^{-1})

Fig. 6

In the next curve we plot the free energy as function of the temperature which clearly exhibits the point at which the free energy of one branch goes below the one of the other branch. The exact solution also gives the density as function of the radius and in Fig.7 we plot the density for various temperatures. Note that above the transition point the density is reasonably homogeneous whereas at the phase transition the ratio between the density in the middle to the density of the surface increases by about five orders of magnitude. Below the phase transition a kind of star develops with a reasonably well defined surface and a very thin atmosphere. This is also emphasized by a plot of the measure of the degeneracy of the Fermi gas (Fig.8).

The atmosphere always remains a Boltzmann whereas the centre becomes degenerate since it is only the zero point pressure which can stop the collapse.

The discussion given here shows the changes in thermodynamics that are going to happen if the gravitational interaction is included. Because of its long range the thermodynamic quantities do not have the usual extensivity property, and energy or entropy do not simply have twice the value of half the system. Because of this the usual thermodynamic arguments, indicating that specific heat and compressibility are positiv, no longer hold. Indeed, there are regions where they become negative.

I would like to thank H.Narnhofer for her permission to use our unpublished results. Furthermore I am grateful to P. Hertel for stimulating discussions and to M.Breitenecker and H.-R.Grümm for help with some calculations.

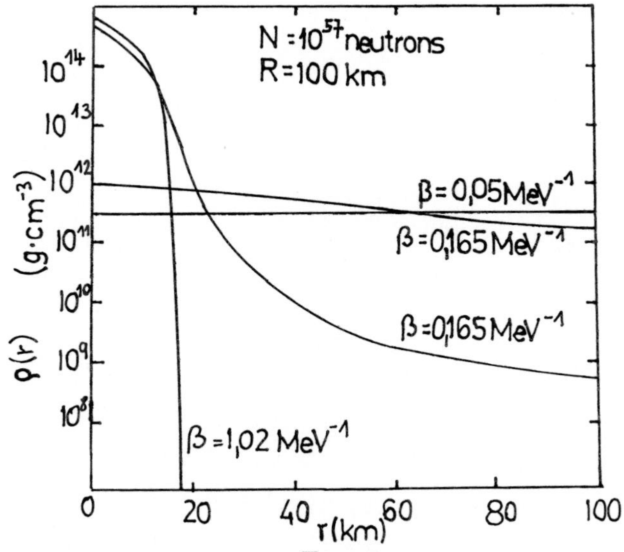
N = 10^{57} neutrons
R = 100 km
10^{14}
10^{13}
10^{12}
10^{11}
10^{10}
10^{9}
10^{8}
ρ(r) (g·cm^{-3})
β = 0,05 MeV^{-1}
β = 0,165 MeV^{-1}
β = 0,165 MeV^{-1}
β = 1,02 MeV^{-1}
0
20
40
60
80
100
r(km)

Fig. 7

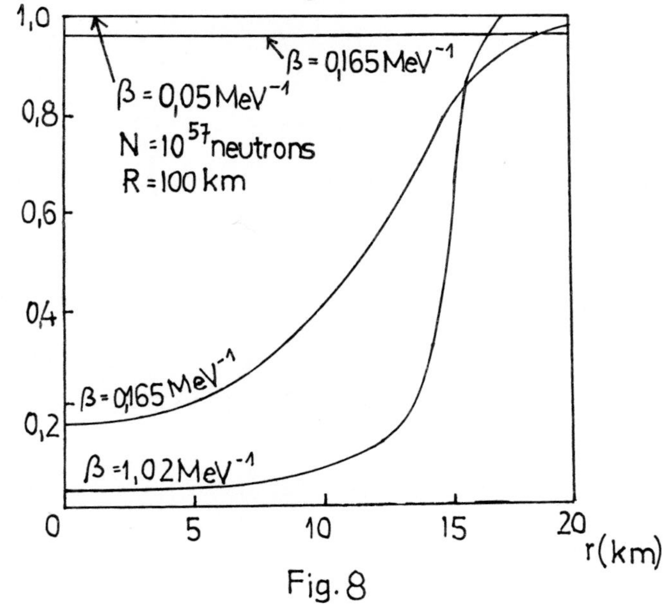
1,0
0,8
0,6
0,4
0,2
β = 0,165 MeV^{-1}
β = 0,05 MeV^{-1}
N = 10^{57} neutrons
R = 100 km
β = 0,165 MeV^{-1}
β = 1,02 MeV^{-1}
0
5
10
15
20
r(km)

Fig. 8

REFERENCES

1. T. Kato, Perturbation Theory for Linear Operators, Berlin, Springer-Verlag 1966.
2. E. Balslev, J. M. Combes, Comm. Math. Phys. 22, 280 (1971).
3. J. D. Dollard, J. Math. Phys. 7, 802 (1966).
4. N. W. Bazley, D. W. Fox, Phys. Rev. 124/2, 483 (1961); P. O. Löwdin, Phys. Rev. 139A, 357 (1965); T. Kinoshita, Phys. Rev. 105/5, 1490 (1957); C. L. Pekeris, Phys. Rev. 112, 1649 (1958).
5. F. J. Dyson, A. Lenard, J. Math. Phys. 8, 423 (1967).
6. F. J. Dyson, In Statistical Physics, Phase Transition and Superfluidity, Brandeis University Summer School in Theoretical Physics, Lecture notes (1966).
7. J. L. Lebowitz, E.H. Lieb, Advances in Mathematics 9, 316 (1972).
8. J. M. Lévy-Leblond, J. Math. Phys. 10, 806 (1969).
9. P. Hertel, H. Narnhofer, W. Thirring, Comm. Math. Phys. 28, 159 (1972).
10. P. Hertel, W. Thirring in Dürr, Quanten und Felder, Braunschweig, Vieweg 1971.
11. A. Wehrl, to be published in Acta Phys. Austr.
12. W. Kohn, Phys. Rev. 71, 635 (1947), H. Narnhofer, W. Thirring, in preparation.
13. S. Okubo, J. Math. Phys. 12, 1123 (1971).
14. M. Reed, B. Simon, Methods of Modern Mathematical Physics, New York and London 1972.
15. B. Simon, J. Math. Phys. 10, 1123 (1969).

16. D. W. Robinson, The Thermodynamic Pressure in Statistical Mechanics, Springer lecture notes (1971).

17. P. Hertel, W. Thirring, Comm. Math. Phys. 24, 22 (1971).

Acta Physica Austriaca, Suppl.XI, 527–564 (1973)

APPROACH AND RETURN TO EQUILIBRIUM* **

BY

HEIDE NARNHOFER
Institute for Theoretical Physics
University of Vienna

CONTENTS

---

*Lecture given at XII. Internationale Universitätswochen für Kernphysik, Schladming, February 5 - 17, 1973

**Work supported in part by the "Fonds zur Förderung der wissenschaftlichen Forschung", Projekt Nr. 1724.

1. INTRODUCTION

In Thermodynamics we work with thermodynamical ensembles, for the infinite system we speak of equilibrium states. For a well behaving system we hope, starting with a state not too wild, that after a sufficiently long time the results of measurements should equal the results of measurements in an equilibrium state. This can mean mathematically either that the state tends weakly to the equilibrium state or that the mean in time of the state exists and that the time of recurrence is small in comparison to the time the measurement takes.

It is the goal of this lecture to show the existence of such properties. Up to now we have only preliminary results: for free systems and systems which are very similar to free ones we can make some statements concerning the approach to equilibrium, i.e. we can start with an only mildly restricted class of states and get a result for the limit $t \to \infty$. For some more general systems we only obtain results regarding the return to equilibrium. We start with an equilibrium state, then disturb it (here there are different possibilities to define a perturbation) and then under some conditions, which need not be fulfilled in reality, we obtain the result, that the disturbed state tends back towards the equilibrium state.

Since we want to deal with systems of unrestricted size we treat the idealization of an infinite system. The definition of the corresponding algebra of observables poses no problem, but the structure of the time evolution may be different from what one is used to for finite systems. We shall show by some examples that the time evolution may or may not be an automorphism of the algebra, in some cases it is not an automorphism of the abstract algebra but is different in different representations.

Our special interest belongs to the states invariant under time evolution and among these to the equilibrium states. They can be obtained as the limit of Gibbs states, but this limit might contain several phases. It may be better, especially in the algebraic treatment, to use the KMS conditions for the definition of the equilibrium states. We hope that for a thermodynamical system all invariant states are linear combinations of these equilibrium states and further that they are the time limit points of all other states.

Useful characterizations of invariant states exist for systems with some further restrictions on the time automorphism. For quasi-free evolutions, including the free evolution, the class of invariant states is much greater than that of the equilibrium states. Nevertheless, we get here the beautiful result that for a great class of initial states the weak time limit exists, though they need not be equilibrium states.

The proof is taylormade for quasi-free evolutions, and so for more general systems we have to look for another method. Results have only been found for systems obtained by perturbation of equilibrium states. The ideas are similar to that of scattering theory and exactly as in the latter theory the remaining main problem is to demonstrate for real physical systems that the endomorphism corresponding to the wave operator in scattering theory really exists and is or is not invertible. In any case the theory can be easily applied to the problem of a system in a thermal bath.

2. MATHEMATICAL FRAMEWORK

i) Algebraic structure [1]

First I want to specify the mathematical input: We start with the axioms of a quasi local C^*-algebra A. They are the ones for a C^*-algebra:

1) It is an algebra over the complex numbers.
2) It has a norm and is complete with respect to this norm
3) $||A_1 A_2|| \leq ||A_1|| \cdot ||A_2||$ (2.1)
4) $\exists$ an involution: $(A^\dagger)^\dagger = A$

$$(A_1 + \lambda A_2)^\dagger = A_1^\dagger + \bar{\lambda} A_2^\dagger$$

$$(A_1 A_2)^\dagger = A_2^\dagger A_1^\dagger$$

$$||A^\dagger A|| = ||A||^2 \rightsquigarrow ||A^\dagger|| = ||A||$$

which allow us some of the usual manipulations. As for locality we assume Λ being a bounded subset of the space R^ν (or of a lattice Z^ν) that

a) If $\Lambda \subset \Lambda'$, then $\exists$ an isomorphism $\alpha_{\Lambda'\Lambda}$ of A_Λ into $A_{\Lambda'}$, and if $\Lambda \subset \Lambda' \subset \Lambda''$, then $\alpha_{\Lambda''\Lambda'}\ \alpha_{\Lambda'\Lambda} = \alpha_{\Lambda''\Lambda}$; we express this by $A_\Lambda \subset A_{\Lambda'}$.

b) $\exists$ automorphism τ_x of space translation and $\tau_x\ A_\Lambda = A_{\Lambda+x}$.

c) If $\Lambda \cap \Lambda' = \Phi$, $A \in A_\Lambda$ and $B \in A_{\Lambda'}$, then $[A,B] = 0$.

d) $A = \overline{\bigcup_\Lambda A_\Lambda}^{\text{norm}}$. (2.2)

The meaning of a) and b) is obvious. Condition c) tells us that measurements in disconnected regions are independent of each other. Finally d) defines A to be the norm closure of the algebras of local observables.

ii) States and Representations [2,3]

I want to recall briefly what we know about states and representations of C^*-algebras.

A state is a normed positive functional over the algebra and by the GNS construction there exists exactly one cyclic representation π_ρ (up to unitary equivalence) such that

$$\rho(A) = (\Omega_\rho | \pi_\rho(A) | \Omega_\rho) \qquad (2.3)$$

and $\pi_\rho(A)\Omega_\rho$ is dense in H_ρ, i.e. Ω_ρ is cyclic. A state is extremal, i.e. cannot be written as the linear combination of two others, if and only if the corresponding representation is irreducible. For classical (= abelian) systems the decomposition of a state into extremal states is always unique. In quantum theory this is no longer the case and this will require a generalization of the concepts of classic ergodic theory.

We define the commutant $\pi_\rho(A)'$ and the weak closure $\pi_\rho(A)''$ of our algebra. From every state ρ we can construct the class of normal states of $\pi_\rho(A)''$, (i.e. states, which are ultraweakly continuous with respect to $\pi_\rho(A)''$). They are the uniform closure of the convex hull of the set of states $\{\rho_A, A \in A\}$ with [2,p.54]

$$\rho_A(B) = \frac{\rho(ABA^\dagger)}{\rho(AA^\dagger)}, \quad \rho(AA^\dagger) > 0. \tag{2.4}$$

If ρ can be written as sum of ρ_1 and ρ_2

$$\rho = \alpha\rho_1 + (1-\alpha)\rho_2 \tag{2.5}$$

then [3,p.35]

$$\rho_i(B) = \frac{(\Omega_\rho | T_i \ \pi_\rho(B) \ T_i^\dagger | \Omega_\rho)}{(\Omega_\rho | T_i T_i^\dagger | \Omega_\rho)}, \quad T_i \in \pi_\rho(A)', \tag{2.6}$$

$$\frac{T_i T_i^\dagger}{||T_i T_i^\dagger||} \quad \text{unique} .$$

Due to the cyclicity of Ω_ρ the ρ_i are normal.

3. TIME EVOLUTION

Quantum mechanics tells us that for finite systems A_Λ the time evolution is governed by a local Hamiltonian H_Λ. What about the time evolution of A? What we measure are the states E over this algebra. Physically it would be sufficient if we could demonstrate the existence of a group of automorphisms over E (it need not be continuous) representing the time evolution. But it is much more convenient to assume that time evolution is represented by an automorphism of A. To get some feeling under which conditions this assumption is satisfied we shall look at typical examples to which we will return later on.

i) Take free fermions [4,p.305] . The algebra we deal with is the even part A^e of the algebra A built up by the creation and annihilation operators a(f) and $a^\dagger(g)$ with

$$[a(f), a^\dagger(g)]_+ = (f,g) \quad , \quad f,g \in L^2(R^3) \ . \tag{3.1}$$

Here $\tau_t\, a(f) = a(\widehat{e^{-ip^2 t}\, \hat{f}(p)})$ and we have an automorphism.

ii) We treat now lattice systems which are easier to handle because the local algebras are algebras over a finite Hilbert space, so that a_x, $a_y^\dagger$ are really operators of the algebra.

Then we know [1,p.192]: If $H_\Lambda = \sum_{X\subset\Lambda} \Phi(X)$, (3.2)

where $\Phi(X)$ is determined recursively and if further

$$||\Phi||_1 = \sum_{X\ni 0} ||\Phi(X)|| \ \exp(N(X) - 1) < \infty \tag{3.3}$$

is fulfilled (e.g for the Heisenberg ferromagnet), then

$$\tau_t(A) = \lim_{\Lambda\to\infty} e^{iH_\Lambda t} \ A \ e^{-iH_\Lambda t}$$

exists in norm topology. The proof is algebraic using the formula

$$e^{iH_\Lambda t} A e^{-iH_\Lambda t} = \sum_{n=0}^{\infty} i^n \frac{t^n}{n!} [H_\Lambda, A]^{(n)} \tag{3.4}$$

with

$$[H_\Lambda, A]^{(n)} = [H_\Lambda, [H_\Lambda, A]^{(n-1)}]$$

and demonstrating that the series converges in the norm uniformly in Λ.

An example is the quasi-free evolution [5] which results from a Hamiltonian $H_\Lambda = \sum_{x,y\varepsilon\Lambda} \Phi(x-y)\, a_x^\dagger a_y$. Then

$$\tau_t\, a(f) = \tau_t \sum_x f(x)\, a_x = a\left(\widehat{e^{-i\hat{\Phi}(\xi)} \hat{f}(\xi)}\right) . \tag{3.5}$$

This automorphism is called quasi-free, because the essential point of the free automorphism, that creation and annihilation operators remain creation and annihilation operators holds again.

iii) We have also a counter example where a time automorphism does not exist: the BCS-model. Here we can examine how this may happen and what the other possibilities are to represent time evolution [6,7,8,9].
We work with a Hamiltonian

$$H_N = \varepsilon \sum_{p=1}^{N} (1 - \sigma_p^z) - \frac{2T_o}{N} \sum_{p,p'=1}^{N} \sigma_p^- \sigma_{p'}^+ , \quad \frac{\varepsilon}{T_o} \leq 1 . \tag{3.6}$$

The interaction is independent from the distance, so

we can enumerate the points instead of speaking of finite regions. Notice, that the Hamiltonian depends also on a parameter N. Therefore, it is not of the type discussed in (ii). Actually $\tau_N(t)A$ does not converge in norm as $N \to \infty$. But we can hope that it may converge in weaker topologies and here we have three different results:

a) We take a representation defined by a state $\rho_{\vec{n}}$ where

$$\frac{1}{N} \sum_{p=1}^{N} \pi_\rho (\vec{\sigma}_p) \underset{\text{strongly}}{\to} \eta\vec{n}\, 1_{\pi_\rho} \quad . \tag{3.7}$$

Then $s - \lim_{N\to\infty} e^{i\pi(H_N)t}\, \pi(\vec{\sigma}_p)\, e^{-i\pi(H_N)t}$ exists and belongs to $\pi(A)$. The time evolution is a combined rotation around the $\vec{z}$-axis and the $\vec{n}$-axis and obviously representation dependent. The algebras A_p do not change at different points. Since our algebra is simple and hence its nontrivial representations are faithful we can project the automorphism of $\pi(A)$ on the algebra A itself. Notice that the automorphism in $\pi(A)$ is not strongly continuous for fixed t (it changes in (3.7) one side and not the other), except if $\vec{n}$ coincides with the $\vec{z}$-axis or if the gap equation $n(z) = \varepsilon/T_o$ holds. In these cases the angular velocity of rotation around the $\vec{z}$-axis vanishes. Otherwise $\vec{n}$ changes in time while rotating around the $\vec{z}$-axis. Therefore the automorphism is not unitarily implementable.

b) Something else happens in the case of Gibbs states, i.e. in the case of the states

$$\Phi_\beta(A) = \lim_{N\to\infty} \Phi_\beta^{(N)}(A) = \lim_{N\to\infty} \frac{\mathrm{Tr}\, e^{-\beta H_N} A}{\mathrm{Tr}\, e^{-\beta H_N}} \quad . \tag{3.8}$$

Above a critical temperature $T_o = \varepsilon(\mathrm{Arth}\ \varepsilon/T_c)^{-1}$ they are factorstates and

$$\Phi_\beta(\sigma_{p_1}^{(i_1)} \ldots \sigma_{p_m}^{(i_m)}) = \eta(\beta)^m\, n^{(i_1)} \ldots n^{(i_m)} \qquad (3.9)$$

with $\vec{n} = (0,0,1)$, $\eta = \mathrm{Th}\ \varepsilon\beta$

$$\frac{1}{N} \sum_{p=1}^{N} \pi_\beta(\vec{\sigma}_p) \to \eta\vec{n}\ 1_{\pi_\beta} \quad .$$

The automorphism of $\pi_\beta(A)$ is a rotation around $\vec{n} = \vec{z}$-axis. Hence the automorphism can be continously extended on the weak closure $\Pi(A)''$ and, since this is a factor III, it is unitarily implementable. Below the critical temperature the Gibbs state is not any more a factorstate but an integral over them:

$$\Phi_\beta(A) = \int_0^{2\pi} d\phi\ \Phi_{\beta,\phi}(A) \qquad (3.10)$$

with

$$\Phi_{\beta,\phi}(\sigma_{p_1}^{(i_1)} \ldots \sigma_{p_m}^{(i_m)}) = \eta(\beta)^m\, n_\phi^{(i_1)} \ldots n_\phi^{(i_m)} \quad ;$$

$$\eta(\beta) = \mathrm{Th}\ \eta(\beta)\ T_c\ \beta$$

with corresponding representation

$$\pi_\beta(A) = \int_+ d\phi\ \pi_{\beta,\phi}(A)$$

$$\vec{n}_\phi = (\alpha\cos\phi, \alpha\sin\phi, \sqrt{1-\alpha^2}).$$

$\alpha = \alpha(T)$ satisfies the gap equation. For $0 < T \leq T_c \to$

$\sqrt{1 - \frac{\varepsilon^2}{T_o^2}} \geq \alpha \geq 0$. The time automorphism reads

$$\tau_t^{\beta}\, \pi_{\beta}(A) = \int_o^{2\pi} d\phi \;\; \tau_t^{\beta,\phi} \pi_{\beta,\phi}(A) \qquad (3.11)$$

where $\tau^{\beta,\phi}$ is only a rotation around $\vec{n}_{\phi}$ since the gap equation is fulfilled (cf. a)). It is unitarily implementable by

$$e^{iH_{\phi}t} = w - \lim_{+} \int \pi_{\beta,\phi}(e^{iH_N t})\, d\phi \quad .$$

Correspondingly the automorphism of $\pi(A)''$ is unitarily implementable, but it does not map $\pi(A)$ onto $\pi(A)$ and therefore cannot be projected into A.

c) It would be nice if we could construct a representation depending automorphism for all representations, maybe only of the weak closure. But here we fail. Take

$$\rho(A) = \frac{1}{2}\, \rho_{\vec{n}_1}(A) + \frac{1}{2}\, \rho_{\vec{n}_2}(A) \qquad n_i^{(z)} \neq \varepsilon/T_o \quad .$$

Now we cannot avoid the trouble that we have different automorphisms by going to the weak closure because new representations arise. Thus in these cases there is neither an automorphism of $\pi(A)$ nor of $\pi(A)''$.

iv) Dubin and Sewell [10] wanted to generalize the theory for equilibrium states for systems without time automorphism (in the norm limit). So they started with two conditions:

$$\lim_{N\to\infty} \Phi_{\beta}^{(N)}(\tau_{t_1}^{(N)} A_1 \ldots \tau_{t_m}^{(N)} A_m) \quad \exists \qquad (3.12)$$

$$\lim_{M\to\infty} \lim_{N\to\infty} \Phi_\beta^{(N)} (\tau_{t_1}^{(N)}(A_1)\ldots\tau_{t_k}^{(N)}(A_k)\ \tau_{t_{k+1}}^{(M)}(A_{k+1})\ldots\tau_{t_\ell}^{(M)}(A_\ell))$$

$$= \lim_{N\to\infty} \Phi_\beta^{(N)} (\tau_{t_1}^{(N)}(A_1)\ldots\tau_{t_\ell}^{(N)}(A_\ell)) \quad . \qquad (3.13)$$

They proved that, if these conditions are met, then there exists an automorphism of $\pi_\beta(A)''$ which represents time evolution. So again we have a representation dependent automorphism, though unfortunately this automorphism cannot in general be projected on the algebra itself.

v) As far as arbitrary continuous systems are concerned no results are known to me. It was only shown by Ruelle [11] that the limits of the Green's functions exist for a dilute Fermi gas with pair potential under some restrictions, excluding especially the Coulomb potential because of its behaviour at $r = \infty$. So condition (3.12) of Dubin and Sewell is satisfied, but as far as condition (3.13) is concerned this remains an open problem. In reality the Coulomb potential is relevant and because of its long range one has to anticipate some similarity with the BCS-model. However, in the following we shall adopt the optimistic point of view that there is some automorphism of a suitable algebra.

4. INVARIANT STATES AND EQUILIBRIUM STATES

i) Extremal invariant states

Before starting with the problem of examining the time evolution of states it is useful to get some information about the invariant states and the equilibrium states.

The invariant states form a linear subset of E. They are the generalization of density matrices commuting with H. So extremal invariant states which cannot be written as linear combinations of two other invariant states are well defined. They generalize to the eigenstates of H*. We ask whether an arbitrary invariant state can be written uniquely as sum of extremal invariant states. That is not true in general (take as automorphism the trivial one), but if we make further assumptions on the time automorphism we will obtain a positive answer.

The significant mathematical facts used in the considerations are the following:

1) If a state is invariant, then

$$\pi_\rho(\tau_t A) = e^{iHt}\, \pi_\rho(A)\, e^{-iHt} \quad , \tag{4.1}$$

and if we further want $e^{-iHt}|\Omega_\rho) = |\Omega_\rho)$ then H is unique and, of course, selfadjoint. Thus the automorphism is spatial.

2) Every commutative W^*-algebra is isomorphic to the algebra of essentially bounded functions on some locally compact measure space.

Using this fact one can prove the following theorem:

Theorem: Given A, ρ, $B \subset \pi_\rho(A)'$ with $B \subset B'$. Then $\exists!$ measure μ over the states with resultant ρ ($\int d\mu(\sigma)\sigma = \rho$) and $B \simeq L^\infty(E,\mu)$ such that $\forall\, B \in B$ $\; B \simeq \int f(\sigma)\, 1_\sigma d\mu(\sigma)$.

---

*But in contradiction to finite systems extremal invariant $\not\to$ extremal.

The measure can be constructed in the following way:

$$B_1, \ldots, B_n > 0, \ \varepsilon \ B \ , \qquad \sum_{i=1}^{n} B_i = 1_\pi \tag{4.2}$$

$$\sigma_i = \frac{(\Omega_\rho | B_i \pi_\rho (A) | \Omega_\rho)}{(\Omega_\rho | B_i | \Omega_\rho)}$$

$$= 0 \qquad \text{if } (\Omega_\rho | B_i | \Omega_\rho) = 0$$

$$\mu_n = \sum_{i=1}^{n} (\Omega_\rho | B_i | \Omega_\rho) \ \delta_{\sigma_i} \quad .$$

If $n \to \infty$ then it can be shown that $\mu_n \rightharpoonup \mu$ independent of the chosen $\{B_i\}$. If $B_1 \subset B_2$, then the decomposition corresponding to B_2 is finer than that corresponding to B_1. From these results we conclude, if ρ_1 of (2.6) is time invariant, then $e^{iHt} T^\dagger T e^{-iHt}$ has to be equal to $T^\dagger T$. Therefore, if $\pi(A)' \cap \{e^{iHt}\}' = B$ is abelian, the theorem gives a unique and extremal decomposition into extremal invariant states. In this case we call the system (A,τ) G-abelian. Of course, we need other criteria whether or not our system is G-abelian. The most useful one reads:

Lemma: (A,τ) is G-abelian, if and only if for all self-adjoint $A_1, A_2 \in A$ and all invariant states $\rho \in E \cap L_G{\perp}$

$$\lim_{T\to\infty} \frac{1}{2T} \int_{-T}^{+T} dt \rho([\tau_t A_1, A_2]) = 0 \tag{4.3}$$

and ρ is extremal invariant if

$$\lim_{T\to\infty} \frac{1}{2T} \int_{-T}^{+T} dt \rho(\tau_t A_1 . A_2) = \rho(A_1) . \rho(A_2) \ . \tag{4.4}$$

Let us take as example our BCS-model. We have already seen how the equilibrium states look like (3.6, 3.7) and now we will concentrate on those above the critical temperature where we have an automorphism of A. The corresponding time-automorphism is a rotation around the $\vec{z}$-axis, such that σ_z is invariant and the expectation values of σ_x and σ_y in invariant states = 0. It is easily checked that the system is G-abelian with respect to this automorphism, but also, that the equilibrium states are not extremal invariant, because $\rho(\sigma_z.\sigma_z) \neq \rho(\sigma_z).\rho(\sigma_z)$. Extremal invariant are those with

$$\bar{\rho}(\sigma_{p_1}^{(i_1)} \ldots \sigma_{p_n}^{(i_n)}) = \bar{\rho}(\sigma_{p_1}^{(i_1)}) \ldots \bar{\rho}(\sigma_{p_n}^{(i_n)})$$

and

$$|\bar{\rho}(\sigma_p^{(i)})| = \delta_{i3} \; .$$

The same experience can be made in a finite system, which is not expected to be thermodynamical. A = M_2 = complex 2×2 matrices with $U(t) = e^{it\sigma_z}$. The invariant states are those with $\rho(\sigma_i) = \eta \; \delta_{i3}$ and the extremal invariant ones those with $\eta = \pm 1$. Of course, in a finite system we cannot expect a state to converge in time, but it makes sense to examine the invariant mean of a state (which is here, of course, unique) and we recognize that in this case every invariant state is the mean of some extremal state. So it is physically as relevant as an extremal invariant one.

ii) Abelianess of the time automorphism

Later on we will need some stronger assumptions on the

time automorphism than G-abelianess and these are: η-abelianess:

The system (A,τ_t) is η-abelian, if η is some invariant mean and for all $\rho \in E$ and all $A,B \in A$

$$\eta\ \rho([\tau_t\ A,B]) = 0 \tag{4.5}$$

holds. For G-abelianess we required (4.5) only on invariant states and there all invariant means coincide. Thus η-abelianess is the stronger assumption. The advantage of this definition is the following: We can demonstrate that

$$w - \lim \frac{1}{2T} \int_{-T}^{+T} dt \quad \tau_t\ \pi_\rho(A) \quad \doteq \eta\ \pi_\rho(A) \in \pi_\rho(A)' \cap \pi_\rho(A)''.$$

Using that Ω_ρ is cyclic, one can demonstrate after some calculation that $\pi_\rho(A)' \cap \{e^{iHt}\}' \subset Z$. Thus the decomposition into factorstates is finer than the central decomposition into extremal invariant states. For weak asymptotic abelianess we require that

$$\lim_{t\to\infty} \rho([\tau_t\ A,B]) = 0 \qquad \forall\ \rho \in E;\ A,B \in A \tag{4.6}$$

such that we can really concentrate on the time limit and need not worry about the recurrence time.

Strong asymptotic abelianess is defined by

$$\lim_{t\to\infty} ||[\tau_t\ A,B]|| = 0 \qquad \forall\ A,B \in A \tag{4.7}$$

so that we can concentrate on the algebra and need not work with states. Obviously (4.7) $\to$ (4.6) $\to$ (4.5) $\to$ (4.3). We ask under which circumstances these assumptions are satisfied in reality. We return to our already examined automrophisms. For the BCS-model we have already seen that $(A,\tau_\beta(t))$ is G-abelian, but not η-abelian (factors are not extremal invariant). For free fermions we have strong abelianess, also for the quasi-free evolution of a lattice system, if the potential $\Phi(x)$ in (3.5) vanishes exponentially at $\pm\infty$ [5]. One has to use the Riemann Lebesgue lemma for

$$\lim_{t\to\infty} \| [\tau_t\, a(f), a^{(+)}(g)]_+ \| = \lim_{t\to\infty} \left| \int e^{i\hat{\Phi}(\xi)t} f(\xi) g^{(*)}(\xi) d^\nu \xi \right. \tag{4.8}$$

and therefore it must be quaranteed that the transformation $d\,\xi_i = (\partial\hat{\Phi}/\partial\xi_i)^{-1} d\Phi$ is possible almost everywhere for some i. But conversely one can construct a counter example with two point interactions where $\hat{\Phi}(\xi)$ is constant in some region and where the system is even not G-abelian

iii) KMS-states [1,p.194]
It remains to characterize the equilibrium states. They are the limit of Gibbs states (3.8) and obey the following KMS-condition:

Let $\tilde{A}$ be the norm dense subalgebra of A of operators, where time automorphism can analytically be continued to imaginary times. Then for $A,B \in \tilde{A}$ $\rho(A\ \tau_t\ B)$ is an analytic function of t in the strip $\{t: 0 < \text{Im } t < \beta\}$, continuous at the boundary and

$$\rho(\tau_t A\ B) = \rho(B\ \tau_{t+i\beta}\ A) \quad . \tag{4.9}$$

I have to remark that the condition also holds if we have

the automorphism only defined in the sense of Dubin and Sewell (DS) [10].

KMS-states are defined to be faithful states which obey the KMS-condition and for finite systems they are the Gibbs-states. KMS-states of some temperature form a linear set and again it can be demonstrated that there exists a unique decomposition into extremal KMS-states. One has the important result that it coincides with the central decomposition (= decomposition into factors). The considerations remain true in the case of DS. But here in the various factor representations the automorphisms will differ. (4.9) will only hold with the corresponding automorphisms. We summarize the various decomposition and cluster properties in the following table (p.17).

iv) Average states

We have already seen in a two dimensional example, that states being the mean of an extremal state need not be extremal invariant ones. Let us call these states average states. We ask whether there exists a unique decomposition into these average states and whether they coincide with extremal equilibrium states. Here, of course, great mathematical difficulties arise. In practical examples we do not have any explicit construction for the extremal states on our algebra. Moreover we do not know, even if we assume a time automorphism, whether a unique invariant mean exists and specifying some invariant mean is physically very unsatisfying.

But let us assume that

$$\lim_{T\to\infty} \frac{1}{2T} \int_{-T}^{+T} dt\ \sigma(A\tau_t BC) = \sigma(AC)\cdot\eta\sigma(B) \qquad \forall \sigma\varepsilon E_{\text{pure}},\ A,B,C\varepsilon\mathcal{A}. \qquad (4.10)$$

This condition is stronger than η-abelianess and weaker than strong asymptotic abelianess.

(4.10) $\rightarrow$ (4.5)

$$\eta\ \sigma([\tau_t A, B]) = \eta\ \sigma(\tau_t A\ B) - \eta\ \sigma(B \tau_t A) =$$

$$= \eta\ \sigma(A)\ \sigma(B) - \sigma(B)\ \eta\ \sigma(A) = 0$$

(4.7) $\rightarrow$ (4.10)

$$\eta\ \sigma(A \tau_t BC) = \eta\ \sigma([A, \tau_t B]C) + \eta\ \sigma(\tau_t BAC) =$$

$$= \eta\ \sigma(B)\ \sigma(AC)$$

because in this case, σ being a factor state [1,p.165]

$$\lim_{t \to \infty} [\sigma(\tau_t B\ A) - \sigma(\tau_t B) \cdot \sigma(A)] = 0 \quad .$$

Let ρ be extremal invariant and one decomposition into pure states read

$$\rho = \int d\mu(\sigma) \quad \sigma \quad . \tag{4.11}$$

Then

$$\rho(A) = \frac{1}{2T} \int_{-T}^{+T} dt\ \rho(\tau_t A) = \int d\mu(\sigma) \quad \sigma\left(\frac{1}{2T} \int_{-T}^{+T} \tau_t(A) dt\right)$$

$$\underset{T \to \infty}{\rightarrow} \int d\mu(\sigma) \quad \eta\sigma(A)\ ; \tag{4.12}$$

since μ is a probability measure and $\left|\sigma\left(\frac{1}{2T} \int_{-T}^{+T} \tau_t A dt\right)\right|$ $\leq ||A||$ we can exchange limit and integration because of Lebesgue's lemma of dominated convergence.

Table 1: DECOMPOSITION AND CLUSTER PROPERTIES

	Decomposition with B	Extremal points	Cluster properties of the extremal points
	Maximal abelian sub algebra $B = \pi(A)' \cap B'$ not unique	extremal states (if A separable)	none
ρ G invariant	$B = \pi(A)' \cap \{U_G\}'$	extremal invariant	$\lim_{T\to\infty} \frac{1}{2T} \int_{-T}^{+T} \rho(\tau_t A_1 \cdot A_2)dt = \rho(A_1)\rho(A_2)$
η abelian	$B =$ " $\subset Z$	"	"
w.a.a.	"	"	$\lim_{t\to\infty} [\rho(\tau_t A_1 \cdot A_2) - \rho(\tau_t A_1)\rho(A_2)] = 0$ holds also for all factorstates
st.a.a.	"	"	"
ρ KMS-state	$B = Z$	extremal KMS-states	
KMS-state + η-abelian	$B = Z = \pi(A)' \cap \{U_G\}'$	extremal KMS $\leftrightarrow$ extremal invariant + KMS	as for η-abelian

$\eta\sigma$ is invariant and therefore equals ρ almost everywhere. Therefore extremal invariant $\rightarrow$ average.

In the contrary, take $\rho(A) = \eta\,\sigma(A)$. We see immediately that

$$\left|\frac{1}{2T}\int_{-T}^{+T} dt\ \rho(A\tau_t BC) - \frac{1}{2T}\int_{-T}^{+T} dt\ \frac{1}{2U}\int_{-U}^{+U} du\ \sigma(\tau_u A\tau_{u+t} B\tau_u C)\right.$$

$$\left. + \frac{1}{2U}\int_{-U}^{+U} du\ \sigma(\tau_u A\tau_u C)\ \sigma(\eta B) - \rho(AC)\rho(B)\right| < \varepsilon$$

so that $\eta\,\pi_\rho(B) = \rho(B)\,1_{\pi_\rho}$. Taking into account that for η-abelian systems $\pi(A)' \cap \{U\}' \subset \pi(A)'' \cap \{U\}'$, we conclude that ρ is extremal invariant.

Therefore extremal invariant and average states coincide. If we take into account that for an η-abelian system the decomposition into extremal invariant states is the same as into extremal KMS-states, namely the central decomposition, the result is rather satisfying. For systems which obey (4.10): Factor $\leftarrow$ extremal KMS $\rightarrow$ extremal invariant $\leftrightarrow$ average.

This generalizes the classical idea that the energy surface is covered by a simple orbit. Thus f.i. for the infinite free fermion system this is actually the case (we shall later give explicitly for each temperature an extremal state whose average is the KMS-state). This is all the more remarkable since in classical theory it is notoriously hard to find examples with an orbit dense on the energy surface. But in quantum theory the converse: average $\overset{?}{\rightarrow}$ KMS is much harder to answer and therefore we have only a partial answer to the problem of classical ergodic theory. It is amusing to observe that for the two-dimensional example we have indeed the implication

$$\text{average} \leftrightarrow \text{invariant} \leftrightarrow \text{KMS} .$$

It is not η-abelian and average $\not\rightarrow$ extremal invariant. For a classical system this implication cannot hold.

5. APPROACH TO EQUILIBRIUM

i) Quasi-free automorphism for a lattice system

We turn now to the concrete results which have yet been obtained, and these exist of course for systems with automorphisms which are well defined and easy to handle, namely the quasi-free systems.

I start with the easiest example, that of a fermi lattice system with a quasi-free evolution (3.5). [12]. To the invariant states belong all quasi-free translation invariant and gauge invariant states, i.e. states for which

$$\rho(a_x^\dagger a_y) = m(x - y) \tag{5.1}$$

and

$$\rho(a_{x_1}^\dagger \dots a_{x_n}^\dagger a_{y_m} \dots a_{y_1}) = \delta_{nm} \sum_{p,p'} (-1)^{p+p'} \prod_i \rho(a_{x_{p_i}} a_{y_{p_{i'}}}) .$$

These states are well examined. We know they are factor-states, disjoint (i.e. they have no common subrepresentations) and for a given temperature there exists exactly one KMS-state, namely the one with

$$\hat{m}(\xi) = \frac{1}{1 + \exp \beta\hat{\Phi}(\xi)} . \tag{5.2}$$

Let us start with an arbitrary state. We want to know what happens to it in the limit $t \to \infty$. We are only interested in the weak limit, so it is sufficient to examine the functions $\rho(a^\dagger_{x_1} \ldots a^\dagger_{x_n} a_{y_m} \ldots a_{y_1})$. Starting with $\rho(a^\dagger_x a_y) \doteq \rho(x,x-y)$ we assume that it is L^1 as function of x-y and that an average exists as function of x in the sense that its Fourier transform reads

$$\hat{\rho}(\xi,\eta) = (2\pi)^\nu \quad \delta(\xi)\ \hat{m}(\eta) + \hat{\rho}_1(\xi,\eta) \tag{5.3}$$

and for each $\varepsilon > 0$ $\exists$ δ such that

$$\int_{-\delta}^{+\delta} \hat{\rho}_1(\xi,\eta)\ d\xi_i < \varepsilon \quad . \tag{5.4}$$

So

$$\lim_{t\to\infty} \rho(\tau_t[a^\dagger(f)a(g)]) = \frac{1}{(2\pi)^\nu} \int d\eta\ \hat{m}(\eta)\ \hat{f}^*(\eta)\ \hat{g}(\eta) +$$

$$+ \lim_{t\to\infty} \frac{1}{(2\pi)^\nu} \int d\xi\ d\eta\ \hat{\rho}_1(\xi,\eta)\ \hat{f}^*(\xi+\eta)\ \hat{g}(\eta)\ \cdot$$

$$\cdot\ \exp\ \{i[\hat{\Phi}(\xi+\eta) - \hat{\Phi}(\eta)]t\} \quad . \tag{5.5}$$

The method is the same as that for proving asymptotic abelianess. Again we use a slightly generalized Riemann Lebesgue lemma. Therefore the second expression tends to zero if $\hat{\Phi}(\xi+\eta) - \hat{\Phi}(\eta)$ is not constant in some region in the sense indicated after (4.7). Notice, that this condition is stronger than that for asymptotic abelianess. We will compare them with another restriction arising for return to equilibrium a little bit later.

The further steps for higher correlation functions are similar and no new ideas are needed for the proof.

Expectation values of higher polynomials are again arranged as sum of functions which in some variables have an average in space, in the others are L^1, e.g.

$$\rho(a^\dagger_{x_1}\, a^\dagger_{x_2}\, a_{y_2}\, a_{y_1}) = \rho_1(x_1,\, x_2,\, x_1-y_1, x_2-y_2)$$
$$+ \rho_2(x_1, x_2, x_1-y_2, x_2-y_1)$$
$$+ \rho_3(x_1, x_1-x_2, y_1, y_1-y_2) \qquad (5.6)$$

and for every term we have an exponential function $\exp \sum \pm \hat{\Phi}(\xi_i)\, t$ to use Riemann Lebesgue lemma. For instance the contribution of ρ_3 is zero and that of ρ_1 and ρ_2 such that we obtain a quasi-free state determined by $\hat{m}(\eta)$. Therefore for these states the weak limits exist, are quasi-free but not necessarily KMS.

ii) Free fermions [13]

About the same methods can be used for free fermions and nearly the same results are obtained, with one exception. Since $a^\dagger_x$, a_y are not operators of the algebra we come into some technical difficulties and cannot write the expectation value corresponding to (5.1) so nicely as function of two variables. This difficulty was revolved by starting with a stronger assumption, namely, that the state is already translation invariant. This assumption is not essential and with some mathematical effort it should be avoidable. The proof is similar. First you show that the so called two point functions $\rho(a^{(\dagger)}(f), a^{(\dagger)}(g))$ have the right behaviour, then, that the truncated functions, i.e. the difference between the expectation values and the values of the quasi-free state, constructed by the two point functions, vanish.

Similarly we can obtain the desired results for bosons, though we have to take more mathematical care due to the more clumsy form of the algebra.

iii) Discussion

Let us summarize the physical results. We start with a state with some restrictions: it should not be too inhomogeneous under space translations. That is reasonable. Take, for instance, the system divided into two infinite parts, both parts in thermal equilibrium, but with different temperature. Then a thermal current will result, which will become constant in time and the system will never become reflection invariant. Notice, that nevertheless in the weak limit (concentrate on operators in finite regions) an invariant state will appear, a fact, which is not contained in our result [14]. The second restriction is that of the clustering implied by $f(x) = \rho(\tau_x A.B) \in L^1$. This means, that measurements far away from each other are independent. The assumption to start with such a state is not bad, why should we have an ordering in the beginning, but I do not see any good reason why such an ordering should alter behaviour in time. I only want to remark, that this restriction might be not strong at all, since our system is asymptotically abelian under space translations, for factorstates an only somewhat weaker property is automatically fulfilled ($\lim_{x\to\infty}[\rho(\tau_x A.B) - \rho(\tau_x A)\rho(B)] = 0$ instead of being a summable function) and we want to start especially with irreducible and therefore a fortiori factor-states.

The question arises, whether there exists an extremal state, such that our invariant state is the limit (remember, it is quasi-free, therefore factorial, extremal invariant and since the whole system is asymptotic abelian, should be an average state). In fact this extremal state can easily be constructed, namely the quasi-free, not gauge invariant state with

$$\rho(a^{\dagger}(f)a(g)) = \int d\xi \ \hat{m}(\xi) \ \hat{f}^{*}(\xi) \ \hat{g}(\xi) \ \frac{1}{(2\pi)^{\nu}}$$

$$\rho(a(f)a(g)) = \int d\xi \ \sigma(\xi) \ \hat{f}(-\xi) \ \hat{g}(\xi) \ \frac{1}{(2\pi)^{\nu}} \tag{5.7}$$

where $|\sigma(\xi)|^2 = \hat{m}(\xi)[1 - \hat{m}(-\xi)]$ [15].

We have seen before that in quantum theory it is in most cases easy to find extremal states whose average is the Gibbs state. Indeed, even for finite systems with discrete energy values ε_i and eigenfunctions ϕ_i the average of the state corresponding to the vector

$$\sum_i \phi_i \, e^{-\beta\varepsilon_i/2} \Big/ \sum_i e^{-\beta\varepsilon_i}$$

has as average the Gibbs state provided the ε_i are mutually irrational. In the infinite system we find not only that the KMS-state is not only the average but even the limit of an extremal state. This result cannot be imagined for classical systems where the extremal states correspond to a single point in phase-space.

We see, that the result is rather satisfying. Of course, quasi-free systems have not a thermodynamically good behaviour in the sense that the class of extremal invariant states is much greater than that of equilibrium states and we see that equilibrium states do not play a distinguished role. So one should make efforts to get results for a system with a Hamiltonian which can be argued to satisfy thermodynamical requirements. Finally I want to mention that for an Ising model

$$(H_\Lambda = - \sum_{x,y\varepsilon\Lambda} \Phi(x-y)\sigma_x^z \sigma_y^z - \sum_{x\varepsilon\Lambda} B(x)\sigma_x^z)$$

it was shown by Radin, that for all states the invariant mean exists independent whether the system is G-abelian or not. For a class of states (namely $\rho(\sigma^A) = 0$ if $A_3 \neq \Phi$) this invariant mean coincides with the limit in time. But also a counterexample can be constructed, where the state remains periodic in time. But there still lacks a complete discription of the time invariant states and we do not know whether the equilibrium states are the only average states and whether for a greater class of systems the invariant mean can be replaced by the limit.

6. RETURN TO EQUILIBRIUM

As far these are the results for approach to equilibrium. Not seeing a way to get informations yet for arbitrary systems we become more modest and concentrate on a smaller class of initial states, which result by perturbation of equilibrium states and should, of course, tend back to it.

i) Normal states

The first result is rather well known and is due to Araki [17]. Assume we have a perturbation during a finite time, represented by a spatial automorphism. So the resulting state ρ is normal with respect to the initial one $\bar{\rho}$ which we assume to be extremal KMS (2.4). Now the perturbation is switched off and if the time automorphism is asymptotically abelian:

$$\lim_{t\to\infty} \rho(\tau_t A) = \lim_{t\to\infty} \frac{\bar{\rho}(B\tau_t AB^\dagger)}{\bar{\rho}(BB^\dagger)} .$$

Now we know that $\bar{\rho}$ as extremal KMS-state is extremal invariant so (see table 1)

$$= \frac{\bar{\rho}(BB^\dagger)\bar{\rho}(A)}{\bar{\rho}(BB^\dagger)} = \bar{\rho}(A) \qquad \text{q.e.d.} \tag{6.1}$$

Of course, we should ask when such a perturbation is represented by a spatial automorphism. This is true if it is weakly continuous, so that it can be enlarged to the weak closure of $\pi_\rho(A)$, because here every automorphism is spatial. (These are usually I_∞ or III factors and there every automorphism is spatial).

ii) Perturbed automorphisms [18]
On the other hand you can really perturb the automorphism. Take $P \in A$ and define

$$\tau_t^P(A) = \tau_t(A) + \sum_{n\geq 1} i^n \int_{0\leq s_1 \leq \cdots \leq s_n \leq t} ds_1 \ldots ds_n \; .$$

$$\cdot \; [\tau_{s_1}(P)\;[\ldots[\tau_{s_n}(P), \tau_t(A)]\ldots] \; . \tag{6.2}$$

This is well defined, because the series converges in norm. If

$$\pi(\tau_t A) = e^{iHt}\,\pi(A)\,e^{-iHt}$$

then

$$\pi(\tau_t^P A) = e^{i[H+\pi(P)]t}\,\pi(A)\,e^{-i[H+\pi(P)]t} \qquad (6.3)$$

But now we ask whether we are allowed to let the system be perturbed during an infinitely long time, such that we may assume to arrive at a state which is invariant or KMS for the perturbed system.

We have already seen in our previous calculation concerning approach to equilibrium that we can start with a pure state and end with a factor III state [19], which is not normal with respect to the initial state. Other examples of this type, that in the limit $t \to \infty$ normality is broken, have been constructed by Hepp [20].

So we ask, can the theory of Araki be generalized to states which are not any longer normal but invariant or KMS for a perturbed Hamiltonian. We obtain some results, if the Hamiltonians only differ by a bounded operator (6.2) such that we can use the formula

$$\frac{d}{dt}\,\tau_t^P\,\tau_{-t}(A) = i\,\tau_t^P\,\tau_{-t}([\tau_t(P),A])$$

or

$$\tau_{t_2}^P\,\tau_{-t_2}(A) - \tau_{t_1}^P\,\tau_{-t_1}(A) = i\int_{t_1}^{t_2} dt\,\tau_t^P\,\tau_{-t}([\tau_t^P(P),A]) \,. \qquad (6.4)$$

There is some symmetry between τ_t^P and τ_t, namely $(\tau_t^P)^{-P} = \tau_t$. But for instance asymptotically abelianess τ_t need not imply the same for τ_t^P. Look at a lattice system, such that the local Hamiltonian is a bounded operator and choose $P = -\sum_{X>\Lambda_o} \Phi(X)$ (H_Λ as in (3.2)).

Then A_{Λ_o} is invariant under τ_t^P and therefore the system cannot be asymptotically abelian.

We assume now, that (A,τ_t) is asymptotically integrable with respect to P, by which we mean, that $f(t) = \|[\tau_t(P),A]\|$ is a function integrable to infinity for all A in a uniformly dense subset of A. We might ask, in which relation stands this to asymptotical abelianess. It may be true for some P without asymptotical abelianess, but asymptotic abelianess is not sufficient, we should need a theorem, when the commutator vanishes rapidly enough for $t \to \pm\infty$. Again τ_t^P and τ_t need not be interchangeable in this relation. Let us examine again our quasi-free evolution. The condition is fulfilled for all P, if roughly

$$\int_{-T}^{+T} dt \int d\xi \; e^{i\hat{\Phi}(\xi)t} \; \hat{f}(\xi)\; \hat{g}(\xi)$$

$$\to \sum_{\substack{\xi_o \\ \hat{\Phi}(\xi_o)=0}} \frac{\hat{f}(\xi_o)\;\hat{g}(\xi_o)}{\hat{\Phi}'(\xi_o)} \neq \pm\infty \qquad . \tag{6.5}$$

Compare this with the condition we needed for approach to equilibrium. ($\hat{\Phi}(\xi+\eta) - \hat{\Phi}(\eta) \neq$ const in any region). Both conditions are stronger than asymptotic abelianess, but have no relation between each other. The advantage of asymptotic integrability is that it guarantees the existence of

$$\gamma_{\pm}(A) = u - \lim_{t\to\infty} \tau_{-t}^{P}\; \tau_t(A) \tag{6.6}$$

and defines an endomorphism. The proof is evident:

$$||\tau^P_{-t_2}\,\tau_{t_2}(A) - \tau^P_{-t_1}\,\tau_{t_1}(A)|| \leq \int_{t_1}^{t_2} ds||[P,\tau_s(A)]|| \quad .$$

If both (A,τ) and (A,τ^P) are asymptotically integrable, then the endomorphisms $\gamma_\pm$ are invertible and therefore automorphisms. But in general, even though γ is the uniform limit of automorphisms and therefore norm conserving the range need not be dense and therefore it need not be invertible.

The result similar to that for normal states reads: Let ω be a τ^P-extremal invariant (τ^P weakly mixing, τ^P strongly clustering) state, then

$$\omega(\gamma_\pm A) = \lim_{t\to\pm\infty} \omega(\tau_t\, A) \tag{6.7}$$

exists and is τ-extremal invariant (τ weakly mixing, τ strongly clustering). Weakly mixing requires that

$$\lim_{T\to\infty} \frac{1}{2T} \int_{-T}^{+T} dt[\omega(A\;\tau^P_t(B)) - \omega(A)\;\omega(B)] \quad = \quad 0$$

and strongly clustering

$$\lim_{t\to\infty} \omega(A\;\tau^P_t(B)) = \omega(A)\;\omega(B) \quad .$$

For instance, for η abelian systems extremal invariance implies weakly mixing, whereas for asymptotic abelian systems this implies strongly clustering.

The existence and invariance of the state is evident, noticing that $\gamma_\pm\,\tau_t = \tau^P_t\,\gamma_\pm$. The clustering properties can be proved using the integral relations (6.3) between the automorphisms, similar as in the proof of the existence of $\gamma_\pm$.

Thus starting with an extremal invariant state with respect to τ the perturbation will lead after a sufficiently long time into an extremal invariant state with respect to τ^P. Switching off the perturbation, we will finally end in the extremal invariant state we started with, provided that both τ_t and τ_t^P are asymptotically integrable with respect to P.

I want to remind you on the approach of equilibrium for free and quasi-free particles. There we treated lim $\tau_t A$. But it can be easily checked that this limit only exists weakly in the representations, thus the present theory does not apply. I have to mention one further result concerning KMS-states. For those we need in our proof that the set for which time automorphism is analytic remains invariant under perturbation. This is guaranteed if P itself belongs to this set (6.2). Now we want τ_{t_P} to be asymptotic abelian and demonstrate that ω being τ-KMS implies that all weak limit points in t of $\omega \circ \tau_t$ are KMS with respect to τ. The existence of limit points is guaranteed by the weak compactness of E. We use again the integral relations between the two automorphisms which lead to

$$|\omega(\tau_t(\tau_{i\beta}A.B)) - \omega(\tau_t(B.A)| =$$

$$|\omega(\tau_{t+i\beta}\, A\, \tau_t\, B) - \omega(\tau^P_{i\beta}\, \tau_t\, A\, \tau_t\, B)| \tag{6.8}$$

$$= \left|\int_0^\beta ds\; \omega(\tau^P_{is}([P, \tau_{t+i(\beta-s)}\, A])\, \tau_t\, B)\right|$$

$$\leq \int_0^\beta ds \, \|[P,\tau_{t+i(\beta-s)}\, A]\| \;\; \|\tau^P_{-is}\, \tau_t\, B\| \, .$$

$\rightarrow 0$ for $t\rightarrow\infty$ remains bounded for all t.

What we miss here is a proof that the limit really exists unless we know already that there exists exactly one KMS-state with respect to τ_t and the given temperature or if we can apply (6.6) and (6.7).

iii) A system in a thermal bath

Up to now we considered the problem: we have an infinite thermodynamical system and want to know how it behaves during time corresponding to some perturbation. But the same method can be applied to an other situation. Suppose we have a system of finite volume Λ in an infinite thermal bath. What happens to the system? Does it become a KMS-state with the temperature of its surroundings? Of course, here are new problems involved. Certainly we need some coupling between the system and its surroundings. Then we can hope using the results already obtained that the complete system will become KMS. What remains to examine is, how does then the reduced state of the finite system look like, how does it depend on the coupling parameter. One would guess that for the coupling tending to zero it should tend to the KMS-state corresponding to the Hamiltonian of the finite system.

The theorem and ideas of (ii) can be used to show under some assumptions the weak continuity of the KMS-state on the coupling parameter.

We want the coupling operator C bounded and such that $||[C,\tau_t^{\alpha C}(A)]|| \leq f(t)$ with f(t) integrable. Let ω_α be the KMS-state for the automorphism $\tau_t^{\alpha C}$, then $\lim_{\alpha\to 0} \omega_\alpha(A)$ exists and

$$\lim_{\alpha\to 0} \omega_\alpha(A) = \lim_{\alpha\to 0} \lim_{t\to\infty} \omega_\alpha(\tau_t^{\alpha C} A)$$

$$= \lim_{t\to\infty} \lim_{\alpha\to 0} \omega(\tau_t^{\alpha C} A) = \lim_{t\to\infty} \omega_o(\tau_t A) = \omega_o(A) \qquad (6.9)$$

where the condition above was needed to allow us to exchange the limits, whereas $\lim_{\alpha\to 0} ||\tau_t^{\alpha C} A - \tau_t A|| = 0$ results immediately from the definition of $\tau_t^{\alpha C}$.

We return to our finite system in a thermal bath. What information does our theory of return to equilibrium give? We concentrate on $A_\Lambda \subset A$, A our quasi-local algebra. Then $A = A_\Lambda \otimes A_{\Lambda^\perp}$. Since we have to assume that the various time automorphisms differ only by bounded operators, we restrict ourselves to lattice systems. Our inital state is assumed to be of the form $\rho = \rho_\Lambda \otimes \tilde{\rho}_\beta$. Here ρ_Λ is arbitrary, $\tilde{\rho}_\beta$ is KMS for the restricted automorphism of $A_{\Lambda^\perp}$, i.e. a quasi-free evolution obtained by a quadratic Hamiltonian with $\Phi(X) = 0$ if $X \cap \Lambda \neq \Phi$. We further assume for simplicity that it has finite range. We can construct an operator Q in A_Λ, which gives the KMS-automorphism for the given β and ρ_Λ ($\rho_\Lambda(A) = \mathrm{Tr}\, e^{-\beta Q} A/\mathrm{Tr}\, e^{-\beta Q}$). So the whole state is KMS for the time automorphism τ_o^P with τ_o the quasi-free evolution for the whole system and $P = Q - \sum_{X, X\cap\Lambda\neq\Phi} \Phi(X)$. τ_o is such that a strictly local perturbation is asymptotically integrable. Therefore $\gamma(A) = \lim_{t\to\infty} \tau^P_{o-t}\, \tau_{ot}(A)$ exists in norm. Now we turn to the actual time automorphism $\tau_o^{\bar{H}}$. It is true that we can write it in this way for some $\bar{H} = H_\Lambda + C - \sum_{X, X\cap\Lambda\neq\Phi} \Phi(X)$.

But here we need the stronger assumption that $\bar{H}$ is asymptotically integrable with respect to $\tau_o^{\bar{H}}$ (for τ_o it is guaranteed if C is strictly local). This is primarily a condition on C, if e.g. C = 0, then A_Λ remains fixed and therefore $\tau_o^{\bar{H}}$ cannot be asymptotically abelian and $\rho(A) = \lim_{t\to\infty} \tau_{o-t}^{\bar{H}}\, \tau_{ot}(A)$ cannot be invertible. But that we need, because only in this case

$$||\tau_{o-t}^{P}\, \tau_{ot}\, \tau_{o-t}\, \tau_{ot}^{\bar{H}}(A) - \gamma\, \rho^{-1}(A)|| \leq$$

$$\leq ||\tau_{o-t}\, \tau_{ot}^{\bar{H}}(A) - \rho^{-1}(A)|| + ||\tau_{o-t}^{P}\, \tau_{ot}\, \rho^{-1}(A)$$

$$-\gamma\, \rho^{-1}(A)|| \leq 2\varepsilon \qquad .$$

So $\bar{\omega}(A) = \lim_{t\to\infty} \omega(\tau_{ot}^{\bar{H}} A)$ exists and, since $\tau_{ot}^{\bar{H}}$ is asymptotically abelian, is KMS. Now, if we swithch off the coupling, as we have shown in (6.9) we come to the KMS-state of the finite system with the corresponding temperature.

7. SUMMARY

To summarize the results: we know that quasi-free systems turn to an invariant state corresponding to the spreading out of the wave functions though not necessarily to an equilibrium state. We know, that systems whose time automorphism is asymptotically abelian or has similar properties, equilibrium or invariant states are stable under perturbations. But we know further that our mathematical assumptions do not yet describe reality so that all considerations only give us some feeling about what might happen and encourage us to weaken the mathematical assumptions. Up to now there is no reason why the results should then become weaker though certainly they will require more effort. I only want to remind you on the BCS-model. Here we do not even have an automorphism but beautiful equilibrium states with many properties we want. This might encourage us to stay in one representation but as a consequence of our considerations we have noticed that an essential property of thermodynamical behaviour is the change of a representation in the limit $t \to \infty$. Therefore algebraic methods seem to me to be a necessary tool to get some insight in statistical mechanics.

ACKNOWLEDGMENT

The author wants to express her gratitude to Professor W. Thirring for many helpful and critical discussions on this subject.

REFERENCES

1. D. Ruelle, Statistical Mechanics, Benjamin 1969.
2. J. Dixmier, Les Algèbres d'Operateurs dans l'Espace Hilbertien, Gauthier Villars 1957.
3. J. Dixmier, Les C^*-Algèbres, Gauthier Villars 1964.
4. G. Emch, Algebraic Methods in Statistical Mechanics and Quantum Field Theory, New York 1972.
5. H. Narnhofer, Acta Phys. Austr. 31, 349 (1970).
6. W. Thirring and A. Wehrl, Comm. Math. Phys. 4, 303 (1967).
7. W. Thirring, Comm. Math. Phys. 7, 181 (1968).
8. F. Jelinek, Comm. Math. Phys. 9, 169 (1968).

9. W. Thirring in L.M. Garrido, A. Cruz, T.W. Preist, The Many Body Problem, Plenum Press, London 1969.

10. D. Dubin, G. L. Sewell, J. Math. Phys. 11, 2990 (1970).

11. D. Ruelle, Helv. Phys. Acta 45, 215 (1972).

12. H. Narnhofer, Acta Phys. Austr. 36, 217 (1972).

13. O. E. Lanford, D. W. Robinson, Comm. Math. Phys. 24, 193 (1972).

14. H. Narnhofer, W. Thirring and R. Sexl, Ann. of Phys. 57/2, 351 (1970).

15. J. Manuceau, F. Rocca, D. Testard, Comm. Math. Phys. 12, 43 (1969).

16. C. Radin, J. Math. Phys. 11, 2945 (1970).

17. H. Araki, H. Mijata, Publ. Res. Inst. Math. Sci., Kyoto Univ., ser. A 4, 373 (1968).

18. D. W. Robinson, preprint, Marseille 1972.

19. H. Porta, J. T. Schwartz, Comm. on Pure and Appl. Math., 20, 457 (1967).

20. K. Hepp, Helv. Phys. Acta 45, 237 (1972).

21. A. Lôpez, Z. f. Physik 192/1, 63 (1966).

22. E. Presutti, E. Scacciatelli, G. L. Sewell, F. Wanderlingh, J. Math. Phys. 13, 1085 (1972).

23. C. Radin, Comm. Math. Phys. 21, 291 (1971).

LIST OF SYMBOLS

A, B	algebra
A_Λ	local algebra
$A_{\Lambda\perp} = \overline{\bigcup_{M\cap\Lambda=\Phi} A_M}$	norm closure of the union
τ_t, τ_x, τ_t^P	automorphisms
E	states over A
$E \cap L_{G\perp}$	states invariant under G-translations
ρ, ω, σ	states
Φ_β	Gibbs state
Ω, Ω_ρ, ϕ_i	vector
π, π_ρ	representation
$[\ ]$	commutator
$[\ ]_+$	anticommutator
u-lim	limit in norm
$\rightarrow$	converges strongly
$\rightharpoonup$	converges weakly
η	invariant mean
$\exists$	it exists
$\exists !$	there exists exactly one
$A^\dagger$	adjoint
f^*, $\bar{\lambda}$	complex conjugate
$\hat{f}(\xi)$	Fourier transform

Acta Physica Austriaca, Suppl.XI, 565–595 (1973)

MAGNETIC CHARGES AND STRINGS AS BUILDING BLOCKS OF HADRONS - CONNECTION BETWEEN ELECTROMAGNETIC AND DUAL STRINGS*

BY

A.O. BARUT†
International Centre for Theoretical Physics,
Trieste, Italy

I. INTRODUCTION

The dynamics of particles having both electric and magnetic charges leads to infinite multiplets which can be used to describe the structure of hadrons. Both the infinite multiplets and the magnetic charges have close mathematical and physical connections to the spectrum of the dual models, to the underlying string models and to saturation of current algebras.

The purpose of this investigation is to point out these connections and to show how one can obtain, starting from a very simple physical model involving magnetic charges, the internal dynamical properties of hadrons, in particular of the proton, such as the form

---

*Seminar given at XII. Internationale Universitätswochen für Kernphysik, Schladming, February 5 - 17, 1973.

†On leave of absence from the University of Colorado, Boulder, Colo., USA.

factors (elastic and inelastic), mass spectra and inelastic structure functions.

I shall assume familiarity with the general concept and classical properties of magnetic charges. I should like to describe only four recent results about the physics of magnetic charges:

a) electromagnetic origin of spin,
b) parity and superselection rule,
c) group structure and exact solutions,
d) string between the magnetic charges.

From the group structure of the composite systems built from the magnetic charges, a connection can be established to infinite multiplets which in turn are used to describe the structure of hadrons. I shall describe quantitative results about form factors, mass spectra and inelastic structure functions.

II. ELECTROMAGNETIC ORIGIN OF SPIN

The bound state problem of two spin-zero magnetic charges provides a new mechanism to obtain a half-odd integer spin from the electromagnetic field. The electromagnetic field produced by two oppositely charged magnetic charges is singular along a string connecting the two charges. Although the exact position of the string is immaterial up to suitable performed gauge transformations (to be explained below), there is energy momentum and angular momentum carried by the singular line. Thus the string acquires a physical meaning and carries, at the end points, electric charges and masses. The existence of electric charges is essential for the occurrence of half-odd integer spins. This is also connected with the topological properties of the space and string.

We begin with the following facts:

a) The electromagnetic field produced by a single magnetic charge g at P is the same as that of a string of infinitesimal magnets from P to infinity (i.e. an infinitely thin solenoid).

b) The contribution of a closed singularity line to the vector potential is gradient of a scalar Λ determined up to an additive multiple of $4\pi g$ $(\Lambda + 4\pi gn)$. For such a change of the string the wave function undergoes a gauge transformation $e^{i\Lambda}$.[1]

c) In order, therefore, to make the physics independent of the position of the string (not its existence), we have to compactify the space R^3. The simplest compactification is given by the map

$$R^3 \longrightarrow S^3 \quad .$$

Thus the proper and necessary framework to describe magnetic charges is the space S^3. Physically, this means that the opposite magnetic charge is situated at one point, i.e. at the north pole N of S^3.

d) If we now have two opposite magnetic charges +g and -g, we can connect their strings to N in such a way that the non-vanishing part lies between +g and -g only. Then any deformation of the string corresponds to a gauge transformation.

For a test charge (e_1, g_1) in the field of another

charge (e_2, g_2) situated at the origin, the non-relativistic Hamiltonian has the form

$$H = \frac{1}{2m} (\vec{p} - \mu D(\hat{n},\vec{r}))^2 - \frac{\alpha}{r} \quad . \tag{1}$$

Here α and μ are charge products invariant under rotations in the two-dimensional charge space (e, g):

$$\begin{aligned} \alpha &= e_1 e_2 + g_1 g_2 \\ \mu &= e_1 g_2 - e_2 g_1 \quad . \end{aligned} \tag{2}$$

The vector potential $D(\hat{n},\vec{r})$ is given by

$$\vec{D} = \frac{\vec{r} \times \hat{n} \; (\hat{r}\cdot\hat{n})}{r(1 - (\hat{r}\cdot\hat{n})^2)} \tag{3}$$

and is singular along the direction $\hat{n}$ (string).

The conserved angular momentum operators are

$$\vec{J} = \vec{r} \times (\vec{p} - \mu\vec{D}) \tag{4}$$

and commute with the Hamiltonian (1). The rotational invariance of the system (1) to (4) is guaranteed only if we formulate the problem in S^3. This is because two different choices of the vector $\hat{n}$ are not compensated by a gauge transformation unless the space R^3 is compactified to S^3. In other words, in R^3 it is easy to see from (3) that the difference

$$\vec{D}(\hat{n}_1) - \vec{D}(\hat{n}_2)$$

cannot be written as $\nabla\Lambda(\hat{n}_1,\hat{n}_2)$. On general grounds, in S^3 the choice of $\hat{n}$ can always be compensated by a gauge transformation.

It is well known that the angular momentum operator for a spinless particle can have only integer values when represented on R^3. However, the space S^3 is the group space of SU(2) and it is also known, perhaps not so well, that angular momentum represented on S^3 allows half-odd integer values. Thus, both the proper rotational invariance and the occurrence of half-odd integer values of spin is justified by the compactification $R^3 \to S^3$. That spin and extended invariant structures occur in sufficiently non-linear theories of bosons and that these are connected with the global topological structures of space is known on general grounds; these phenomena are sometimes called kinks[2]. Thus, the magnetic charge model seems to give an explicit realization.

Mathematical details

S^3 can be parametrized by 4 co-ordinates (or fields) U_i such that (Fig.1)

$$\sum U_i^2 = 1 \ .$$

We consider the map $F : R^3 \to S^3$ with the boundary condition

$$f(\vec{x}) \longrightarrow (0001) \equiv N, \quad \text{as } |\vec{x}| \to \infty \quad .$$

Let F be the set of all such maps: $F = \{f(x)\}$. The set

F may be divided into a collection of pathwise connected components $F_1, F_2, F_3, \ldots$ called homotopy classes. Members in any homotopy class can be connected by a path. The collection of homotopy classes form the third homotopy group of S^3 (because of R^3); $\pi_3(S^3,N) \equiv \pi_3(S^3)$ (it is independent of the point N). The order of the homotopy group is a characteristic invariant feature of the map.

It is known that $\pi_3(S^3) = Z$ = additive group of integers. (In general $\pi_n(S^n) = Z$.) Thus each homotopy class may be characterized by an integer F_{-2}, F_{-1}, F_0, F_1, $F_2, \ldots$ (This integer is interpreted physically as the number of particles or kinks.)[3]

The existence of spin is related to the first homotopy group of SO(n):

n	$\pi_1(SO(n))$	
1	0	
2	Z	SO(2) ∞-connected
≥3	Z_2	SO(≥3) doubly connected.

III. QUANTIZATION CONDITION

The value of the magnetic charge g is determined by the value e of the electric charge or vice versa. The generalized electrodynamics has only one fundamental coupling constant. There are several ways to derive the relation between the two charges. The angular momentum quantization gives from Eq. (4),

$$e_1 g_2 - e_2 g_1 = \mu = 0, \pm \frac{1}{2}, \pm 1, \ldots \qquad (5)$$

As we have seen, both integer and half-odd integer representations of SU(2) occur and all these representations are integrable.

The quantization condition (5) can also be related to the quantization of magnetic flux. For a single pole (with the singularity line going to infinity), or for two opposite magnetic charges, the flux in the string is $4\pi g$ and is not detectable when it is quantized in units of $\frac{hc}{e}$. This gives $4\pi g = n\,\frac{hc}{e}$ or $\frac{eg}{\hbar c} = \frac{n}{2}$.

IV. CONSERVATION OF PARITY

Having discussed the rotational invariance of the theory, we now come to invariance under parity. Magnetic charge is an axial charge. Hence, if we had eigenstates of magnetic charge $|g\rangle$, we would have under parity

$$P|g\rangle = |-g\rangle \quad . \tag{6}$$

The Hamiltonian (1) violates parity if the parameter is treated as a number. The way to obtain a parity conserving theory is to double the Hilbert space and allow, together with states $|g\rangle$, also states $|-g\rangle$.[4] In this case the parameter μ can be viewed as a dichotomic variable and the Hamiltonian becomes

$$H = \begin{bmatrix} H(\mu) & 0 \\ 0 & H(-\mu) \end{bmatrix} \quad , \tag{7}$$

with the parity operator

$$P = \begin{pmatrix} 0 & 1 \\ 1 & 0 \end{pmatrix} . \qquad (8)$$

The diagonalization of this system leads to parity eigenstates

$$A_{\pm} = |g> \pm |-g> \qquad (9)$$

and the fact that the eigenvalues of H depend only on g^2.

The expectation valuesof the magnetic charge in parity eigenstates vanish. Some time ago I made the hypothesis that magnetic charges occur in parity eigenstates and not in eigenstates of magnetic charge,[4] contrary to the direct extrapolation from the classical theory. There are of course many other examples in physics where a charge need not be in diagonalized form, e.g. hypercharge for neutral kaons

$$|K_0> + |\bar{K}_0> = |K_{1,2}>$$

i.e. neutral K mesons occur in superpositions of positive and negative hypercharge.

V. SUPERSELECTION RULE

For the electric charge the following situation occurs. If we <u>assume</u> that we have eigenstates of the electric charge (according to obvious experience), then one can show that the superpositions

$$|e\rangle \pm |-e\rangle$$

lead, in external electromagnetic fields, to an undetermined relative phase between the two states. Consequently such superpositions do not belong to a coherent subspace (or sector) of the Hilbert space. One says that there exists a superselection rule for the electric charge. Such superposition states are not realizable, in the sense that we cannot measure in a Stern-Gerlach type experiment the relative amplitudes of each component, i.e. split the beam $a|e\rangle \pm b|-e\rangle$ in an electric field into $|e\rangle$ and $|-e\rangle$.

In the case of the magnetic charge, having postulated the coherent states $A_{\pm} = |g\rangle \pm |-g\rangle$ (in agreement with the fact that states $|g\rangle$ and $|-g\rangle$ are not seen), the question arises whether we can separate these states in an external magnetic field into its components. The answer is no.[5] Notice that in the case of electric charge there was the tacit assumption that eigenstates $|e\rangle$, $|-e\rangle$ exist. Now our initial hypothesis is that the states $A_{\pm} = |g\rangle \pm |-g\rangle$ exist. In Maxwell's equations we must treat, as we stated above, g as a 2×2 matrix. For example,

$$\text{div } B = \begin{pmatrix} g & \\ & -g \end{pmatrix} .$$

Now we apply the argument with the external field to the states

$$|A_+\rangle \pm |A_-\rangle \quad .$$

There are undetermined phases between $|A_+\rangle$ and $|A_-\rangle$ in an external electromagnetic field. Hence such superpositions are not realizable and, consequently, we cannot obtain pure states $|g\rangle$ or $|-g\rangle$ by superposing $|A_+\rangle$ with $|A_-\rangle$.

Mathematical details

The existence of superselection relations can be traced to the situation when the identity element of a group is represented quantum-mechanically not by 1 but by a phase. The well-known examples are the superselection rules for mass in the Galilei group and the electric charge. In the former case, the commutation relations between momenta P_i and Galilei boosters M_{0j}, instead of giving zero as might be expected geometrically, is

$$[P_i, M_{0j}] = -M\delta_{ij} \ , \qquad (10)$$

where M is the mass operator. We can then find a combination of boosts and translations equivalent to identity, $B_{-\vec{v}}\, T_{\vec{a}}\, B_{\vec{v}}\, T_{\vec{a}} = I$, which, however, is represented on the space of wave functions by the phase $e^{-im\vec{a}\cdot\vec{v}}$, as a result of (10). Consequently the relative phase between two eigenstates of mass in the superposition $|m_1\rangle + |m_2\rangle$ is undetermined.

In the case of the electric charge, the commutation relations of the translation generators π_μ of a particle in the presence of an external c-number field $F_{\mu\nu}$ are

$$[\pi_\mu, \pi_\nu] = -i\, F_{\mu\nu}\, Q \ , \qquad (11)$$

where $\pi_\mu = p_\mu - e A_\mu$, $F_{\mu\nu} = A_{\nu,\mu} - A_{\mu,\nu}$ and Q is the charge operator. The identity element $T_a\ T_b\ T_{-a}\ T_{-b}$ is again represented by a phase $e^{i\phi Q}$, where $\phi = \phi(F_{\mu\nu})$ is a function of $F_{\mu\nu}$. Thus, if we have eigenstates of Q, it follows that the relative phase in the superposition $|e_1> \pm |e_2>$ is not observable. In the case of the magnetic charge, we write Eq. (11) as follows:

$$[\pi_\mu, \pi_\nu] = -i\ F^*_{\mu\nu}\ A\ , \qquad (12)$$

where $A = UQ^*U$, Q^* being the magnetic charge operator and U is the 2×2 matrix such that the eigenvalues of A are $|g> \pm |-g>$.

According to our hypothesis, there exist only eigenstates of A. It follows then from (12) that the superpositions

$$|A_+> + |A_-> = |g>$$

$$|A_+> - |A_-> = |-g>$$

are not realizable. Consequently, the splitting of the A states in external fields is not measurable.

VI. GROUP STRUCTURE

The group describing the spectrum of the atom formed from two dyons (called dyonium) is SO(4,2), the same as the ordinary H atom, except that a new representation characterized by a number μ is realized. This μ is the same as the invariant μ in Eqs. (1) and

(2)[6]. The representation for the H atom has $\mu = 0$. For each μ we get a different dyonium. For the proton tower we shall take $\mu = \frac{1}{2}$ and for the π^0 tower $\mu = 0$. The weight diagram is shown in Fig.2. The invariant μ also determines the lowest spin $|\mu| = j_{min}$.

Exact solutions

The Hamiltonian (1) as well as the corresponding Hamiltonians in the Klein-Gordon and Dirac form are exactly soluble. One indeed exhibits explicitly the spinor form of the wave functions which are also exactly soluble.[6] There are other exactly soluble models with spin and O(4) symmetry, which are obtained from the Dirac Hamiltonian by approximation to order α^2.[7]

VII. CHARGE-SPIN-STABILITY RELATIONS

We take dyonium states to be magnetically neutral. Hence we have the two relations between the charges,

$$g_1 + g_2 = 0$$

$$e_1 g_2 - e_2 g_1 = j_{min} \cdot$$

Thus, for mesons ($j_{min} = 0$), $e = e_1 + e_2$ must be zero. For $j = \frac{1}{2}$ (baryons), $e = e_1 + e_2 \neq 0$. This is in agreement with the lowest states π^0 and p, for mesons and baryons, respectively. The "π^0 state" must then consist of dyon-antidyon (with 2γ decay), but the proton state of two different dyons ((e,g) and (0,-g), for example).[8]

VIII. THE ELECTROMAGNETIC STRING

Fig. 3 shows the two dyons moving round each other with a singularity line between them which then sweeps a two-dimensional world sheet $z_\mu(\sigma,\tau)$. The electric charge is at one of the end points and is responsible for the total spin of the system. There are also masses at the end points.

The variables of the singularity sheet must be introduced into the action principle as dynamical variables, or as gauge co-ordinates, or as gauge fields. The electromagnetic field outside the sheet is regular, but we have another part of the field on the sheet which is a distribution. We may distinguish between the orbital motion of the end points with the singularity line being in the "ground state" (for electromagnetic interactions), plus the excited states of the string.

As mentioned above, the electromagnetic potentials must be singular in the presence of magnetic charges, otherwise, from $F_{\mu\nu} = A_{\nu,\mu} - A_{\mu,\nu}$ it would follow that $\tilde{F}_{\mu\nu}{}^{,\nu} = 0$, i.e. no magnetic charge. One possibility to exhibit the singularity is to separate $F_{\mu\nu}$ into a regular and a singular part:[9]

$$F_{\mu\nu} = A_{\nu,\mu} - A_{\mu,\nu} + \lambda_{\mu\nu} \quad , \tag{13}$$

where $\lambda_{\mu\nu}$ is different from zero only on the singularity sheet

$$\lambda_{\mu\nu}(x) = \varepsilon_{\mu\nu\lambda\rho} \quad g \int d\sigma \; d\tau \; \delta^4(x-z) \; [\frac{\partial z^\lambda}{\partial\tau} \frac{\partial z^\rho}{d\sigma} - \frac{\partial z^\lambda}{\partial\sigma} \frac{\partial z^\rho}{\partial\tau}] \; . \tag{14}$$

It follows from (13) and (14) that

$$\tilde{F}_{\mu\nu}{}^{,\nu} = \tilde{\lambda}_{\mu\nu}{}^{,\nu} = g \int ds\ \delta(x-z)\ \frac{dz_\mu}{ds}\ , \qquad (15)$$

where we have put $z_\mu(\tau,0) = z_\mu(s)$ (end point of the string), i.e. we have a magnetic charge as a source with the world line z_μ .

There exists an action principle for interacting electric and magnetic charges, which can be written in various forms. A symmetric form is[10]

$$L = \sum m\int ds+\tfrac{1}{2}F_{\mu\nu}\ F^{\mu\nu} - F_{\mu\nu}\ A^{[\nu,\mu]}+\tfrac{1}{2}G_{\mu\nu}\ G^{\mu\nu} - G_{\mu\nu}\ B^{[\nu,\mu]}+$$

$$+\ j_\mu\ A^\mu + k_\mu\ B^\mu + \lambda_{\mu\nu}[F^{\mu\nu} + \varepsilon^{\mu\nu\lambda\sigma}\ G_{\lambda\sigma}]\ , \qquad (16)$$

where $A^{[\nu,\mu]} = A_{\nu,\mu} - A_{\mu,\nu}$, etc., and we have introduced two fields, $F_{\mu\nu}$ and $G^{\mu\nu}$, potentials A^μ, B^μ and Lagrange multipliers $\lambda_{\mu\nu}$. From (16) we obtain

$$F_{\mu\nu} = A_{[\nu,\mu]} - \lambda_{\mu\nu} \qquad \text{(i.e. Eq. (13))},$$

$$G_{\mu\nu} = B_{[\nu,\mu]} - \varepsilon_{\mu\nu\lambda\sigma}\ \lambda^{\lambda\sigma}$$

$$F^{\nu\mu}{}_{,\nu} = j^\mu$$

$$G^{\mu\nu}{}_{,\nu} = k^\mu$$

$$F^{\mu\nu} = -\ \varepsilon^{\mu\nu\lambda\sigma}\ G_{\lambda\sigma}\ . \qquad (17)$$

We see that through $\lambda_{\mu\nu}(x)$ the string variables enter into the theory as dynamical variables. Dirac considers the string variables as "unphysical" and tries to eliminate them at the end. However, this is not entirely possible, because at the end he arrives at a boundary condition that charges can never cross the strings, showing that strings cannot be completely eliminated. Furthermore, because field quantities (consequently the energy momentum and spin densities) exist on the string, we must recognize the dynamical role of the string. The dynamical role of the string variables is restricted by the freedom of choosing the string variables, i.e. gauge transformations, but cannot be erased completely. All these aspects are in remarkably close correspondence with the theory of the relativistic string.[11]

IX. INFINITE-COMPONENT WAVE EQUATIONS FOR NUCLEONS

Contact with phenomenology is established by postulating a relativistic wave equation based on the spectrum and dynamical group of the problem (see Fig.2).

The solutions of the wave equation are equivalent to Bethe-Salpeter amplitudes in the case of a two-particle structure. The wave function is a c-number non-perturbative solution (already including all internal diagrams) of internal dynamics and the system couples minimally to the external electromagnetic field. It is of the form

$$\psi_{njm}(P)\ , \quad \text{in a discrete basis}$$

or

$$\psi(q,P)\ , \quad \text{in a continuous basis,}$$

where P is the total momentum of the system, and njm or

q are internal variables. The wave equation is modeled after the Dirac equation and the relativistic H atom. For the latter the equations are

$$(\gamma_\mu P^\mu - K)\ \psi(P) = 0 \tag{18}$$

and

$$(J_\mu P^\mu - K)\ \psi(P) = 0,$$

$$J_\mu = \alpha_1 \Gamma_\mu + \alpha_2 P_\mu + \alpha_3 P_\mu \Gamma_4,$$

$$K = \beta \Gamma_4 + \gamma , \tag{19}$$

respectively.

Here Γ_μ, Γ_4 are the generators of the dynamical group SO(4,2) mentioned in Sec.VI. Note that the Lie algebra of the γ matrices and that of the Γ matrices are the same, only the dimensionality is different: 4 in the former case, infinity in the latter.

The general features (independent of the actual values of the parameters) of the description of the proton as an "atom" by the relativistic wave equation (19) are the following:

a) An increasing mass spectrum linear in n for small values of n (Fig.4).

b) Elastic form factors of the form

$$G^M(t) = 1/(1 + at)^{2|\mu|+1}, \quad |\mu| = 0, \tfrac{1}{2}, 1, \ldots \tag{20}$$

where $|\mu|$ is the lowest spin, as before, characterizing the O(4,2) representation.

This equation selects the particular dynamical group and its representation. Note that for wave equations using the dynamical group SO(3,1) the form factors behave like $1/(1 + at)^{3/2}$.

c) The existence of a new quantum number n (the analogue of the principal quantum number) which distinguishes states with the same spin and parity, e.g., $\frac{1}{2}^+(N)$, $\frac{1}{2}^+(1470)$, $\frac{1}{2}^+(1780)$.

d) Approximately the same behaviour of all inelastic form factors given by $\langle n,p|J_\mu|n',p\rangle$.

For more quantitative results we must fit the parameters α_i, β, γ. This has been done in the literature with the addition of the current $J'_\mu = i\alpha_4 L_{\mu\nu} q^\nu$ which is automatically conserved. What is important is that once these parameters are fitted by normalizations and ground state properties, the complete functional forms of the mass spectrum and form factors are predicted.[12] We also obtain a relationship between the slope of the mass spectrum (i.e. the so-called Regge slope parameter) and the slope of the form factor, as might be expected from an atomic model.

Structure of the proton

For static properties of the proton one might use a Dirac equation and treat it as a spin-$\frac{1}{2}$ elementary particle. What we want to show here is that in a large region of energy and momentum transfer we can describe the proton as a relativistic composite particle, as an atom, with all its excited states forming an entity. This picture too will probably have to be changed at very high energies when pair production and other phenomena will become important. For example, an H atom ceases to be an H atom, when pair production becomes dominant. We now apply this picture to determine the inelastic structure functions of the proton. In the inelastic electron proton scattering, we may assume that the proton will be excited (including discrete and continuous spectrum) and then will decay producing any number of particles (Fig.5). The importance of the wave function in this problem lies in the fact that we can treat all the intermediate states n (including the continuum) as single-particle states, i.e. as solutions of the wave equation.

The relevant formulae are as follows. The two structure functions W_1 and W_2 are related to the components of the imaginary part of the Compton amplitude $W_{\mu\nu}$ by

$$W_1 = W_{11}$$

$$W_2 = -\frac{M^2}{p_3^2}\frac{(1 + p_3^2/q^2)}{(1 + M_\nu/q^2)} W_{11} + \frac{M^2}{p_3^2}\frac{W_{33}}{(1 + M_\nu/q^2)} . \tag{21}$$

Here q^2 and ν are the usual invariants and p_3 the momentum of the nucleon in the rest frame of the intermediate state n. The Compton amplitude in turn is given by

$$W_{11} = \frac{\pi}{|N|^2}\sum_n |M_1(n)|^2 \delta(s - M_n^2) . \tag{22}$$

where N is a known normalization constant and

$$M_i(n) = \langle n| e^{i\Theta_n T} \; e^{-i\vec{\xi}\cdot\vec{M}} \; J_i \; e^{-i\Theta_i T} |1\rangle \qquad (23)$$

are the form factors between the ground state $|1\rangle$ and the state $|n\rangle$; T and M are the generators of SO(4,2), J_μ the current in the wave equation (19) and $\vec{\xi}$ the Lorentz boost parameter.

The so-called scaling limit corresponds to

$$M_n \to \infty , \qquad n \to 0 \text{ (continuum)} .$$

In this limit the structure functions W_1 and W_2 do become functions of only the scaling variable $\xi = -\frac{q^2}{2M\nu}$. In this limit we can drop terms containing n alone, $n^2p_3^2$, etc., and keep only terms of the type np_3^2, np_3q_0 and nq_0^2. The final form of the structure functions in the scaling limit is given by

$$F_1(\xi) = \frac{M}{4}\, \alpha_4^2 \cosh^2\Theta_1 \; f_1(\xi) \; (1-\xi)^4$$

$$F_2(\xi) = 2M\xi \; [\tfrac{1}{4}\, \alpha_4^2 \cosh^2\Theta_1 \; f_1(\xi)(1-\xi)^4 + 4f_2(\xi)\frac{\xi}{(2\xi-1)^2}(1-\xi)^3] ,$$

(24)

where $f_1(\xi)$ and $f_2(\xi)$ are definite functions given elsewhere. The results (24) incorporate the Drell-Yan rule,

$$F_2(\xi) \sim (1-\xi)^3 \text{ at threshold,} \qquad (25)$$

and the Callan-Gross rule,

$$F_2(\xi) \simeq 2\xi \; F_1(\xi) \;, \quad \text{if} \quad f_2 \ll f_1 \;. \qquad (26)$$

The terms with $(1-\xi)^3$ and $(1-\xi)^4$ are precisely the phenomenological terms used to fit the experimental data.

In order to understand the intuitive meaning of the structure function, we show schematically in Fig.6, the behaviour of $F_2(\xi)$ for several different structures.[13] For an elementary particle, $F_2(\xi)$ is a δ-function situated at $\xi = 1$. On the other end of the scale, the H atom is a loosely bound system whose structure function is peaked forward, very high and narrow around $\xi = 0$. The Majorana type infinite multiplets have some structure, but are close to the EP. For this system the internal motion is two-dimensional and there is no principal quantum number n. Finally the proton seems to be an intermediate structure between H atom and Majorana particles.

The theory can be fitted to the experiment, except perhaps at very high energies and momentum transfers, i.e. $\xi \to 0$.[14] In this limit, the wave function gives $F_2(\xi) \to 0$. If the experiment shows a non-vanishing $F_2(0)$ at $\xi \to 0$, as it seems to do, this is an indication that processes other than those indicated in Fig. 5 are coming into play, such as the so-called diffractive processes.

Up to this limit of diffractive processes, however, we can describe all the internal properties of the proton in a consistent and unified way by the wave equation which then gives us an overall picture of the proton as an "atom".

Excitation of the string

So far we have considered by the wave equation the orbital excitations of the underlying dyonium model. Because the electric charge is at one end of the string, the electromagnetic field interacts locally and with one of the constituents of dyonium. This is in complete agreement with the fact that the wave equation is so constructed that the E-M field interacts only with one of the particles. For instance, in the H atom case, the wave equation describes the interactions of the electron; those of the proton must be considered separately. Thus, as far as electromagnetic interactions are concerned, the excitations of the string do not seem to come into play, at least at the present energies and, as we have seen, there are good indications that the proton behaves like a two-body system.

Because of the dynamical significance of the string, as we have discussed, the string excitations must also enter into the hadron-hadron interactions. It is here that we expect to see the real connection with the dual models.

ACKNOWLEDGMENTS

The author would like to thank Prof. Abdus Salam, the International Atomic Energy Agency and UNESCO for hospitality at the International Centre for Theoretical Physics, Trieste.

APPENDIX

BOHR PICTURE, QUARK PICTURE AND PARTON PICTURE FOR H ATOM, AND DUAL PICTURE FOR POSITRONIUM - POSITRONIUM SCATTERING

I owe an explanation as to why I consider a two-body "atomic" model for the proton besides the generally used quark and parton models. I would like to show, in the example of the H atom, that, in general, a composite structure admits several seemingly different pictures and an observed multiplet structure of levels does not necessarily imply that the corresponding "quarks" are the physical constituents.

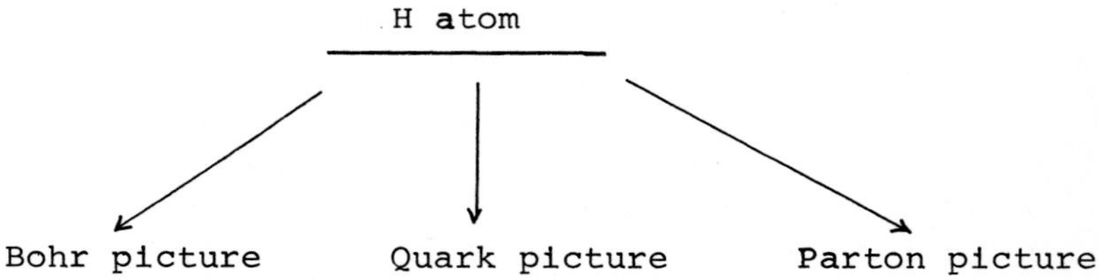

The Bohr picture consists of the physical constituents e and p interacting via the Coulomb potential. The quark picture of the atom results from the observed SO(4) symmetry of the levels. Every state of the atom may be constructed out of four creation operators a_i^+, i = 1,2,3,4, in the form

$$\frac{1}{N} a_1^{+n_1} a_2^{+n_2} a_3^{+n_3} a_4^{+n_4} |0\rangle \quad . \qquad (A.1)$$

These "quarks" carry spin $\frac{1}{2}$ but obey Bose statistics;

thus we have the same "spin statistic problem" as in the quark model of baryons. The four boson operators arise from the diagonalization of the Hamiltonian with interaction. We have, namely, the correspondence

$$\frac{p^2}{2m} - \frac{\alpha}{r} \leftrightarrow \sum_{i=1}^{4} (a_i^+ a_i + c) . \qquad \text{(A.2)}$$

The relation (A.2) is not an equality but a correspondence (the equality is more complicated; it is not necessary here to write it out explicitly). Thus the boson (or quark) excitations enter exactly as normal modes of the Coulomb system with the interaction already included. In (A.2) there are no potentials between the quarks. However, if we include quark potentials, we get a four-body problem which would be different from the two-body problem we are studying. One could, however, pose the problem as to what the potentials between the four quarks would be if we want to get the same states (A.1) from four quarks only, e.g. in the quark language without the so-called "exotic" states.

The parton picture of the H atom is obtained if we identify $|\psi_n(p_L)|^2$, where $\psi_n(p)$ is the wave function of the electron in momentum space (p_L = longitudinal momentum), with the probability function f(x) of finding the n-th parton with the longitudinal momentum fraction x. Indeed, the external photon interacts locally with the electron, and the inelastic structure functions of the atom calculated by the wave function picture and quark picture are the same.

It is also instructive to mention that there is even a dual picture in the interaction between to atoms. To take a symmetric case, consider the elastic scattering of two positroniums. The exchange of positronium states in the t channel would produce an amplitude with positronium poles in the s channel, or vice versa. The usual dual "quark" diagrams apply here to the e^+, e^- constituents of the positroniums.

REFERENCES

1. P. Jordan, Ann. Phys. (NY) 32, 66 (1938).
   B. Grönblom, Z. Physik 98, 283 (1935).
2. D. Finkelstein, J. Math. Phys. 7, 1218 (1965).
   D. Finkelstein and J. Rubinstein, J. Math. Phys. 9, 1762 (1968).

3) J. G. Williams, J. Math. Phys. 11, 2611 (1972).
   T.H.R. Skyrme, J. Math. Phys. 12, 1735 (1971).

4) A. O. Barut, Phys. Letters 38B, 97 (1972).

5) A. O. Barut, ICTP, Trieste, Internal Report IC/73/33.

6) A. O. Barut and G. Bornzin, J. Math. Phys. 12, 841 (1971).

7) A. O. Barut and G. Bornzin, to appear in Phys. Rev. May 1973.

8) A. O. Barut, in Structure of Matter, Proc. Rutherford Centenary Symposium, Ed. B. Wybourne (Univ. of Canterbury Press, 1972).

9) P. A. M. Dirac, Phys. Rev. 74, 817 (1948).

10) A. O. Barut and G. Bornzin, to be published.

11) T. Goto, Progr. Theoret. Phys. (Kyoto) 46, 1560 (1971).

L. N. Chang and F. Mansouri, Phys. Rev. D5, 2535 (1972).
F. Mansouri and Y. Nambu, Phys. Letters 39B, 375 (1972).
P. Goddard, J. Goldstone, C. Rebbi and C. B. Thorn, CERN preprint TH. 1578 (1972).
H. B. Nielsen and P. Oleson, Copenhagen preprint, 1972.

12. A. O. Barut, D. Corrigan and H. Kleinert, Phys. Rev. Letters 20, 167 (1968).
13. H. Bebié, V. Georgé and H. Leutwyler, Ann. Phys. (NY) 74, 524 (1972).
H. Kleinert, Nucl. Phys. B43, 301 (1972).
14. A. Baiquni and A. O. Barut, ICTP, Trieste, preprint IC/72/160.

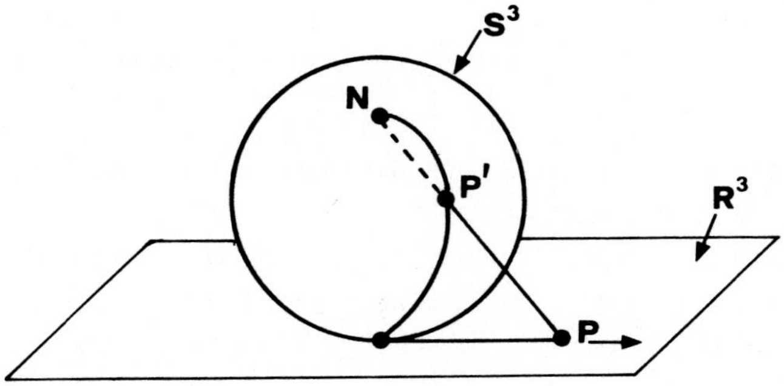

Fig.1

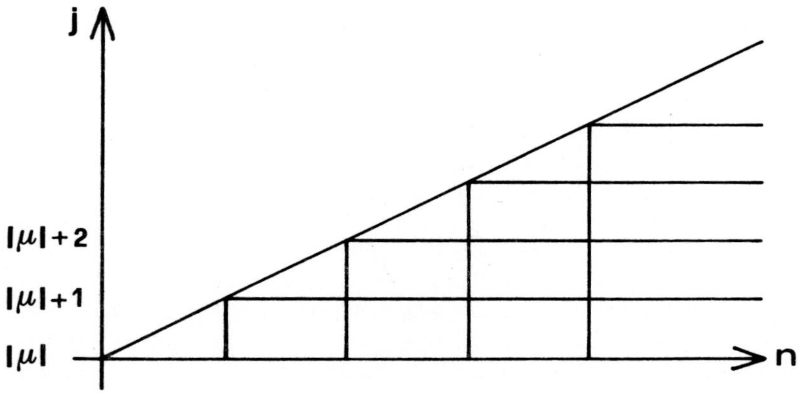

Fig. 2

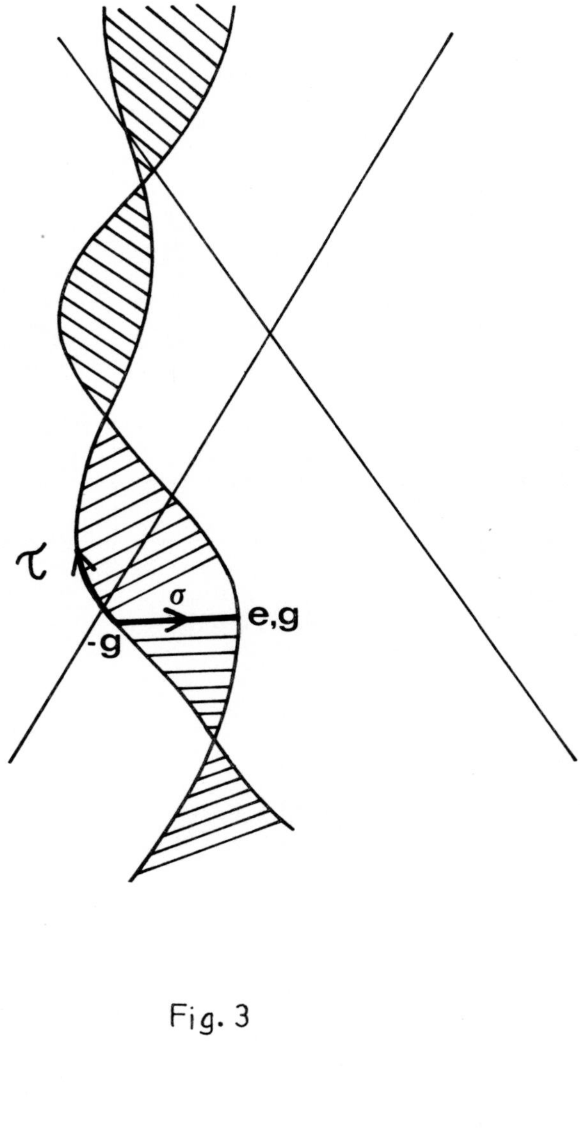

Fig. 3

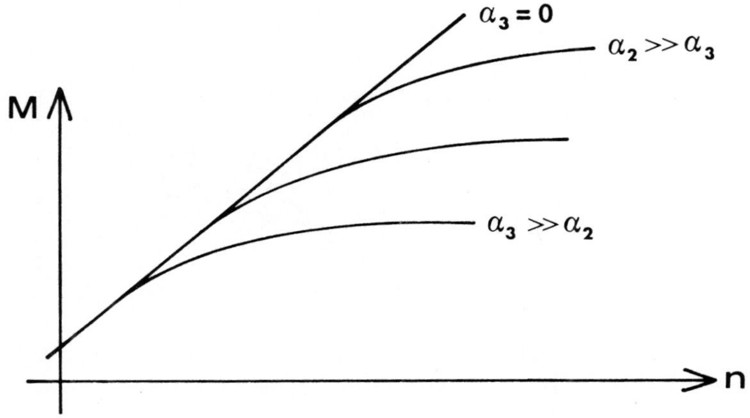

Fig. 4

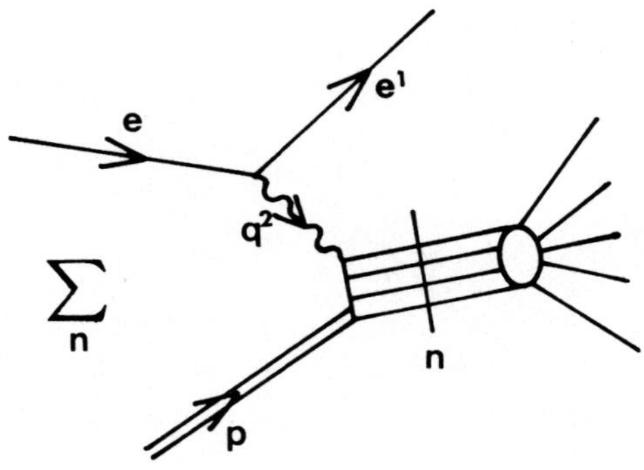

Fig. 5

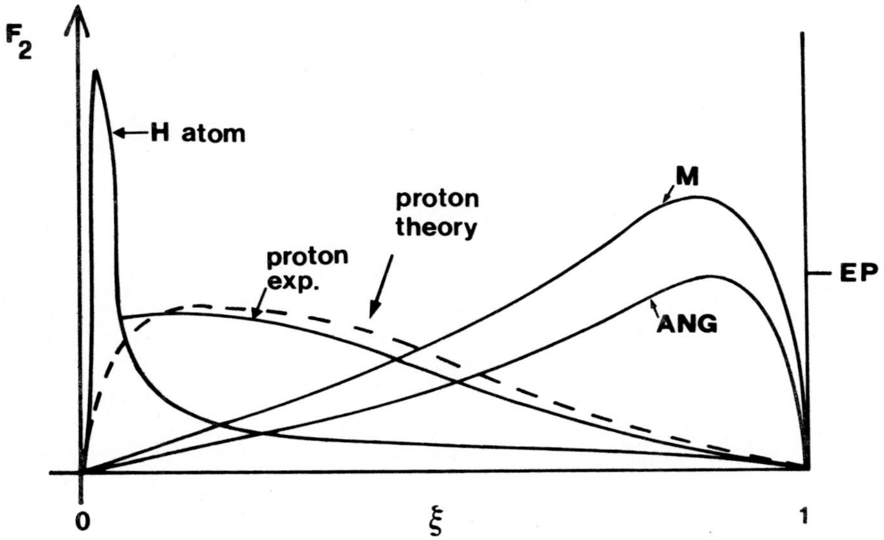

Fig. 6

Acta Physica Austriaca, Suppl.XI, 597–604 (1973)

RECENT DEVELOPMENTS IN MATHEMATICAL PHYSICS

SUMMARY - FIRST WEEK*

BY

L. STREIT† x
Institut für Theoretische Physik
Universität Graz

Theoretical physics, as one observes it in Schladming and at other similar schools, comes in three distinct phases. These phases can be exhibited diagrammatically by plotting the lecturers versus the various topics that they cover.

Mathematical physics in particular has been in the "ivory tower" phase (Fig.1) most of the time in the past decades.

The other predominant situation that one encounters is the "Hula hoop" phase (Fig.2), where "Hula hoop" stands for fashions like the Veneziano model, Regge poles, peratization and the like.

The transition from the "ivory tower" to the "Hula hoop" phase tends to occur spontaneously, pointing

---

*Summary given at XII. Internationale Universitätswochen für Kernphysik, Schladming, February 5 - 17, 1973.

†On leave of absence from Fakultät für Physik, Universität Bielefeld, Germany.

xDrawings by W. Plessas.

to a structural instability arising from the social conditions under which physics research is done.

Ideally the physics scene should be reflected in a plot as ridly populated as Fig.3. This one would call the strongly interacting or Ideal Phase, and there can be no doubt that at this point in time mathematical physics is more enjoyable than it has been for quite a while, with rapid progress on a broad front through a rigorous interaction of various approaches.

In particular, Euclidean quantum field theory presents a case of Sleeping Beauty not only kissed awake but turning violently promiscuous right on the bedside.

Barry Simon has shown us the fruit it has borne with constructive quantum field theory, most remarkably via the lattice approximation, the link between Ising model and constructive quantum field theory, with $(a\phi^4 + b\phi^2 - \mu\phi)_2$ emerging as a Wightman theory.

Another area of recent progress, and again one in which Euclidean quantum field theory has become an important tool, is the realm of massless theories.

The exact solution of such a theory was presented by Q. F. Dell'Antonio, while I. Todorov exploited systematically the information on Green's functions that emanates from conformal invariance.

Interest in scale invariant models stems from their possible occurrence as asymptotic theories. Such high momentum and zero-mass limits were discussed by K. Symanzik, as well as the problem of introducing masses perturbatively into massless theories.

One of the most pleasing aspects of the present

"Ideal Phase" of mathematical physics is how quantum field theory, i.e. dynamical concepts like anomalous dimensions etc. become accessible to and relevant for high energy experiments, as shown to us in H.Leutwyler's talk.

Loosely speaking the advantage of massless models in the present context is that they are more manageable because of their simpler structure. However, theories with zero mass particles also have extra complications through their energy spectrum. This is illustrated by the fact that constructive, i.e. Hamiltonian quantum field theory has a hard time with such models. It also means, as you all know, that one has to think harder about the concept of scattering. Simple models like e.g. the Pauli-Fierz model as further pursued by Ph. Blanchard show that the $t \to \pm \infty$ notion of scattering can be an over-idealization, linked to that of exact energy measurements. O. Steinmann shows that full-fledged quantum electrodynamics also has finite predictions for realistic, finite time measurements of charged particle scattering. One might feel stimulated to think about a link to the work of Kulish and Faddeev and on the practical side, to the usual cookbook prescriptions for the elimination of infrared divergences.

There were two talks in the first week that stick out from the rest. The first one is that of John Klauder, if only because unlike the others it was basically of a speculative nature.

His observation was that the free field measure $d\mu_F$ given e.g. by the characteristic functional

$$\int e^{i(f,\chi)} \, d\mu_F(\chi) = \langle e^{i\phi(f)} \rangle_F$$

picks up an extra factor

$$d\mu_g = e^{-gV(\phi)} \, d\mu_F \, .$$

So, formally, if $V(\phi) = \infty$ on a set A of non-zero μ_F-measure, we can throw in the characteristic function of A as a non-trivial factor

$$d\mu_g = e^{-gV(\phi)} \quad \chi_A(\phi) \, d\mu_F$$

which, however, does not go away in the limit $g \to 0$, with the result that

$$\langle e^{i\phi(f)} \rangle_o \quad \neq \quad \langle e^{i\phi(f)} \rangle_F$$

and the "pseudofree" $\langle\rangle_o$ presumably a better starting point for such theories than $\langle\rangle_F$.

To meet the challenge of this speculation I would like to point out that the ϕ^2-model provides an example of this situation if the cut-off function g in the interaction term

$$\int g(x) \; : \phi^2(x) : \; dx$$

is chosen not to be smooth but instead sufficiently singular. This model has the nice feature that it is exactly solvable and that the measures in question are fairly easy to control.

K. Hepp's lectures on phase transitions in open systems were rather more related to the topics dealt with in the second week. It is too bad that he was too modest to show us a photograph of his laser (Fig.4; there is also an infinite dog somewhere in the background).

Despite the couple of abstractions that went into the construction of this little engine, Hepp's contribution may well be considered as the one with the most far reaching implications. In the face of the tremendously small probability for the origin of life through a fluctuation in thermal equilibrium, one looks to phase transitions in open systems, i.e. far from thermal equilibrium, as a possible explanation. Clearly the first step here is to develop this concept with the help of simple models, and Klaus Hepp demonstrates how indeed spontaneous ordering can occur in open systems when the energy throughput increases to a critical level.

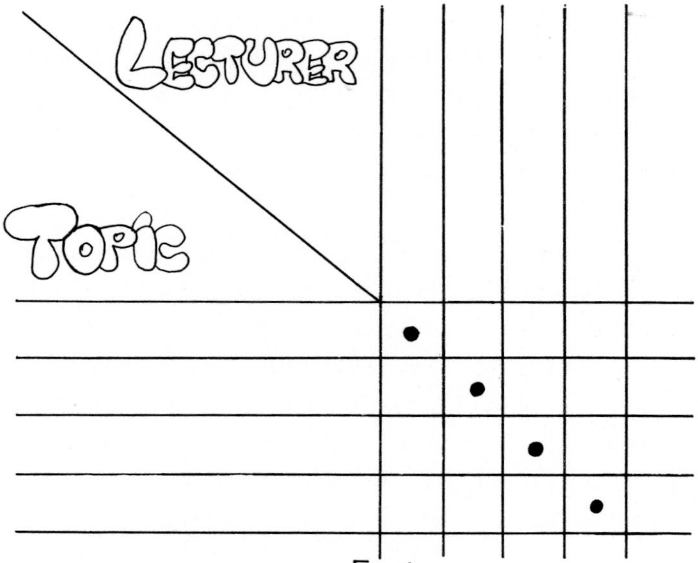

Fig.1

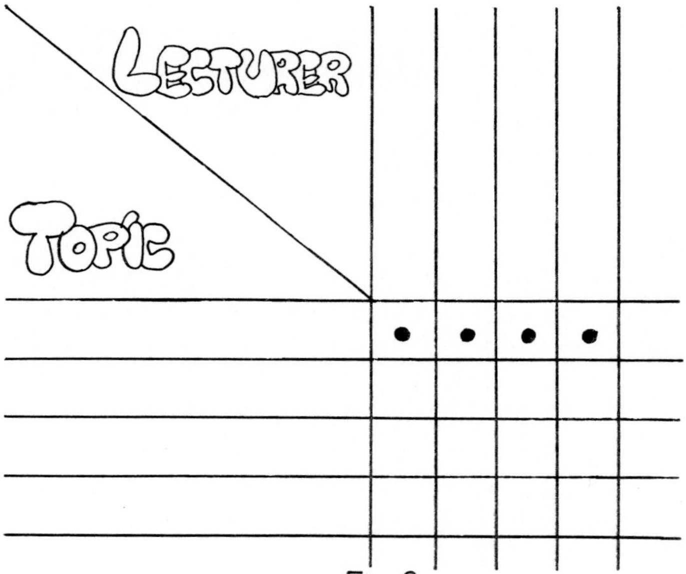

Fig. 2

TOPIC \ LECTURER	DELL'ANTONIO	KLAUDER	LEUTWYLER	SIMON	STEINMANN	SYMANZIK	TODOROV
ANOMALOUS DIM.	•	•	•			•	•
CCR REPRES.	•	•		•			
CURRENTS	•		•				•
DIL. + CONF.	•	•				•	•
EQFT	•			•		•	•
HIGH MOMENTA			•			•	•
LIGHT CONE	•		•				•
MASSLESS THEORIES	•	•			•	•	•
OP. PROD. EXP.	•		•			•	•
QFT + STAT.MECH.	•			•		•	•
RENORMALIZATION	•	•	•	•	•	•	•
s = 1 MODELS	•	•		•			

Fig. 3

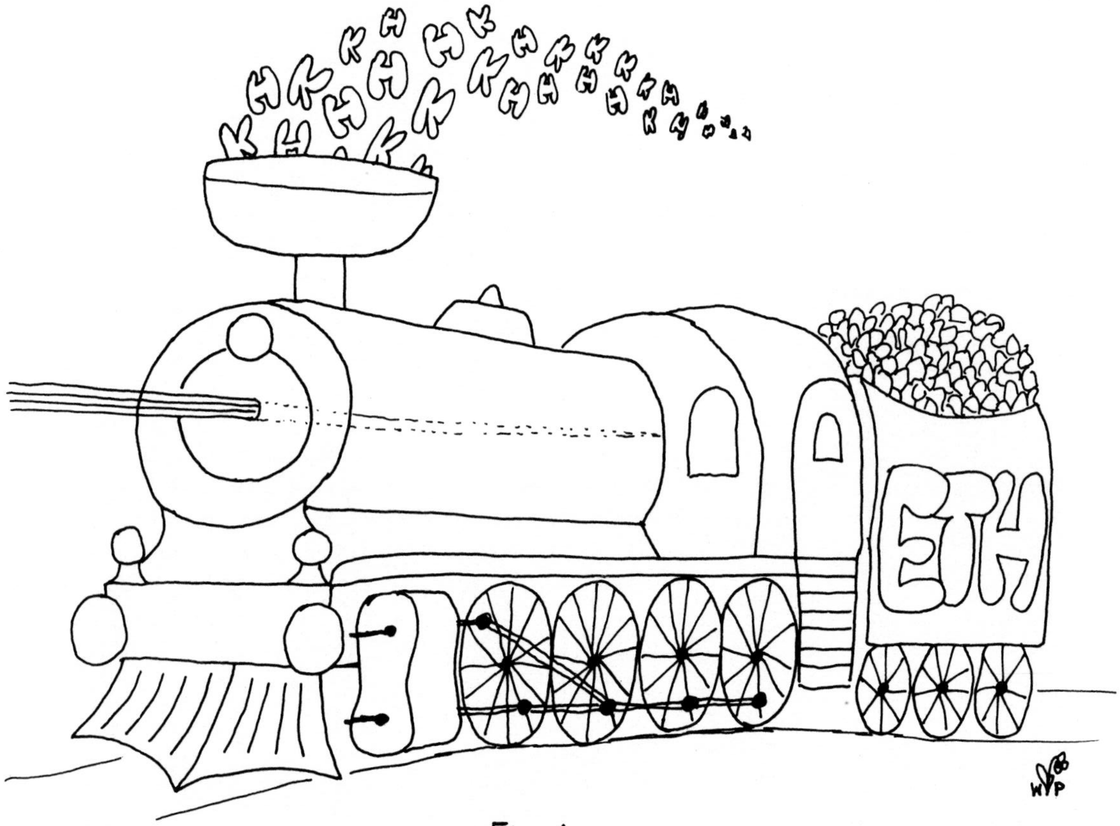

Fig. 4

The ETH Steam Laser

Acta Physica Austriaca, Suppl.XI, 605–610 (1973)

SUMMARY - SECOND WEEK*

BY

I. T. TODOROV
Physical Institute of the Bulgarian Academy
of Science, Sofia

I am for the first time in Schladming and it was, I should say, quite an experience. One thing that I discovered, perhaps too late, was the great persuasive power of Austrians. First, I was induced to go to the ski school, then I found myself at the bowling contest, and eventually I was pushed down the gigantic slalom. Well, you would say, that looks enough for a newcomer. But is was not all! I had to prepare in addition this summary. Persuading me to do it Ludwig Streit said that what I will say could be just some jokes, not necessarily related to the lectures at this school. So I will concentrate on things unrelated to the second week of the meeting.†

One should be careful in telling jokes in high energy physics. Some 10 years ago, at the heights of

---

*Summary given at XII. Internationale Universitätswochen für Kernphysik, Schladming, February 5 - 17, 1973.

†This should be considered as an excuse for all injustices I am going to do to the speakers of the second week.

the Regge pole fashion when theorists used to continue over discrete quantum numbers all over the place, someone was joking that we are going eventually to continue analytically the number 4 of space time dimensions. Well, now we heard from Kenneth Wilson that this is a rather serious matter.

Wilson's lectures were probably the most talked about event during the second week. People were going around and asking each other what did he mean by saying this and that. One should not get exasperated about it. After all Wilson told us that it took him some 5 years to make sense out of his talk in Schladming of 10 years ago.

One thing of Wilson's 1973 series which I believe, one should keep in mind, is his general view of the mountain chain of Lagrangeans. The idea was that it does not matter much where you start "skiing in the mountain" you will always end up on a stationary low point. Some of us were asking why should one make at all random walks in the set of Lagrangeans, using the cut-off as a vehicle, and look for a stationary point? It may be helpful to realize that looking for some "fixed points" in the space of theories could also be justified with arguments lying outside the renormalization group program. A Lagrangean, even if it is formally scale invariant, does not lead, in general, to scale invariant matrix elements because of renormalization. One can sum up some of the appearing logarithms into power thus defining the anomalous dimension of the field. Certain logarithms still remain, however, unless we choose the coupling constant equal to its critical value which satisfies the Gell-Mann-Low equation and the Callan-Symanzik equation. Then we will end up precisely with one of those "stationary theories" we heard here from Wilson.

A derivation of the so called pre-scaling equations, which include the renormalization group and the Callan-Symanzik equations, on the basis of the Zimmermann normal product algorithm was given by B. Schroer. He even dedicated his lectures to K. Wilson, although I am not sure that the two of them do agree on all points.

I had a hard time to tell in either Schroer or Streater lectures when they were taling about quantum field theory (QFT) and when - about statistical mechanics (SM). Fortunately, Streater gave us a clue: whenever you see a β in an exponent, it means that the speaker is concerned with SM. Schroer used the temperature $t = T - T_c$ instead, but the message was the same. As we also heard last week from B. Simon, in the Euclidean formulation QFT and SM differ only in the number of space dimensions, which we vary at will, anyway.

We also had a high-brow mathematical approach to the near equilibrium behaviour in the lectures of Heide Narnhofer.

I had the feeling that an introductory lecture to her, and may be also to Streater's series was missing; it should have explained in some detail the beautiful mathematical technique that Dr. Narnhofer was using all the way.

W. Thirring gave us a first-class example of how mathematical physics should look like. We learned from him how rigorous mathematics can apply to important, down-to-earth, physical problems. Let me remind you, for instance, about the infinite volume and domain subtleties involved in deducing Bogolubov type inequalities for $\frac{\partial^2 E}{\partial \alpha^2}$. Paradoxes of the type

$$0 < \mathrm{Tr}(e^{-\beta p^2}) = \frac{1}{i}\mathrm{Tr}[e^{-\beta p^2}(xp-px)] = 0 ,$$

due to unjustified formal manipulation with canonical commutation relations, exhibited by Prof. Thirring, are a warning that mathematical rigour should not be regarded as a superfluous ornament, when dealing with physical problems. It took a little more than 40 years after the foundations of quantum mechanics were set, before the first important work on the stability of matter was done. Would such a time interval prove sufficient to obtain equally fundamental results in the relativistic quantum theory? The answer to this question is in the hands of the young people.

The same type of problem - search for a dynamical theory of resonances - was attacked with rather different tools in the lectures of H. Joos and M. Krammer on one hand, and H. Lehmann on the other.

Joos and Krammer start with a Bethe-Salpeter (B-S) type of equation with a phenomenological potential for the quark-antiquark system. There are two questions which one may ask about this starting point. The first question is whether one should start with a 4-dimensional B-S type equation or work instead with a 3-dimensional quasipotential equation. Since I have worked on the 3-dimensional formulation of the relativistic 2-body problem, I am prejudiced on this point. I would rather agree with the remark of Eddington that "a proton today and an electron yesterday do not constitute a hydrogen atom". Joos, on the other hand, believes that the presence of the relative time variable in the equation plays an important role in the relativistic dynamics of a strongly bound system. It is nice that Joos and coworkers suggest an experimental check for this controversy. If they are right there will be three ρ-resonances around the same mass value; otherwise there will be at most two. So we shall wait and see. The second problem, which I had in mind, is concerned with the spin structure of the potential. As we learned from M. Krammer, all derivative couplings are neglected in the present study. Of course, one can hardly start with the most general expression, but I would better trust a potential coming from some symmetry principle like approximate chiral invariance. I enjoyed to hear the high-brow argument quoted by Joos using the general ideas of the algebra of observables, which tells us that one should be able to see fractional charges if there are quark fields.

Lehmann had more limited objectives: the description of the $\pi\pi$-system below 1 GeV including the ρ-resonance, but his conclusions were based on a firmer theoretical ground. It is instructive to know that the low-energy behaviour of the $\pi\pi$-system up to order p^4 is determined up to two real parameters from unitarity, ordinary dispersion relations, and chiral invariance. The calculation of the remaining two parameters, discussed so far, uses. however, a non-covariant approximation scheme for superpropagators. It looks preferable to try to work out, as Ecker and Honerkamp had attempted, a manifestly chiral invariant technique in the spirit of the work of Faddeev and Slavnov.

To summarize, here is a picture of what was going on in the lecture hall during the second week:

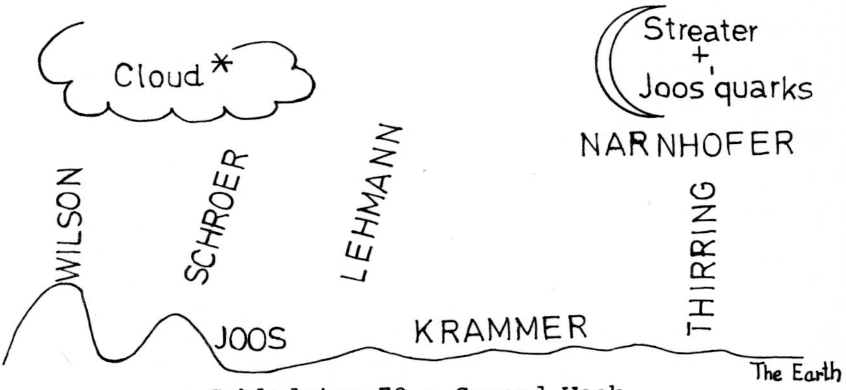

Schladming 73 - Second Week

The sky symbolizes abstract mathematical methods.
*Do not try to pick up unrenormalized theories out of it!

The weather was really unfair to the seminar speakers, especially during the second week. A. Barut, G. Cicuta and J. Nilsson had to complete in the early afternoon with a bright sunshine and all the ski attraction which this beautiful place is offering. Nevertheless, none of them missed his own talk. Moreover, there were interesting and imaginative things to hear about on all three seminars. For instance, Nilsson's talk was the only one in the past two weeks devoted to the fashionable unified theory of electromagnetic and weak interactions, which combines the ideas of gauge fields and spontaneous symmetry breaking.

We all spent two instructive and enjoyable weeks in Schladming. I am sure that I express everyone's feelings when I thank again Professor P. Urban for his kind hospitality during the XII. Internationale Universitätswochen and the members of the Organizing Committee: Prof. L. Streit, Drs. H. Latal, H.J. Faustmann and the secretaries Miss M. Krautilik and Mrs. I. Primschitz for the excellent job they have done. Thank you!